Communications in Computer and Information Science

2793

Series Editors

Gang Li , *School of Information Technology, Deakin University, Burwood, VIC, Australia*
Joaquim Filipe, *Polytechnic Institute of Setúbal, Setúbal, Portugal*
Zhiwei Xu, *Chinese Academy of Sciences, Beijing, China*

Rationale

The CCIS series is devoted to the publication of proceedings of computer science conferences. Its aim is to efficiently disseminate original research results in informatics in printed and electronic form. While the focus is on publication of peer-reviewed full papers presenting mature work, inclusion of reviewed short papers reporting on work in progress is welcome, too. Besides globally relevant meetings with internationally representative program committees guaranteeing a strict peer-reviewing and paper selection process, conferences run by societies or of high regional or national relevance are also considered for publication.

Topics

The topical scope of CCIS spans the entire spectrum of informatics ranging from foundational topics in the theory of computing to information and communications science and technology and a broad variety of interdisciplinary application fields.

Information for Volume Editors and Authors

Publication in CCIS is free of charge. No royalties are paid, however, we offer registered conference participants temporary free access to the online version of the conference proceedings on SpringerLink (http://link.springer.com) by means of an http referrer from the conference website and/or a number of complimentary printed copies, as specified in the official acceptance email of the event.

CCIS proceedings can be published in time for distribution at conferences or as postproceedings, and delivered in the form of printed books and/or electronically as USBs and/or e-content licenses for accessing proceedings at SpringerLink. Furthermore, CCIS proceedings are included in the CCIS electronic book series hosted in the SpringerLink digital library at http://link.springer.com/bookseries/7899. Conferences publishing in CCIS are allowed to use our online conference service (Meteor) for managing the whole proceedings lifecycle (from submission and reviewing to preparing for publication) free of charge.

Publication process

The language of publication is exclusively English. Authors publishing in CCIS have to sign the Springer CCIS copyright transfer form, however, they are free to use their material published in CCIS for substantially changed, more elaborate subsequent publications elsewhere. For the preparation of the camera-ready papers/files, authors have to strictly adhere to the Springer CCIS Authors' Instructions and are strongly encouraged to use the CCIS LaTeX style files or templates.

Abstracting/Indexing

CCIS is abstracted/indexed in DBLP, Google Scholar, EI-Compendex, Mathematical Reviews, SCImago, Scopus. CCIS volumes are also submitted for the inclusion in ISI Proceedings.

How to start

To start the evaluation of your proposal for inclusion in the CCIS series, please send an e-mail to ccis@springer.com

Camelia Chira · Oliviu Matei · Florin Pop ·
Petrică Pop-Sitar
Editors

Innovative Perspectives on Computational Intelligence and Data Science

First International Conference, InnoComp 2025
Cluj-Napoca, Romania, October 22–24, 2025
Proceedings, Part I

 Springer

Editors
Camelia Chira [iD]
Babeş-Bolyai University
Cluj-Napoca, Romania

Oliviu Matei [iD]
Holisun
Baia Mare, Romania

Florin Pop [iD]
National University of Science
and Technology Politehnica Bucharest
Bucharest, Romania

Petrică Pop-Sitar [iD]
Technical University of Cluj-Napoca
Baia Mare, Romania

ISSN 1865-0929   ISSN 1865-0937 (electronic)
Communications in Computer and Information Science
ISBN 978-3-032-12477-7   ISBN 978-3-032-12478-4 (eBook)
https://doi.org/10.1007/978-3-032-12478-4

Preface

These volumes present the refereed proceedings of the First International Conference on Innovative Perspectives on Computational Intelligence and Data Science (InnoComp 2025), held in Cluj-Napoca, Romania, during October 22–24, 2025.

The objective of InnoComp is to establish a multidisciplinary forum for researchers, practitioners, and educators to disseminate original research results, exchange novel ideas, and explore emerging challenges in the areas of computational intelligence, data science, artificial intelligence, and intelligent systems. By integrating theoretical advances with practical applications, the conference aims to contribute to the development of innovative solutions addressing complex scientific and societal problems.

The inaugural edition of InnoComp attracted 117 submissions from diverse geographic regions, reflecting its international scope and relevance. Following a rigorous peer-review process conducted by the International Program Committee, assisted by additional expert reviewers, 47 full papers were accepted for inclusion in this volume. Each submission underwent thorough evaluation with respect to originality, technical quality, and relevance to the conference themes. The editors gratefully acknowledge the efforts of the Program Committee members and reviewers, whose careful assessments ensured the high scientific quality of these proceedings.

The accepted contributions are organized into four topical sections: Computational Intelligence; Data Science; Intelligent Systems; and Applications.

The scientific program of InnoComp 2025 was further enriched by keynote addresses delivered by three distinguished invited speakers: José Antonio Barata de Oliveira, NOVA University of Lisbon, Portugal; Aniello Castiglione, University of Salerno, Italy; and Andrei Păun, University of Bucharest, Romania.

Their lectures provided authoritative perspectives on state-of-the-art research and future directions, stimulating in-depth discussion and contributing to the intellectual breadth of the conference.

The editors wish to extend their gratitude to all authors for contributing their research, to the keynote speakers for sharing their expertise, to the reviewers for their rigorous evaluations, and to the organizing institutions and technical staff for their invaluable support. It is our conviction that this volume will serve as a scholarly reference for the community of researchers and practitioners working in computational intelligence, data science, and related fields. We further hope that InnoComp will continue to evolve into a leading international forum, fostering collaboration, innovation, and excellence in the years to come.

September 2025

Camelia Chira
Oliviu Matei
Florin Pop
Petrică Pop-Sitar

Organization

General Chairs

Camelia Chira	Babeş-Bolyai University, Romania
Oliviu Matei	Holisun, Romania
Florin Pop	Politehnica Bucharest, Romania
Petrică Pop-Sitar	Technical University of Cluj-Napoca, Romania

Program Committee

Lenuţa Alboaie	Alexandru Ioan Cuza University, Romania
Apostolos Ampatzoglou	University of Macedonia, Greece
Anca Andreica	Babeş-Bolyai University, Romania
Costin Badica	University of Craiova, Romania
Ovidiu Bagdasar	University of Derby, UK
Constantin Bala Zamfirescu	Lucian Blaga University of Sibiu, Romania
José Antonio Barata De Oliveira	University Nova de Lisboa, Portugal
Ana Bazzan	Federal University of Rio Grande do Sul, Brazil
Ion Bica	Military Technical Academy, Romania
Mihaela Breabăn	Alexandru Ioan Cuza University, Romania
Alexandros Chatzigeorgiou	University of Macedonia, Greece
Reggie Davidrajuh	University of Stavanger, Norway
Marc Demange	Royal Melbourne Institute of Technology, Australia
Liviu Dinu	University of Bucharest, Romania
Laura Dioşan	Babeş-Bolyai University, Romania
Ciprian Dobre	Politehnica Bucharest, Romania
Enol García González	University of Oviedo, Spain
Dorian Gorgan	Technical University of Cluj-Napoca, Romania
Alvaro Herrero	University of Burgos, Spain
Vasile Manta	Technical University Gheorghe Asachi, Romania
Daniel Mican	Babeş-Bolyai University, Romania
Mihai Micea	Polytechnic University of Timişoara, Romania
Sanaz Nikghadam Hojjati	University Nova de Lisboa, Portugal
Constantin Orasan	University of Surrey, UK
Michele Pagano	University of Pisa, Italy

Vincenzo Pallotta	University of Applied Science and Arts, Switzerland
Andrei Păun	University of Bucharest, Romania
Dana Petcu	West University of Timişoara, Romania
Radu Emil Precup	Polytechnic University of Timişoara, Romania
Hector Quintian	University of A Coruña, Spain
Dimitrios Raptis	Aalborg University, Denmark
Günter Rudolph	Technical University Dortmund, Germany
Stefano Secci	Conservatoire National des Arts et Métiers France
Gabriel Semanišin	Pavol Jozef Šafárik University, Slovakia
Dragan Simić	University of Novi Sad, Serbia
Andrei Sleptchenko	Khalifa University Abu Dhabi, UAE
Ioannis Sorokos	Fraunhofer IESE, Germany
Giovanni Stilo	University of L'Aquila, Italy
Nikolai Velikov	Nikola Vaptsarov Naval Academy, Bulgaria
José Ramón Villar	University of Oviedo, Spain
Michał Woźniak	Wroclaw University of Science and Technology, Poland
Daniela Zaharie	West University of Timişoara, Romania

Additional Reviewers

Anca Avram	Elena Nechita
Anca Mărginean	Cristina Feier
Diana Borza	Emil Paşca
Tudor Mihoc	Bela Genge
Alina Călin	Adrian Petrovan
Gabriela Moise	Mara Hajdu-Măcelaru
Ionela Chereja	Mădălina Răschip
Simona Motogna	Barna Iantovics
Cerasela Crişan	Mihai Suciu

Organizing Committee

Camelia Chira	Babeş-Bolyai University, Romania
Anca Andreica	Babeş-Bolyai University, Romania
Laura Dioşan	Babeş-Bolyai University, Romania
Simona Motogna	Babeş-Bolyai University, Romania
Cristiana Moroz-Dubenco	Babeş-Bolyai University, Romania
Mădălina Dicu	Babeş-Bolyai University, Romania
Raluca Chiş	Babeş-Bolyai University, Romania

Alexandru Manole   Babeş-Bolyai University, Romania
Oskar Picus   Babeş-Bolyai University, Romania
Ciprian Stupinean   Babeş-Bolyai University, Romania
Rudolf Erdei   Technical University of Cluj-Napoca, Romania
Daniela Delinschi   Technical University of Cluj-Napoca, Romania
Laura Andreica   Technical University of Cluj-Napoca, Romania

Contents

Computational Intelligence

Data Science

Computational Intelligence

Analysis of Customer Journey Video Data Using Eye-Tracking and Multimodal AI

Sami Heikkinen$^{(\boxtimes)}$, Jaani Väisänen , Hannu Kaikonen, Sami Makkula, and Petteri Markkanen

LAB University of Applied Sciences, Lahti, Finland
`sami.heikkinen@lab.fi`

Abstract. This study introduces a novel methodological approach for analyzing customer journey data using eye-tracking technology and multimodal AI. Customer journey research faces significant limitations due to reliance on resource-intensive qualitative methods that are difficult to scale. We address this gap by employing vision language models (VLMs) to automate the interpretation of eye-tracking data, eliminating the manual coding bottleneck while enabling analysis of larger and more diverse datasets. Our research evaluates five locally-deployable VLMs to determine the optimal balance between semantic accuracy and computational efficiency. Using data collected with Tobii eye-tracking glasses, we developed an analytical pipeline that synchronizes gaze location data with verbal think-aloud protocols to create a comprehensive multimodal dataset. Results demonstrate that Gemma3 (4B parameters) achieved 100% semantic accuracy on our test set while maintaining reasonable processing efficiency (43.26 s per image). When validated against human coding across the complete dataset, the model achieved a 74.2% recall rate. The integration of eye-tracking and verbal data revealed distinctive attention patterns including "navigational uncertainty," "confirmatory scanning," and "socially-mediated attention" throughout the customer journey. Our approach provides objective behavioral evidence of visual attention that complements traditional self-reported measures, enabling more comprehensive touchpoint analysis while aligning with event-driven perspectives from process mining research. This methodology offers promising applications for service design by identifying discrepancies between reported and actual customer attention patterns and providing a foundation for developing automated behavioral indicators to detect moments of customer confusion, decision-making, or confirmation throughout service journeys.

Keywords: Customer journey analysis · Eye-tracking · Vision language models · Service design · Customer experience

1 Introduction

Halvorsrud et al. [1] observe that the lack of formalism and tools to automatically capture the actual journeys for analytical purposes hinders its operational use, highlighting the need for more systematic and automated approaches.

C. Chira et al. (Eds.): InnoComp 2025, CCIS 2793, pp. 3–22, 2026.
https://doi.org/10.1007/978-3-032-12478-4_1

Eye-tracking technology offers a promising data source for understanding customer journeys, providing objective behavioral evidence of visual attention throughout Customer journeys have gained widespread recognition for portraying customer behaviors and experiences from the perspective of human end-users [2]. The analysis of these journeys provides critical insights for service design, but current methodological approaches face significant limitations. Customer journey research has traditionally relied heavily on qualitative data collection methods and manual coding processes that are resource-intensive and difficult to scale [3].

The challenge of analyzing large volumes of customer journey data has been well-documented in the literature. Bernard and Andritsos [4] note that current research typically limits the number of journeys to be compared to less than ten, making the overall process relatively straightforward, but this approach fails to capture the diversity and complexity of customer experiences at scale. Similarly, service interactions. However, the analysis of eye-tracking data has traditionally required labor-intensive manual annotation of areas of interest (AOIs), creating a resource bottleneck that limits its application in large-scale studies [5]. This constraint has prevented researchers and practitioners from fully leveraging the rich insights that eye-tracking can provide about customer attention patterns across touchpoints.

Recent advances in artificial intelligence, particularly vision language models (VLMs), present an opportunity to overcome these limitations. These models can potentially automate the interpretation of eye-tracking data, eliminating the manual coding bottleneck while enabling the analysis of larger and more diverse datasets. Kobialka et al. [6] highlight the value of systematic, data-driven methods for deriving user journeys, noting the shift from manually constructed journey maps toward automated analysis approaches.

This study aims to give an example of how vision language models can be utilized to automate the analysis of visual attention patterns across customer journey touchpoints. The approach presented here offers potential for scaling up customer journey analysis while providing deeper insights into the visual dimension of customer experience—a dimension that has been challenging to capture systematically with traditional methods.

2 Background

Understanding human visual attention is essential in numerous fields, including computer vision, human-computer interaction, psychology, and customer experience research. Eye movements provide rich information about real-time human visual attention and cognition, offering valuable insights into where people look and how they process visual information [7].

Customer journeys have gained widespread recognition for portraying customer behaviors and experiences from the perspective of human end-users [2]. A customer journey consists of a sequence of steps called touchpoints, each representing an instance of communication or interaction with a service. Due to digitalization, most touchpoints leave digital traces in IT systems, although some touchpoints may remain invisible to service providers, such as face-to-face conversations or interactions with external actors [2]. These actions in the real world require an approach that provides a methodology to help organizations understand their customers' needs and ways of perceiving service offerings.

2.1 Conceptualizing Customer Experience and the Customer Journey

Lemon and Verhoef [8] provide a comprehensive framework for understanding customer experience throughout the customer journey. They define customer experience as "a multidimensional construct focusing on a customer's cognitive, emotional, behavioral, sensorial, and social responses to a firm's offerings during the customer's entire purchase journey" (p. 74). This definition emphasizes the holistic and dynamic nature of experience, which aligns with our multimodal approach to analyzing customer experience data. The customer journey perspective is closely linked to customer experience, with research using customer journeys not only as a means to understand the customer's point of view but also as a way to gain insight into their experiences [3]. This connection reflects the growing recognition that customer experience is shaped during interactions between customers and service providers [9, 10], making the systematic analysis of these interactions crucial for experience management.

Recent literature has emphasized the need for more sophisticated frameworks to analyze customer experience (CX). [11] introduced the Touchpoints, Context, Qualities (TCQ) nomenclature to provide structure and clarity to the customer experience field. While this framework offers conceptual clarity, measuring these components objectively remains challenging. Traditional research methods often rely on self-reported data, which may be subject to recall bias and cannot capture real-time reactions. One particularly challenging aspect is understanding visual attention during customer journeys - specifically, what customers look at during touchpoint interactions on their customer journey. Lemon and Verhoef [8] highlight that customer experience measurement remains fragmented, with no strongly established scales for measuring the overall customer experience across the journey. This measurement gap presents an opportunity for eye-tracking methodologies that can capture objective behavioral data across the customer journey.

2.2 Customer Journey Modeling and Analysis

An important distinction exists between planned journeys (design-time) and actual journeys that materialize during execution (run-time) [2]. Bogale et al. [12] frame this distinction as "patient pathways" representing the planned blueprint of the care process, while "patient journey" denotes an individual's actual experience revealed retrospectively. This conceptualization can be applied to customer journeys in general, distinguishing between the designed service process and the lived customer experience.

Traditional approaches to analyzing customer journeys typically fall into two categories: customer journey mapping (analyzing existing service processes "as is") and customer journey proposition (generative design activities toward possible services "to be") [3]. Both approaches focus heavily on understanding what customers think and feel throughout their service interactions but often rely on self-reported data subject to recall bias.

The Customer Journey Modeling Language (CJML) has emerged as a domain-specific modeling approach that distinguishes between planned journeys (design-time) and actual journeys that materialize during execution (run-time). CJML also differentiates between objective elements that can be observed and logged, and subjective components encompassing personal customer experiences [2].

Beyond descriptive modeling, recent research has explored predictive approaches to customer journey analysis. Hassani and Habets [13] demonstrated that a customer's next touch point can be predicted using machine learning techniques applied to historical interaction data. Knowing what a customer will do next gives the company the ability to proactively provide support to the customer [13], potentially improving customer satisfaction while reducing service costs.

While traditional customer journey mapping typically relies on qualitative methods, recent research has explored more data-driven approaches. Arias et al. [14] demonstrated how process mining techniques can be applied to systematically map customer journeys in healthcare settings. Their approach treats each touchpoint as a process activity and uses event logs to reconstruct the actual paths customers follow, revealing variations that might not be captured through traditional mapping methods.

Process discovery, trace clustering, and decision mining methods have been used for customer journey analysis [15]. Process discovery creates representational graphs that encapsulate actions observed in an event log, capturing the main flow of customer activities during interactions. Trace clustering addresses the complexity challenge by clustering event logs into smaller groups based on similar properties, making the resulting process models more comprehensible. Decision mining enriches these models by identifying decision points and applying machine learning algorithms to understand what factors influence customer choices at each stage of the journey.

The concept of journey mapping extends beyond marketing and into other domains such as healthcare, where similar methodologies are employed to understand patient experiences. Davies et al. [16] define patient journey mapping as 'a patient-oriented project that has been undertaken to better understand barriers, facilitators, experiences, interactions with services and/or outcomes for individuals and/or their carers and family members as they enter, navigate, experience and exit one or more services in a health system by documenting elements of the journey to produce a visual or descriptive map' (p. 84). This definition emphasizes the holistic nature of mapping experiences across multiple touchpoints, which aligns with our multimodal approach to analyzing customer journey data.

Customer journeys have emerged as a significant concept in service design and management, with various approaches developed to analyze and visualize these journeys. Bernard and Andritsos [4] introduced CJM-explorer (CJM-ex), a web-based approach that bridges customer journey mapping with process mining techniques. Recent innovations in customer journey analysis have expanded beyond traditional qualitative approaches. Halvorsrud et al. [1] propose a framework that combines formal representations of user journeys with process mining techniques to automate the capture and analysis of user interactions extending across service systems. Their approach emphasizes creating a semantic journey database from which patterns that optimize user experience can be identified. This aligns with Bernard and Andritsos's [17] genetic approach to customer journey discovery, which considers both sequential activities and contextual data when measuring similarities between journeys. Their method addresses the challenge of summarizing large collections of customer journeys into representative patterns that effectively capture the essential characteristics of the observed behaviors.

Various visual approaches to journey mapping have been identified in recent literature. Davies et al. [16] categorized journey mapping approaches into six primary types: (1) mapping key experiences and recommendations over the duration of a journey; (2) mapping by location; (3) mapping by event; (4) mapping roles, input, and experiences of key stakeholders; (5) mapping from multiple perspectives; and (6) mapping timelines of events.

Recent advances in user journey analysis have expanded beyond traditional qualitative approaches. There is a need for a shift from manually constructed journey maps toward automated, data-driven analysis, addressing the current problem that hinders their operational use in complex services.

2.3 Eye-Tracking in Customer Journey Research

Despite the widespread adoption of customer journey approaches, there are significant methodological limitations in current practice. Most notably, customer journey research relies heavily on self-reported data and subjective assessments that may be subject to recall bias and cannot capture real-time reactions [3]. One particularly challenging aspect is understanding visual attention during customer journeys—specifically, what customers actually look at during interactions with touchpoints.

Eye-tracking technology offers a promising solution to this limitation by providing objective behavioral data across the customer journey. Previous research has demonstrated the value of eye-tracking in retail environments, showing that visual attention metrics such as fixation count and total visit duration can be meaningfully correlated with consumer decision-making [18]. Similarly, Clement [19] used eye-tracking in supermarket settings to reveal distinct phases in visual decision processes, challenging traditional linear models of buying behavior.

The integration of IoT sensors—such as eye-tracking glasses in our context—with business process analysis allows for finer-grained automated monitoring and analysis [20]. This integration is particularly relevant for customer journey analysis, where understanding visual attention at different touchpoints can provide deeper insights into customer behavior and experience. [20] identify event mining from sensors as one of the four primary topics in IoT-BPM integration research. This directly relates to our approach of extracting meaningful patterns from eye-tracking sensor data.

A structured approach for translating low-level sensor data into higher-level process insights is presented by [21]. Their framework for process discovery from sensor data addresses the challenge that, the event data generated during the execution of a process is at a much lower level of abstraction. This mirrors our challenge of interpreting continuous eye-tracking data within the context of customer journeys.

Traditionally, eye tracking research has relied on specialized hardware operated in controlled laboratory environments. This approach, while accurate, presents several limitations: it is expensive, time-consuming, and difficult to scale for larger participant pools [7]. These constraints have historically limited the size of eye tracking datasets compared to those available for other visual recognition tasks, restricting the potential for data-intensive machine learning approaches to gaze analysis.

The challenge of translating low-level sensor data into higher-level, discrete insights is not unique to eye-tracking research. [21] present a framework for process discovery from sensor data that addresses the challenge of lower level abstraction. Their approach discretizes observed sensor data into activities by applying unsupervised learning through clustering techniques. This conceptual approach is relevant to eye-tracking analysis, where continuous gaze data must similarly be translated into meaningful discrete events representing visual attention patterns.

Recent innovations have explored more accessible methods for collecting eye tracking data. For example, [7] developed a webcam-based eye tracking system that supports large-scale, crowdsourced collection of eye gaze data. Their approach demonstrated that carefully designed algorithms and protocols could obtain eye tracking data comparable to traditional lab settings but at significantly lower cost and effort. The researchers employed various techniques to address common challenges in remote eye tracking, including managing lighting conditions, head movements, and ensuring participant attention.

Gaze data has been effectively utilized in various applications including object detection, tracking, and referring expressions in videos. Vasudevan et al. [22] investigated the problem of object referring in videos using both language descriptions and human gaze data. They highlight that humans naturally gaze at objects when issuing referring expressions, and this gaze information can be a valuable cue for object localization. Their research demonstrated that combining multiple modalities—including appearance, motion, gaze, and spatio-temporal context—significantly improved the accuracy of object localization compared to using visual data alone.

Several research groups have developed datasets and analytical frameworks for various eye tracking applications. For example, Judd et al. [23] created a dataset of natural images viewed by 15 subjects, while others have focused on specific domains such as object search [24] or emotional stimuli [25]. These efforts have established important benchmarks for evaluating the performance of computational models of visual attention [7].

Recent innovations in eye-tracking analysis have moved beyond traditional approaches that rely on manual annotation of areas of interest (AOIs). Castner et al. [5] developed a novel approach to gaze scanpath comparison that incorporates convolutional neural networks (CNNs) to process scene information at the fixation level. Their method, called DeepScan, extracts image patches linked to respective fixations as input for a CNN, with the resulting feature vectors providing temporal and spatial gaze information necessary for accurate scanpath similarity comparison. By utilizing deep learning to automatically extract semantic information from gaze fixation points, their method enables more objective analysis of eye-tracking data across varying stimuli. The DeepScan approach is particularly relevant for real-world applications where stimuli vary naturally, such as customer journey analysis.

3 Rationale and Aim of the Study

Our research builds upon these foundations while addressing the specific challenge of analyzing customer journey data through eye-tracking information captured in video recordings. Rather than focusing on static images or controlled visual stimuli, we examine dynamic real-world interactions where gaze patterns must be understood in relation to verbal exchanges and environmental context. This approach requires integrating techniques from video processing, artificial intelligence-based image analysis, natural language processing, and semantic analysis to effectively identify and categorize patterns of visual attention throughout the customer journey.

While our research employs specialized eye-tracking glasses rather than webcam-based solutions, the analytical challenges addressed in previous research remain relevant. The processing and interpretation of gaze data, particularly the identification of fixations (stable gaze on a specific point) versus saccades (rapid eye movements between fixations), requires sophisticated computational approaches [7]. Similarly, the challenge of mapping raw gaze data to meaningful semantic information about what participants are looking at remains a complex problem across all eye tracking methodologies.

While Huddleston et al. [18] focused on static retail displays rather than dynamic customer journey videos, their methodological approach to defining areas of interest and analyzing visual attention patterns provides valuable precedent for our research. Our work extends this capability by applying similar principles to dynamic interactions throughout the customer journey. Particularly relevant to our research is their methodological approach, where they extracted two primary visual metrics—fixation count and total visit duration—to measure attention.

The user-centric perspective highlighted by Kobialka et al. [6] aligns with our multimodal approach to analyzing customer journey video data. While their focus on system logs as their primary data source, they acknowledge the limitations of this approach, noting that journey diagrams are usually generated by hand, and the user perspective is derived from interviews with experts and users. Our methodology extends this work by incorporating eye-tracking data as an additional layer of objective behavioral evidence, providing insights into visual attention that cannot be captured through system logs alone.

This research extends current customer journey analysis methods by applying an event-driven approach to eye-tracking data. While Halvorsrud et al. [1] emphasize that the lack of formalism and tools to automatically capture the actual journeys, our methodology addresses this gap by providing a structured framework for analyzing visual attention throughout the customer journey. Similar to how Berendes et al. [26] propose a domain-specific modeling language to represent complex customer journeys in retail settings, our approach offers a specialized technique for analyzing customer journey data through the lens of visual attention, thereby contributing to the emerging field of data-driven customer journey mapping.

This study addresses the identified gaps by introducing a novel methodological approach that combines three key elements: a) the use of locally-deployable VLMs to automate the analysis of eye-tracking data, which resolves the scalability and resource issues of manual coding while addressing data privacy concerns, b) the integration of multimodal data, by synchronizing eye-tracking data with think-aloud protocols, enabling a deeper understanding of the relationship between visual attention and cognitive processes during service interactions, and c) the generation of objective behavioral data, which complements traditional self-reported measures and establishes a foundation for data-driven customer journey analysis in line with event-driven perspectives from process mining research. The research questions for the current study are:

RQ1: How accurately can locally-run vision language models (VLMs) identify and interpret gaze locations in eye-tracking video data from customer journeys?

RQ2: What are the trade-offs between computational efficiency and semantic accuracy when using different VLMs for automated eye-tracking analysis?

RQ3: How can the integration of eye-tracking data and verbal think-aloud protocols enhance our understanding of customer attention patterns throughout service interactions?

4 Methods

This study employed a computational approach to analyze customer journey data through eye-tracking information captured in video recordings. The methodology integrated video processing, vision language models (VLMs), and semantic analysis to systematically identify and categorize patterns of visual attention in relation to verbal think-aloud data. Our analytical pipeline, developed in R [27], follows a structured workflow: (1) data collection, (2) video pre-processing, (3) automated gaze location analysis, (4) validation of AI interpretations, and (5) integration with verbal data for comprehensive journey analysis.

4.1 Data Collection

Data was collected using Tobii eye-tracking glasses, which simultaneously recorded three data streams: (1) the participant's field of view, (2) a red marker indicating precise gaze location in real time, and (3) audio of participant think-aloud verbalizations. For this pilot study, we recorded one customer journey wherein a participant completed a predefined task within a physical servicescape environment [28]. The participant was instructed to verbalize their thoughts naturally while navigating the environment, providing concurrent verbal data alongside the eye-tracking measurements. This approach aligns with established protocols for capturing multimodal customer experience data [29]. Following the recording session, the audio component was professionally transcribed and time-stamped to enable synchronization with the visual data during analysis.

Our analytical approach shares conceptual similarities with frameworks developed for processing continuous sensor data in other domains. Janssen et al. [30] propose a four-step methodology for translating sensor event data into meaningful process models: (1) event correlation, (2) activity discovery, (3) event abstraction, and (4) process discovery. While developed for different sensor types, this framework offers a useful structure for interpreting continuous eye-tracking data to analyze run-time [2] customer journeys. Our approach similarly aims to transform low-level gaze events into higher-level semantic understanding through algorithmic processing and pattern identification.

4.2   Video Pre-processing

The continuous video footage was decomposed into sequential still images using the av package [31], enabling frame-by-frame analysis of gaze location. After evaluating various sampling rates, we configured the extraction at 10 frames per second (fps), which provides a temporal resolution of 100 ms between images. This sampling rate was selected as optimal because it is sufficient to detect meaningful gaze fixations, which typically require a minimum duration of 200 ms to indicate cognitive processing [18, 25], while maintaining computational efficiency. Higher sampling rates (e.g., 30 fps) would have created redundant data given the established minimum fixation duration threshold, while lower rates might have missed brief but significant fixations. The extraction process yielded 1,770 still images for subsequent analysis.

4.3   LVM Evaluation with Gaze Location Analysis

The core innovation in our approach was the utilization of vision-capable language models to automatically identify gaze focus points in each frame, eliminating the need for manual coding. This automation addresses a significant limitation in eye-tracking research, where manual annotation of areas of interest (AOIs) typically creates a resource-intensive bottleneck [5].

To address RQ1 (accuracy of VLMs in identifying gaze locations) and RQ2 (trade-offs between computational efficiency and semantic accuracy), we conducted a systematic evaluation of multiple vision language models using a controlled testing protocol.

Privacy considerations for customer journey data necessitated using locally-deployable large language models rather than cloud-based solutions. We implemented the Ollama framework [32] to host and run various vision language models (VLMs). To gain a comprehensive understanding of performance trade-offs, five distinct VLMs were selected for this study based on a set of specific criteria. The selection criteria were local deployability, variations in model sizes, architectural diversity, and open availability. To adhere to privacy considerations for customer journey data, all models had to be executable locally within the Ollama framework. The selected models, with parameter counts ranging from 3.8B to 11B, represent different computational scales, allowing for an evaluation of the trade-off between processing efficiency and semantic accuracy. The models represent different model families (such as Llama, Gemma, and Minicpm) to avoid over-reliance on a single architecture. All chosen models were openly available and well-supported. Five models were systematically evaluated: Llama3.2-vision (11B parameters), Minicpm-v (8B parameters), Llava-phi3 (3.8B parameters), Llava-llama3 (8B parameters), and Gemma3 (4B parameters). This evaluation directly addresses RQ2 by examining how model size and architecture affect the balance between processing efficiency and semantic accuracy.

To assess model performance, we randomly selected five diverse still images from the dataset and prompted each model to describe the object at the gaze marker location. Each model was evaluated on two criteria: (1) semantic accuracy – the ability to correctly identify the object at the gaze location, and (2) computational efficiency – the processing time required per image. This evaluation approach follows established methods for assessing VLM performance in specialized visual analysis tasks [33].

4.4   Standardized Prompting Protocol

Based on the evaluation results, Gemma3 was selected as the optimal model for semantic analysis of the still images. We then conducted prompt engineering trials to determine the most effective instructions for consistent gaze location identification. The finalized prompt was: "The red circle shows eye gaze. Describe the exact location using format: [Object] at [position] (e.g., 'Monitor at upper right', 'Patient chart at center')." This structured prompt format was designed to produce consistent, parsable outputs that could be efficiently processed in subsequent analysis stages.

To maximize computational efficiency, we implemented parallel processing via the parallel package, distributing the image analysis tasks across multiple CPU cores. This parallelization approach reduced total processing time by approximately 65% compared to sequential processing.

4.5   Validation of AI Interpretations

To validate the reliability of the VLM-generated interpretations, we conducted a systematic comparison between AI and human interpretations. It is important to clarify that this process was not a formal cross-validation but rather a validation of the model's performance (recall) against a human-coded "ground truth" benchmark. Human annotation is the bedrock for training and evaluating AI models in supervised learning paradigms, providing factual and verified data obtained through direct observation. As the goal of this pilot study was to assess the model's general performance across the dataset rather than optimize its predictive power, validation based on a random sample provided a sufficient and resource-effective approach. Even though discrepancies between the experts' subjective assessment of a knowledge domain and AI's more know-what approach have been reported [34], this bias was mitigated by meticulously training the observers to use shared terminology and similar points of interest.

For each image in this validation set, we compared the AI-generated interpretation with the human coder's interpretation using object identification agreement, whether both identified the same primary object. This validation approach is consistent with recommended practices for evaluating AI visual interpretation systems [35].

For this study involving 1770 still images, the ground truth was determined from a 160-image random sample. The sample size was determined to estimate the likelihood of the AI model correctly identifying an object within the target population with a specified level of confidence and precision. We calculated the required sample size using the standard formula for estimating a population proportion based on a desired confidence level (95%). Due to the preliminary stage of the research, as well as time and resource limitations, the chosen margin of error was determined as 7,5% instead of the more traditional 5%. As there was no reliable prior estimate for the population proportion, we used $p = 0.5$. This conservative value maximizes the required sample size, ensuring the study is adequately powered for the desired precision regardless of the true proportion. Given that our population size was finite at $N = 1770$, the Finite Population Correction was applied, yielding a result of 156.

Since the AI model was not classifying objects into predetermined categories but rather attempting to interpret the content of the gaze point in a still image, it only produced True Positive (correct identification) and False Negative (incorrect identification) results. To evaluate the validity of AI interpretations, the following procedure took place: 1) Ground Truth definition by human verifiers, 2) The generation of AI output, 3) Comparison, if the Ground Truth object was mentioned in AI output, 4) Calculation of the Recall figure as follows: (Number of GT objects mentioned by AI) / (Total number of GT objects).

Recall for the model's performance was 74,2%, meaning that for every 100 Ground Truth objects, the model would successfully recognize 74 of them. To get a better understanding of how well the AI model would rank different categories of images, the Ground Truth images were categorized into the following five categories: 1) a car (n = 19), 2) a human figure (n = 27), 3) a building (n = 30), 4) entrances into spaces. (n = 29), 5) human-made guidance systems, i.e. roads, signs, etc. (n = 49). The reviewers deemed six observations unintelligible and, therefore, omitted from the analysis. A similar procedure to calculate the Recall value was conducted for all categories. The recall value was consistent across all categories, ranging from 68,4% (category 1) to 79,3% (category 4).

One clarification to our validation scheme needs further addressing for transparency. Our human observers might detect several objects within the subject's gaze point when determining the Ground Truth. This would result in Ground Truth depictions, such as "human figure by a door," effectively combining categories 2 and 4. In such cases, the category was determined by which of the categories appeared first in the description. The category decision was based on the assumption of visual salience, where more animate objects in a scene will be produced earlier in a sentence, an approach partially supported by Montag and MacDonald [36].

4.6 Integration and Semantic Analysis

To address RQ3 regarding the integration of eye-tracking and verbal data, we developed a multimodal analysis approach. Transcript data containing time-stamped verbal exchanges was imported and synchronized with the visual gaze data based on temporal alignment. Each gaze description was paired with any verbal data occurring within ± 1 s of the frame timestamp, creating a comprehensive multimodal dataset.

To identify patterns of sustained attention on specific objects across multiple frames, we developed a semantic similarity analysis pipeline with three components. First, we performed lexical normalization with object references were normalized through lemmatization and stop-word removal to reduce surface-level variation (e.g., "the large computer screen" and "computer monitor" would both be normalized to "computer screen"). Second, we recognized synonyms by creating domain-specific synonym groups for common objects in the servicescape environment. This allowed the system to recognize that terms like "display," "monitor," and "screen" often refer to the same physical object. Third, we determined similarity threshold through iterative testing. We established a semantic similarity threshold of 0.75 (on a scale of 0–1) using cosine similarity measures applied to word embeddings. Frame sequences with similarity scores above this threshold were classified as continuous attention on the same object, even when terminology varied slightly across frames.

This semantic analysis approach builds upon methodologies developed for customer journey analytics by Bernard and Andritsos [4], adapted specifically for eye-tracking data. By detecting semantic similarity between consecutive gaze descriptions, our system could identify both sustained attention on single objects and transitions between different objects of interest, mapping the visual journey through the servicescape.

The final output consisted of a structured dataset containing frame numbers, timestamps, gaze focus descriptions, transcript text, speaker identification, and semantic similarity scores between consecutive frames. This integrated dataset enabled both quantitative analysis of attention patterns (e.g., duration of fixation on specific objects, frequency of transitions between objects) and qualitative assessment of the relationship between verbal communication and visual attention.

5   Results

5.1   Model Accuracy Results (RQ1)

Figure 1 shows an example of a still image captured from the eye-tracking video, with the red marker indicating the participant's gaze focus point. Using our standardized prompt on this image and others, we evaluated each model's ability to correctly identify the object being viewed.

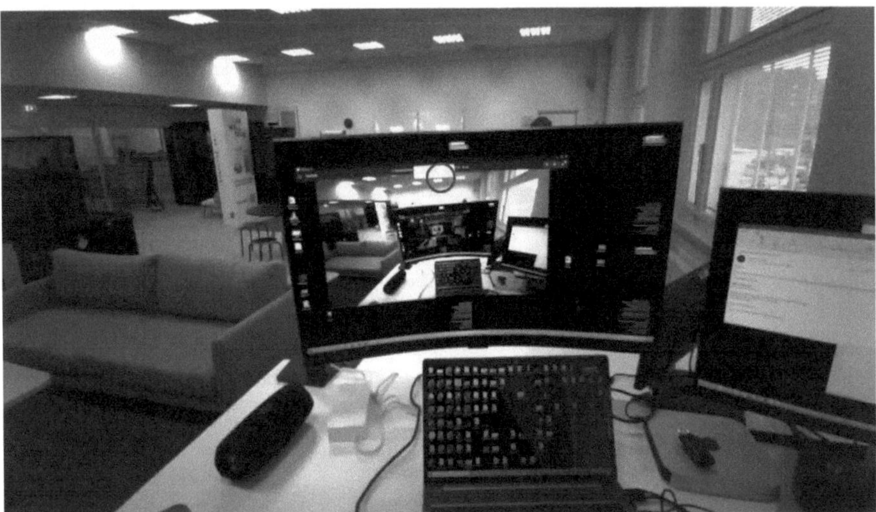

Fig. 1. An example of a still image captured from the video shows the red marker illustrating the focus of the viewer.

Table 1 presents example responses from five different VLMs for the image shown in Fig. 1, along with human evaluation of semantic correctness. As shown in the "Semantically correct" column, four of the five models correctly identified that the gaze was focused on a computer monitor, although with varying levels of descriptive detail.

5.2   Model Efficiency Results (RQ2)

Table 2 summarizes the performance of all five evaluated models across our test set of images, showing both semantic accuracy and processing time requirements.

The relationship between model size, accuracy, and processing time revealed interesting trade-offs. The largest model (Llama3.2-vision:11b) achieved perfect semantic accuracy but required the longest processing time (average 55.48 s per image). In contrast, smaller models like Llava-phi3:3.8b processed images much faster (4.51 s) but with substantially lower accuracy (40%).

Gemma3:4b emerged as the optimal balance point, achieving 100% semantic accuracy on our test set while requiring only 43.26 s per image on average. When scaled to the full dataset of 1,770 images, this represented approximately 21 h of processing time, which our parallel processing implementation reduced to 7.6 h of wall clock time using an 8-core CPU.

Table 1. The comparison between LVMs based on their responses, responses' semantical correctness, and time elapsed getting the responses.

LVM	Response	Semantically correct	Time elapsed (sec)
Llama3.2-vision:11b	The focus of the person is on a computer screen displaying a room with multiple monitors and computers, likely indicating that they are working or interacting with digital content related to the space	yes	84.5
Minicpm-v:8b	The red circle indicates that a person's attention or gaze is focused on another individual who appears to be seated and possibly engaged with equipment, as seen through multiple monitors within what seems like an office environment dedicated to technical work such as video editing or broadcasting control room operations. The exact nature of the interaction between this central figure and others in the scene cannot be determined from the image alone without additional context	no	42.6

(continued)

Table 1. (*continued*)

LVM	Response	Semantically correct	Time elapsed (sec)
Llava-phi3:3.8b	The focus of the person is on a computer screen with a red circle on it	yes	4.3
Llava-llama3:8b	The focus of the person is on the computer monitor, which has a red dot on it	yes	7.8
Gemma3:4b	The image presents a somewhat unsettling and surreal scene. It appears to be a live video feed being displayed on multiple monitors, and the subject of that feed is a chaotic, dimly lit control room. The overall feeling is one of surveillance and potential tension	yes	43.95

Table 2. The overall comparison between LVMs based on the proportion of semantically correct responses and average time elapsed for getting the responses.

LVM	Semantically correct	Time elapsed
Llama3.2-vision:11b	1	55.48
Minicpm-v:8b	0.6	42.49
Llava-phi3:3.8b	0.4	4.51
Llava-llama3:8b	0.4	7.64
Gemma3:4b	1	43.26

5.3 Integration Results (RQ3)

We analyzed the relationship between visual attention patterns and concurrent verbal data throughout the customer journey. Our dataset comprised 1,770 frames of synchronized eye-tracking and verbal data, capturing a rich multimodal representation of the participant's experience through the servicescape environment.

The data revealed distinct patterns in both visual attention and verbal reporting. The contributed utterances on 1,031 instances (58.2% of frames) with the remaining frames (739 instances, 41.8%) containing no verbal data. This differential pattern of verbalization suggests varying levels of engagement throughout the journey, with certain servicescape elements prompting more extensive verbalization than others.

Analysis of eye movements revealed 850 gaze transitions across the dataset, with an average fixation duration of 1.08 frames (approximately 108ms) between transitions.

This relatively short average fixation time suggests an active visual scanning pattern throughout the journey, with the participant's attention frequently shifting between different environmental elements. The longest sustained fixation observed was 24 frames (2.4 s) focused on a road segment, during which the participant made verbal observations about wayfinding considerations.

The top five objects of visual attention were: Building (135 instances), Monitor (128 instances), Person (90 instances), Window (73 instances), and Sign (70 instances). This distribution reflects the environmental context of the journey, with architectural elements and informational displays receiving priority attention.

We identified 979 service interaction patterns where eye gaze on specific objects was accompanied by concurrent verbalization. These patterns represent moments where visual attention and verbal reporting were integrated, potentially indicating heightened cognitive processing at key touchpoints. Analysis of the verbal content revealed that verbalizations were predominantly related to wayfinding and navigational decision-making (e.g., "So I left the car there and now I should go this way."). Interestingly, direct verbal-visual alignment—where the object being viewed was explicitly mentioned in concurrent verbalization—was relatively infrequent. This suggests that participants often verbalized higher-level cognitive processes or contextual considerations rather than simply narrating their views.

The integration of eye-tracking and verbal data revealed several noteworthy patterns in how participants processed and responded to the servicescape environment:

Navigational Uncertainty. Extended visual fixations on signs and directional cues were frequently accompanied by verbalizations expressing uncertainty or decision-making processes (["And from this direction, I don't really know anything other than guessing I'll go this way."]).

Confirmatory Scanning. After verbal statements indicating a decision, participants often exhibited a pattern of brief fixations on environmental elements that could confirm their chosen direction, suggesting an integration of verbal commitment and visual verification.

Socially-Mediated Attention. Following verbal exchanges between speakers, gaze patterns showed responsiveness to suggestions or directions provided by the other speaker, indicating that verbal interaction shapes visual attention allocation.

6 Discussion

6.1 Key Findings

This study demonstrates the viability of using locally deployable VLMs to automate the analysis of eye-tracking data in customer journey research. Our evaluation of five different VLMs revealed significant variations in both semantic accuracy and computational efficiency. Llama3.2-vision (11B parameters) and Gemma3 (4B parameters) both achieved 100% semantic accuracy on our test set, but with substantial differences in processing time. Gemma3 emerged as the optimal model, balancing high accuracy with reasonable processing efficiency (43.26 s per image on average).

When applied to the full dataset, our validation against human coding showed that the Gemma3 model achieved a 74.2% recall rate across all object categories, with relatively consistent performance across different types of objects (ranging from 68.4% for cars to 79.3% for entrances). These results suggest that locally-run VLMs can provide reasonably accurate automated interpretation of gaze locations in eye-tracking videos, though with some limitations in recognition accuracy compared to human coders.

The integration of eye-tracking and verbal data revealed several noteworthy patterns in how participants interacted with the servicescape environment. We identified specific behavioral patterns including "navigational uncertainty" (extended fixations on directional cues accompanied by verbalized uncertainty), "confirmatory scanning" (brief fixations verifying previously stated decisions), and "socially-mediated attention" (gaze patterns responsive to verbal exchanges). These patterns demonstrate how visual attention and verbalization interact throughout the customer journey, providing deeper insights into the cognitive processes underlying service interactions. However, it is important to note that these findings are based on a single-participant pilot study, and the generalizability of the results requires broader validation.

6.2   Comparison with Existing Knowledge

Our findings extend the current body of knowledge in both methodological and theoretical dimensions. Methodologically, our approach addresses limitations identified by Halvorsrud et al. [1]. Whereas they noted that the lack of formalism and tools to automatically capture actual journeys hinders operational use, this study presents a concrete methodological solution that applies VLMs to solve this challenge specifically for the dimension of customer visual attention. By automating visual attention analysis through VLMs, our methodology provides a structured framework for analyzing an important dimension of customer experience that has previously been difficult to capture at scale.

Our results regarding eye movement patterns align and extend with previous eye-tracking research in retail environments. Our work extends the findings of Huddleston et al. [18] and Clement [19] from static retail displays to dynamic, multi-touchpoint customer journeys. Furthermore, by integrating verbal data, we not only identify phases of visual decision-making but also offer explanations for the cognitive processes behind them, such as the "confirmatory scanning" pattern we identified.

While our integration approach shares conceptual similarities with Castner et al.'s [5] methodology in moving beyond manual annotation of AOIs, our work shifts the focus from scanpath comparison for expertise classification to the semantic interpretation of gaze locations and their relationship with verbal data. This enables a content-focused analysis of the customer experience itself, rather than just the classification of behavior.

6.3   Applications for Customer Journey Analysis

The methodology developed in this study offers several promising applications for enhancing customer journey analysis. First, our approach provides objective behavioral data that complements traditional self-reported measures. By capturing what customers actually look at during service interactions, organizations can identify discrepancies between what customers say they notice versus what they actually attend to. For example, a retailer can analyze whether customers actually look at a new promotional sign that they reported noticing in an interview. If their gaze does not land on the sign, its placement or design has likely failed to capture attention effectively. This addresses a significant limitation identified by Følstad & Kvale [3] regarding the reliance on self-reported data in customer journey research.

Second, the integration of eye-tracking and verbal data enables more comprehensive touchpoint analysis. Lemon and Verhoef [8] highlighted that customer experience measurement remains fragmented, with no strongly established scales for measuring overall customer experience across the journey. Our methodology contributes to addressing this gap by providing objective measurements of visual attention that can be quantified and compared across touchpoints.

Third, our approach aligns with the event-driven perspective advocated by process mining researchers [4, 14]. By capturing discrete attention events throughout the customer journey, our methodology enables the analysis of visual attention patterns using similar analytical frameworks to those applied to system-generated event logs. This creates potential for integrating eye-tracking events with other digital trace data to create more comprehensive customer journey models.

Fourth, the patterns we identified (e.g., "navigational uncertainty," "confirmatory scanning") provide a foundation for developing behavioral indicators that could be used to automatically detect moments of customer confusion, decision-making, or confirmation throughout the service journey. For instance, a service provider could use the "navigational uncertainty" indicator to identify locations in a physical servicescape (like a hospital or airport) where signage is unclear. By detecting spots where a customer's gaze wanders between signs for an extended period while they express verbal uncertainty, the provider can implement targeted improvements to wayfinding and measure their impact. These indicators could help service providers identify and address pain points in the physical servicescape design.

6.4 Future Research

Several promising directions for future research emerge from this study. First and foremost, future research must address the primary limitation of this study: the single-participant dataset. The methodology must be validated with a larger and more diverse sample, including multiple participants across different service contexts and with varying behaviors. This is essential to establish the generalizability of the identified attention patterns and to assess the robustness of the VLM-based interpretation approach.

Second, refinement of VLM prompting strategies could improve semantic accuracy. Our current prompt engineering focused on structured output format, but more sophisticated prompting approaches that incorporate domain-specific knowledge about servicescape environments might yield higher accuracy rates.

Third, integration with other sensor data would provide a more comprehensive understanding of customer experience. Combining eye-tracking with other physiological measures (e.g., heart rate variability, electrodermal activity) could reveal emotional responses alongside visual attention patterns, addressing the multidimensional nature of customer experience defined by Lemon and Verhoef (2016).

Fourth, development of more accessible technological approaches would enable wider adoption. Future research should explore whether similar analysis could be conducted using webcam-based eye-tracking or even smartphone-based solutions, reducing the hardware barriers to implementation.

Fifth, longitudinal studies could reveal how visual attention patterns evolve across multiple service interactions. Understanding how customers' visual attention changes as they become more familiar with a service environment could provide insights into learning processes and habit formation throughout the customer lifecycle.

Finally, comparative studies across different VLMs as the technology evolves will be valuable. The field of vision language models is rapidly advancing, and newer models may offer improved semantic accuracy with lower computational requirements, potentially making this approach more accessible and efficient for service providers.

7 Conclusion

In conclusion, this research demonstrates the potential of integrating eye-tracking data and VLM-based analysis to provide deeper insights into customer journeys. While technical and methodological challenges remain, this approach offers a promising path toward more objective, comprehensive understanding of customer experience across service touchpoints.

References

1. Halvorsrud, R., Mannhardt, F., Johnsen, E.B., Tapia Tarifa, S.L.: Smart journey mining for improved service quality. In: 2021 IEEE International Conference on Services Computing (SCC), pp. 367–369. IEEE (2021)
2. Halvorsrud, R., Mannhardt, F., Prillard, O., Boletsis, C.: Customer journeys and process mining – challenges and opportunities. ITM Web Conf. **62**, 05002 (2024)
3. Følstad, A., Kvale, K.: Customer journeys: a systematic literature review. J. Serv. Theory Pract. **28**, 196–227 (2018)
4. Bernard, G., Andritsos, P.: CJM-ex: goal-oriented exploration of customer journey maps using event logs and data analytics. Bus. Process Manag. (2017)
5. Castner, N., et al.: Deep semantic gaze embedding and scanpath comparison for expertise classification during OPT viewing (2020)
6. Kobialka, P., Lizeth Tapia Tarifa, S., Bergersen, G.R., Johnsen, E.B.: User journey games: automating user-centric analysis. Softw. Syst. Model. **23**, 605–624 (2024)
7. Xu, P., Ehinger, K.A., Zhang, Y., Finkelstein, A., Kulkarni, S.R., Xiao, J.: TurkerGaze: crowdsourcing saliency with webcam based eye tracking (2015). http://arxiv.org/abs/1504.06755
8. Lemon, K.N., Verhoef, P.C.: Understanding customer experience throughout the customer journey. J. Mark. **80**, 69–96 (2016)
9. Berry, L.L., Wall, E.A., Carbone, L.P.: Service clues and customer assessment of the service experience: Lessons from marketing. Acad. Manag. Perspect. **20**, 43–57 (2006)
10. Berry, L., Carbone, L., Haeckel, S.H.: Managing the total customer experience. MIT Sloan Manag. Rev. **43**, 85–89 (2002)

11. De Keyser, A., Verleye, K., Lemon, K.N., Keiningham, T.L., Klaus, P.: Moving the customer experience field forward: Introducing the touchpoints, context, qualities (TCQ) nomenclature. J. Serv. Res. **23**, 433–455 (2020)
12. Bogale, B., Vesinurm, M., Lillrank, P., Celius, E.G., Halvorsrud, R.: Visual modeling languages in patient pathways: scoping review. Interact. J. Med. Res. **13**, e55865 (2024)
13. Hassani, M., Habets, S.: Predicting next touch point in a customer journey: a use case in telecommunication. In: European Conference on Modelling Simulation, pp. 48–54 (2021)
14. Arias, M., et al.: Mapping the patient's journey in healthcare through process mining. Int. J. Environ. Res. Public Health **17**, 6586 (2020)
15. Nguyen Chan, N., et al.: Design and deployment of a customer journey management system: the CJMA approach. In: The 5th International Conference on Future Networks & Distributed Systems. ACM, New York (2021)
16. Davies, E.L., et al.: Reporting and conducting patient journey mapping research in healthcare: a scoping review. J. Adv. Nurs. **79**, 83–100 (2023)
17. Bernard, G., Andritsos, P.: Contextual and behavioral customer journey discovery using a genetic approach. In: Advances in Databases and Information Systems. pp. 251–266. Springer, Cham (2019)
18. Huddleston, P., Behe, B.K., Minahan, S., Fernandez, R.T.: Seeking attention: an eye tracking study of in-store merchandise displays. Int. J. Retail Distrib. Manag. **43**, 561–574 (2015)
19. Clement, J.: Visual influence on in-store buying decisions: an eye-track experiment on the visual influence of packaging design. J. Mark. Manag. **23**, 917–928 (2007)
20. De Luzi, F., Leotta, F., Marrella, A., Mecella, M.: On the interplay between business process management and internet-of-things: a systematic literature review. Bus. Inf. Syst. Eng. **67**, 265–288 (2025)
21. Koschmider, A., Janssen, D., Mannhardt, F.: Framework for process discovery from sensor data. In: EMISA, pp. 32–38 (2020)
22. Vasudevan, A.B., Dai, D., Van Gool, L.: Object referring in videos with language and human gaze. In: 2018 IEEE/CVF Conference on Computer Vision and Pattern Recognition, pp. 4129–4138. IEEE (2018)
23. Judd, T., Ehinger, K., Durand, F., Torralba, A.: Learning to predict where humans look. In: 2009 IEEE 12th International Conference on Computer Vision, pp. 2106–2113. IEEE (2009)
24. Papadopoulos, D.P., Clarke, A.D.F., Keller, F., Ferrari, V.: Training object class detectors from eye tracking data. In: Computer Vision – ECCV 2014, pp. 361–376. Springer, Cham (2014)
25. Ramanathan, S., Katti, H., Sebe, N., Kankanhalli, M., Chua, T.-S.: An eye fixation database for saliency detection in images. In: Lecture Notes in Computer Science, pp. 30–43. Springer, Heidelberg (2010)
26. Berendes, C.I., Bartelheimer, C., Betzing, J.H., Beverungen, D.: Data-driven customer journey mapping in local high streets: a domain-specific modeling language. In: Thirty Ninth International Conference on Information Systems. aisel.aisnet.org (2018)
27. R Core Team, R., et al: R: A language and environment for statistical computing (2013)
28. Rosenbaum, M.S., Massiah, C.: An expanded servicescape perspective. J. Serv. Manag. **22**, 471–490 (2011)
29. Holmqvist, K., Nyström, M., Andersson, R., Dewhurst, R., Jarodzka, H., de Weijer J., V.: Eye tracking: a comprehensive guide to methods and measures (2011)
30. Janssen, D., Mannhardt, F., Koschmider, A., van Zelst, S.J.: Process model discovery from sensor event data. In: Lecture Notes in Business Information Processing, pp. 69–81. Springer, Cham (2021)
31. Ooms, J.: av: Working with Audio and Video in R. (2025)
32. Ollama: Run large language models locally. https://ollama.com/. Accessed 14 May 2025

33. Denner, S., Bujotzek, M.R., Bounias, D., Zimmerer, D., Stock, R., Maier-Hein, K.: Visual prompt engineering for vision language models in radiology. In: Medical Imaging with Deep Learning (2025)
34. Lebovitz, S., Levina, N., Lifshitz-Assa, H.: Is AI ground truth really true? The dangers of training and evaluating AI tools based on experts' know-what. MIS Q. 45, 1501–1526 (2021)
35. Steyvers, M., Griffiths, T.: Probabilistic topic models. In: Handbook of Latent Semantic Analysis (2007)
36. Montag, J.L., MacDonald, M.C.: Visual salience modulates structure choice in relative clause production. Lang. Speech 57, 163–180 (2014)

Fine-Tuned Vision Transformers for Mammographic Breast Cancer Classification: Evaluating the Impact of Preprocessing Techniques

Bianca Iacob[(✉)] [ID] and Laura Diosan [ID]

Faculty of Mathematics and Computer Science, University Babes Bolyai,
Cluj-Napoca, Romania
{andreea.lixandru,laura.diosan}@ubbcluj.ro

Abstract. This study examines the comparative performance of Vision Transformers in classifying breast cancer images under various pre-processing conditions. The analysis was conducted on raw breast cancer mammograms, as well as images enhanced using Gamma correction and Contrast Limited Adaptive Histogram Equalization. Specifically, five dataset configurations were considered: (1) original images, (2) Gamma-corrected images only, (3) original images combined with Gamma-corrected versions, (4) Contrast Limited Adaptive Histogram Equalization-enhanced images only, (5) original images combined with Contrast Limited Adaptive Histogram Equalization-enhanced versions.

Additionally, the analysis was stratified by lesion types and mammographic view, comparing model performance for calcification and mass lesions on Mediolateral Oblique and Craniocaudal projections. The results reveal distinct performance patterns in the vision transformer model across these preprocessing scenarios and imaging perspectives, highlighting the impact of image enhancement techniques and view types on model performance. This study provides valuable insights into the effectiveness of pre-processing methods in optimising the performance of advanced neural network architectures for breast cancer image classification.

The model demonstrates strong performance across both datasets, achieving its highest results on MIAS dataset with an accuracy, sensitivity, and specificity of 0.99 and on DDSM dataset with an accuracy of 0.95, sensitivity of 0.96 and specificity of 0.95.

Keywords: ViT · CLAHE · Gamma correction · breast cancer · mammograms

1   Introduction

In recent years, Vision Transformers (ViTs) have emerged as a powerful alternative to traditional convolutional neural networks (CNNs) for image analysis,

C. Chira et al. (Eds.): InnoComp 2025, CCIS 2793, pp. 23–36, 2026.
https://doi.org/10.1007/978-3-032-12478-4_2

including applications in biomedical imaging [10]. ViTs have the potential to outperform CNNs in mammographic image analysis because they excel at capturing long-range dependencies and global context within images. Unlike CNNs, which rely on local receptive fields and hierarchical feature extraction, ViTs process an image as a sequence of patches, allowing them to model relationships across the entire image directly. This capability is especially important in mammography, where subtle patterns and spatial interactions across different regions can be critical for accurate diagnosis. Additionally, ViTs' self-attention mechanism enables more flexible feature learning without the inductive biases inherent in CNNs, potentially leading to improved performance when fine-tuned on domain-specific mammography datasets.

For the detection and diagnosis of breast cancer, automated analysis systems based on various types of imaging, such as mammography, ultrasound, MRI or histopathological images, can be developed. However, this paper focuses on the use of mammography, as it plays an essential role in the early stages of detection and is frequently used in population screening programs. In addition, compared to other imaging methods, mammography is more accessible and less expensive, which makes it suitable for large-scale implementation.

In mammogram classification literature, ViT is often used either as a feature extractor [1,12] or as a classifier [4–6], but rarely in a unified framework. Our proposal applies ViT end-to-end, combining feature extraction and classification. Unlike prior work using only pretrained ViTs (e.g., [7]), we fine-tune ViT on a domain-specific mammography dataset for better adaptation. We also explore how preprocessing synergizes with fine-tuning to improve breast cancer classification. Therefore, the major contributions of this paper are:

- end-to-end use of ViT for mammogram classification
- development of fine-tuning for ViT using various datasets (original or preprocessed) to be able to study the influence of pre-processors
- conducting a stratified evaluation of ViT performance across lesion types (masses vs. calcifications) and mammographic views (craniocaudal - CC vs. mediolateral oblique - MLO), offering deeper insights into classifier behaviour

Through this paper, we want to respond to the following questions:

- **RQ1:** How do preprocessing steps impact the performance of ViT in detecting breast cancer from mammograms?
- **RQ2:** How does the type of tumor abnormality influence breast cancer detection in the context of ViT?

The next section of this paper describes how ViT is currently used in the literature. After that, we continue with the third section which presents the datasets used in this paper. The fourth section details the model involved in the study. The results are presented in the sixth section of the article. The paper ends with the last section where the conclusions are detailed.

2   Related Work

The use of Transformers in medical imaging has advanced breast cancer detection across modalities like mammographies, histopathology, and ultrasound. In mammography, Abd Elaziz *et al.* [1] proposed a Cross Vision Transformer with a Growth Optimizer for IoMT-based detection. Al-Tam *et al.* [7] introduced a hybrid model using residual convolutional transformer encoders for digital X-ray classification. Al Mansour *et al.* [5] developed MammoViT, a custom vision transformer for accurate BIRADS[1] classification, highlighting the versatility of Transformers in mammographic analysis.

In histopathology, Transformers enhance breast cancer diagnosis by improving image classification accuracy and speed. Abimouloud *et al.* [2] used token vision transformers for efficient histopathology image classification. Jahan *et al.* [13] combined deep learning with vision transformers to identify breast cancer subtypes, showing their potential in detailed cancer pathology. These works highlight the impact of vision transformers in early and precise histopathological analysis.

In ultrasound imaging, Transformers improve breast cancer classification accuracy. Gheflati and Rivaz [11] applied vision transformers to handle challenges like image variability. Wang *et al.* [23] proposed a semi-supervised vision transformer with adaptive token sampling, enabling learning from limited labeled data. These studies demonstrate the adaptability and effectiveness of Transformers in ultrasound-based diagnostics.

Abd Elaziz *et al.* [1] proposed CrossViT, a CNN-ViT hybrid, for classifying breast cancer using INbreast [16] and MIAS [22] mammograms. Images are processed to extract features, refined via a Modified Growth Optimizer (MGO), and classified using K-Nearest Neighbors (KNN) into normal or abnormal. This method shows the effectiveness of combining CNN and ViT for improved mammographic analysis [1].

Al Mansour *et al.* [5] proposed MammoViT, a custom Vision Transformer for classifying mammograms into four BIRADS categories (1, 3, 4, 5) using the KAU-BCMD dataset [8]. Features extracted via ResNet are augmented with SMOTE [9] to address class imbalance, then processed by the ViT for final classification. This approach shows the potential of combining ResNet and ViT to enhance BIRADS classification accuracy [5].

Gutierrez *et al.* [12] present a binary classification method for breast cancer using DDSM [19] and INbreast [16] mammograms. A ViT pretrained on ImageNet extracts features, which are reduced via PCA [14], then classified using an MLP or SVM. This method demonstrates the value of pretrained ViT and dimensionality reduction for accurate breast cancer diagnosis [12].

Muthukrishnan *et al.* [17] propose an efficient binary classification method using DDSM [19] and MIAS [22] mammograms. Texture features are extracted

---

[1] The BI-RADS (Breast Imaging Reporting and Data System) is a standardized scoring system developed by the American College of Radiology to categorize breast imaging findings and guide clinical management [20].

via GLCM, then processed by a Lightweight Multihead attention Gannet CNN (LMGCNN), which uses multihead attention to enhance classification. This approach shows the effectiveness of combining GLCM and lightweight networks for accurate, fast diagnosis [17].

Al *et al.* explore a multimodal breast cancer detection and classification method using MIAS [22], INBreast [16], and BUSI [3] datasets. The approach uses YOLOv8 for detection, ResNet for feature extraction, and ViT for classification, focusing on malignancy identification and BI-RADS scoring. Experiments show that combining these models improves accuracy and reliability in breast cancer diagnostics [6].

Al *et al.* [4] propose a method for classifying breast lesions (normal, benign, malignant) using cropped INbreast [16] images. An ensemble of CNNs extracts features, which are then refined by a ViT for final classification. This CNN-ViT integration enhances diagnostic accuracy and reliability in breast cancer detection [4].

Table 1. Comparison of datasets, preprocessing, and augmentation methods applied specifically to mammography-based breast cancer detection, including our approach.

Ref	Data		
	name	ROI and preprocessing	augmentation
[1]	MIAS, INbreast, MiniDDSM		vertical and horizontal flips, cropping, and color jittering
[5]	BIRADS	normalization	SMOTE
[12]	DDSM, INbreast		resize, normalization
[6]	MIAS, INbreast, private KAU-BCMD, DDSM	removing artifacts, lesion segmentation	based on image/ROI, CLAHE, mosaic, rotaiting
[4]	INbreast	intensity normalization, lesion segmentation	augmentation (not mentioned how)
our	MIAS, DDSM	removing artifcats	CLAHE, Gamma correction

Table 2. Comparative overview of classification algorithm and the predicted classes utilised in mammography-based studies, including our proposed approach.

Ref	Classification	
	classes	algorithm
[1]	INbreast - normal, abnormal MIAS - normal, abnormal MiniDDSM - benign, cancer, normal	kNN
[5]	BIRADS 1, 3, 4 or 5	ViT
[12]	DDSM - benign, malign	MLP, SVM
[6]	INbreast - benign, malign	SVM, RF, kNN, NB, LR, DT, YOLO, ViT
[4]	INbreast - normal, abnormal	ViT
our	MIAS - benign, malign, DDSM - benign, malign	ViT

To provide a comprehensive overview of mammography-based breast cancer classification, we present comparative analyses of key methodologies. Table 1 compares datasets, ROI/preprocessing, and augmentation methods across studies including ours. Table 3 details feature computation and selection, while Table 2 summarizes predicted outputs and classification algorithms. These comparisons showcase the advances of our approach within the context of prior research.

The performance metrics used in the literature [21] are shown in Table 4. Accuracy measures overall correctness; precision indicates the quality of positive predictions; recall (sensitivity) assesses true positive detection; F1 score combines precision and recall as their harmonic mean, useful for imbalanced classes. Specificity measures true negative identification, while the Matthews Correlation Coefficient (MCC) provides a balanced summary considering all confusion matrix components, especially effective for imbalanced datasets.

Table 3. Comparative overview of ML-based discovered features utilised in mammography-based studies, including our proposed approach.

Ref	Features	
	Learning model	selection
[1]	Cross ViT	MGO
[5]	ResNet	–
[12]	ViT	PCA
[6]	ResNet, ViT	–
[4]	DenseNet201, VGG16, InceptionResNetV2	–
our	ViT	–

Table 4. Comparative overview of performance metrics involved in mammography-based studies, including our proposed approach.

Ref	Performance					
	Acc	Prec	Recall	F1	Specificity	MCC
[1]	x		x		x	
[5]	x		x	x		
[12]	x	x	x	x		
[6]	x	x	x	x	x	
[4]	x		x	x	x	x
our	x		x		x	

In class-imbalanced scenarios, relying solely on accuracy can be misleading, as a model predicting only the majority class may still show high accuracy without detecting important cases. Recall is crucial for identifying actual positives, minimizing missed diagnoses. Specificity is also important to avoid misclassifying healthy individuals, preventing unnecessary anxiety and follow-ups.

3   Experimental Study

This section presents a comprehensive overview of the datasets utilized in the experiments, the preprocessing techniques applied, and the machine learning model employed. The primary objective of the model is to investigate the influence of various preprocessing strategies on the performance of the ViT architecture. Additionally, the study aims to examine how different types of tumor abnormalities affect the effectiveness of breast cancer detection when using ViT. By analyzing these factors, the experiments seek to provide insights into optimizing preprocessing pipelines and understanding the sensitivity of ViT to specific pathological features in breast imaging data.

3.1   Overview

The proposed pipeline starts with the acquisition of the original images, which are subsequently subjected to a cleaning step aimed at removing noise and irrelevant information. Next, the cleaned data is pre-processed to meet the input requirements of the ViT model, thus ensuring its consistency and quality. The next step consists of training and validating the ViT model on the processed dataset, in order to optimize the generalization capacity of the network. Finally, the model performance is evaluated through relevant metrics, providing a quantitative insight into the efficiency of the proposed pipeline.

In the breast cancer case, the malignant tumours often have distinct, irregular shapes and higher density compared to benign lesions. Benign lesions typically have smoother, more regular shapes and lower density. By using a nonlinear operation to modify the brightness and contrast of digital images, Gamma correction can help in emphasising the breast lesions [18]. Because of its ability to normalise brightness and contrast across many images, Gamma correction plays a crucial role in avoiding overfitting. On the order side, Contrast Limited Adaptive Histogram Equalization (CLAHE) [24] is especially helpful in bringing out subtle details in medical images while maintaining overall image quality. CLAHE helps in reducing the noise while still improving the contrast, adjusting the contrast locally rather than globally, leading to clearer images.

The ViT model applies the transformer architecture from language processing to image analysis [10]. Unlike traditional methods that separate feature extraction and classification, ViT works end-to-end, handling everything in one model. It splits images into patches, converts them to vectors, and uses self-attention layers to learn important spatial and semantic features. Fine-tuning a pre-trained ViT adapts it to specific tasks, improving performance by adjusting the model based on the target dataset instead of training from scratch.

3.2   Datasets

MIAS Dataset. The MIAS dataset consists of images that contain normal tissue and tumors, with explicit classes malignant, benign, and normal [22]. The dataset contains 335 PNG images, partitioned into 68 benign, 51 malignant and 216 normal. All images have the size 1024 × 1024 pixels.

We used in our experiments a subset of the MIAS dataset which contains only the crops of benign and malignant images. Thus, the dataset consists of 119 images and two classes: benign and malignant.

DDSM Dataset. Curated Breast Imaging Subset DDSM (DDSM) is a dataset that contains mammograms from 6775 patients [19]. The images are categorized into two primary classes: calcification and mass. Each of these classes is further subdivided based on the view, either craniocaudal (CC) or mediolateral oblique (MLO). Subsequently, these categories are further classified into benign, malignant, and benign without callback.

For our approach, we decided not to include the mammograms that are from class benign without callback, as they do not require further immediate follow-up or additional diagnostic procedures, and the ones that contain both a benign and malign tumor, as they are not clear in which class to be included. After the exclusion of these cases, we have the number of images from each class as described in Table 5. For the experiments performed, we used the Region of Interest (ROI) from the images, which are already present in the dataset and their dimension them is variadic.

Table 5. DDSM - Number of images from each class after triage.

Abnormality	View	Tumor type	No of img
calcification	CC	benign	362
		malign	281
	MLO	benign	176
		malign	164
mass	CC	benign	347
		malign	342
	MLO	benign	554
		malign	552

3.3   Pre-processing

We explore the impact of preprocessing techniques on image classification performance in detecting breast cancer using state-of-the-art vision transformers. We
utilize two distinct datasets, referred to as the MIAS dataset and DDSM dataset.
Each dataset undergoes a structured preprocessing pipeline, involving the previously described preprocessing steps, denoted as Contrast Limited Adaptive
Histogram Equalization [24] and Gamma correction [18]. These preprocessing
techniques aim to enhance image quality and feature representation, thereby
improving classification model performance.

To systematically analyze the effects of preprocessing, we generate multiple
configurations for each dataset as in Fig. 1. The first configuration corresponds
to the original images. The second configuration consists of original images
enhanced by Gamma correction. The third configuration includes both the original images and the same images after the application of Gamma pre-processing
(it is the reunion of the first two subsets). Similarly, the fourth configuration consists of original images after CLAHE was applied, while the fifth configuration
extends the first by incorporating images pre-processed by CLAHE.

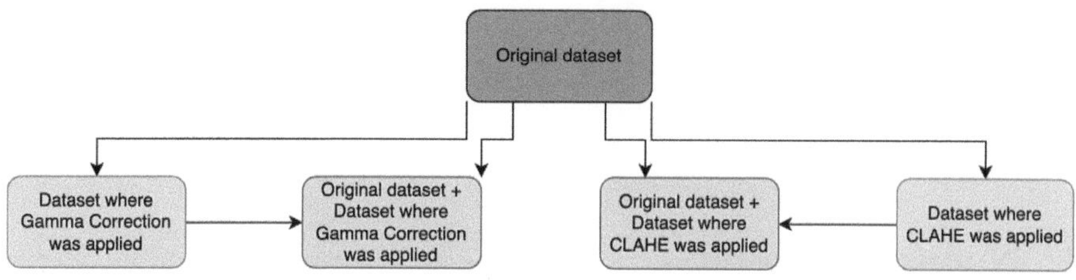

Fig. 1. Subsets creation for the dataset.

3.4   Experimental Setup

Following the dataset preprocessing and configuration creation, we employ a ViT model (vit_base_patch16_224) [10], which was pretrained on the extensive ImageNet dataset and fine-tuned on each configuration of a dataset, to assess its classification performance. We evaluate and compare model performance across the different configurations to determine the impact of preprocessing techniques CLAHE and Gamma correction. The results of our analysis provide insights into the effectiveness of data preprocessing strategies and augmentation of images in enhancing deep learning-based mammogram classification.

In the context of fine-tuning, the training process involves updating the ViT model's parameters to adapt it to a new dataset. This is achieved using an Adam optimiser [15], which is responsible for adjusting the weights of the model based on the computed gradients during backpropagation. By setting a learning rate, the optimiser determines the step size for each update, allowing the model to gradually learn the specific features and patterns of the new dataset. Fine-tuning is particularly effective when leveraging a pretrained model, as it starts with weights that have already captured general features from a large dataset, such as ImageNet. To ensure robust evaluation, a 5-fold cross-validation strategy was applied. The data was initially split into 80% for training and 20% for testing, and all procedures involving data shuffling and model initialization were conducted with controlled random seeds to ensure reproducibility. This approach enables the model to quickly converge to a solution that is well-suited for the target task, improving performance and reducing the time required for training compared to training a model from scratch. Figure 2 describes the pipeline used in our experiments.

After running multiple experiments with different learning rates, we considered that the learning rate for all experiments run is 0.00001 as it was giving good results. The number of epochs is variable for each experiment run as we used a mechanism of early stopping.

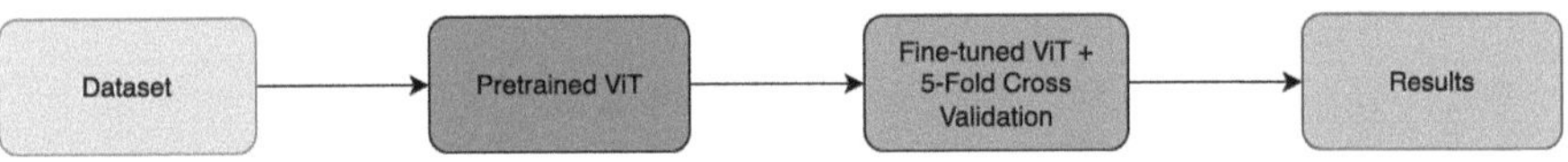

Fig. 2. Overview of the experimental pipeline.

4   Numerical Results

Our experiments aim to evaluate the performance of a ViT model on mammogram classification tasks using two datasets: MIAS and DDSM, with a focus on understanding the impact of dataset composition and clinical grouping. In this study, the malign class was considered the positive class, and the benign class was treated as negative. The dataset was split 80% for training and 20% for validation, and all results presented in the table are obtained on the validation set.

We first evaluate the performance of fine-tuned Vision Transformers on the full dataset, establishing a baseline for model effectiveness in a real-world, heterogeneous setting. This experiment provides a holistic view of ViT capabilities when exposed to the full spectrum of mammographic variability, without stratification by clinical attributes. It serves as a reference point for understanding how well the model generalises across diverse imaging conditions and lesion types.

Secondly, we analyse how fine-tuned Vision Transformers perform across clinically relevant subgroups, revealing differential sensitivity to lesion characteristics and imaging perspectives. This work provides a granular assessment of ViT-based classification, highlighting the importance of lesion-specific and view-specific performance in breast cancer imaging.

4.1   Results Obtained on MIAS Dataset

The results obtained for the MIAS dataset using a pretrained ViT model can be seen in Table 6.

Table 6. Performance of the ViT model on five configurations of the MIAS dataset

Pre-processing/Augmentation	ACC	SEN	SPC
Original	0.58	0.48	0.69
Gamma Correction	0.63	0.56	0.69
Original + Gamma Correction	0.99	0.99	0.99
CLAHE	0.54	0.50	0.59
Original + CLAHE	0.83	0.77	0.88

The baseline performance without any image pre-processing shows moderate specificity but lower accuracy and sensitivity. Applying Gamma correction improves both accuracy and sensitivity compared to no pre-processing, while specificity remains the same. Combining original images with Gamma-corrected images results in almost perfect accuracy, sensitivity, and specificity. This suggests that the diversity and the enlarged data volume introduced by combining both types of images significantly enhance the model's ability to correctly classify all samples.

Using CLAHE alone results in lower performance across all metrics compared to Gamma correction. This indicates that CLAHE may not be as effective in enhancing the features needed for accurate classification in this context. Combining original images with CLAHE-processed images improves performance significantly compared to using CLAHE alone. This suggests that the combination provides a beneficial diversity that enhances the model's ability to classify images correctly.

4.2   Results Obtained on DDSM Dataset

For the whole dataset, without pre-processing, the accuracy, sensitivity and specificity obtained are 0.7, 0.66 and 0.73. We want to dig deeper with our analysis and discover how the type of tumour abnormality and mammographic view influences breast cancer detection.

The results obtained in the experiments using a pretrained ViT and fine-tuned on the **DDSM calc CC subset** can be seen in Table 7. The baseline performance without any pre-processing shows moderate accuracy and specificity, but relatively low sensitivity. Applying Gamma correction improves all metrics slightly, with a notable increase in specificity. Combining original images with Gamma-corrected images significantly boosts all metrics, achieving high accuracy, sensitivity, and specificity. This indicates that the combination enhances the model's ability to classify images correctly across the board, making it highly effective. Using CLAHE results in a slight improvement in sensitivity compared to no pre-processing, with accuracy and specificity remaining similar. Combining original images with CLAHE-processed images results in very high performance across all metrics, similar to the combination with Gamma correction. This indicates that the diversity introduced by combining both types of images significantly enhances the model's classification capabilities.

Table 7. Performance of the ViT model on five configurations of the DDSM calc CC subset.

Pre-processing/ Augmentation	ACC	SEN	SPC
–	0.69	0.57	0.79
Gamma Correction	0.72	0.60	0.81
– + Gamma Correction	0.95	0.95	0.95
CLAHE	0.71	0.61	0.78
– + CLAHE	0.95	0.94	0.96

Table 8. Performance of the ViT model on five configurations of the DDSM calc MLO subset.

Pre-processing/ Augmentation	ACC	SEN	SPC
–	0.58	0.51	0.65
Gamma Correction	0.61	0.60	0.62
– + Gamma Correction	0.91	0.90	0.92
CLAHE	0.58	0.52	0.63
– + CLAHE	0.91	0.90	0.93

The results obtained by a pre-trained ViT and fine-tuned on the **DDSM calc MLO subset** are described in Table 8. The baseline performance without any pre-processing shows relatively low accuracy and sensitivity, with slightly better specificity. This indicates that the model struggles to correctly identify true positives, while it performs moderately better at identifying true negatives. Applying Gamma correction improves accuracy and sensitivity, bringing them closer to specificity. Combining original images with Gamma-corrected images significantly boosts all metrics, achieving high accuracy, sensitivity, and specificity. Using CLAHE results in similar performance to no pre-processing, with a slight improvement in sensitivity. Combining original images with CLAHE-processed images results in very high performance across all metrics, similar to the combination with Gamma correction. This indicates that the diversity introduced by combining both types of images significantly enhances the model's classification capabilities.

Table 9 presents the results obtained using the pretrained ViT model and fine-tuned on the **DDSM mass CC subset**. The baseline performance without any image enhancement shows moderate accuracy, sensitivity, and specificity. Applying Gamma correction results in a decrease in accuracy and sensitivity but an increase in specificity. Combining original images with Gamma-corrected images significantly boosts all metrics, achieving high accuracy, sensitivity, and specificity. This indicates that the combination enhances the model's ability to classify images correctly across the board. Using CLAHE results in slightly improved accuracy and specificity compared to no pre-processing. Combining original images with CLAHE-processed images results in high performance across all metrics, similar to the combination with Gamma correction. This indicates that the diversity introduced by combining both types of images significantly enhances the model's classification capabilities.

Table 9. Performance of the ViT model on five configurations of the DDSM mass CC subset.

Pre-processing/ Augmentation	ACC	SEN	SPC
–	0.74	0.71	0.76
Gamma Correction	0.71	0.60	0.82
– + Gamma Correction	0.95	0.96	0.95
CLAHE	0.75	0.70	0.80
– + CLAHE	0.95	0.96	0.94

Table 10. Performance of the ViT model on five configurations of the DDSM mass MLO subset.

Pre-processing/ Augmentation	ACC	SEN	SPC
–	0.77	0.75	0.77
Gamma Correction	0.72	0.73	0.70
– + Gamma Correction	0.95	0.94	0.95
CLAHE	0.72	0.74	0.72
– + CLAHE	0.95	0.94	0.95

DDSM calc MLO results using ViT are described in Table 10. The baseline performance without any pre-processing shows relatively balanced accuracy, sensitivity, and specificity. Applying Gamma correction results in a slight decrease in accuracy and specificity, with sensitivity remaining similar. Combining original images with Gamma-corrected images significantly boosts all metrics, achieving high accuracy, sensitivity, and specificity. This indicates that the combination enhances the model's ability to classify images correctly across the board, making it highly effective. Using CLAHE results in similar accuracy and specificity compared to Gamma correction, with a slight improvement in sensitivity. Combining original images with CLAHE-processed images results in very high performance across all metrics, similar to the combination with Gamma correction.

This indicates that the diversity introduced by combining both types of images significantly enhances the model's classification capabilities.

5   Conclusions

To answer **RQ1**, we notice that both Gamma and CLAHE pre-processings help especially in MIAS and DDSM calc cases in terms of accuracy and sensitivity. Furthermore, CLAHE pre-processing also helps for the DDSM mass CC case in terms of accuracy, contributing to **RQ1 RQ2**. Gamma pre-processing favours a better ViT performance than CLAHE for DDSM-calc, both CC and MLO views, and DDSM-mass, MLO view, compared with pre-processed subsets Gamma or CLAHE vs. original subsets (**RQ1, RQ2**). If we compare results obtained on Gamma-subsets with those obtained on CLAHE-subsets, a greater adaptability to intricate patterns is favoured by Gamma pre-processing (**RQ1**). ViTs are made to recognise intricate patterns and distant relationships in input images. ViTs may more easily learn and categorize the images thanks to Gamma correction, which intensifies these patterns without adding noticeable artefacts. The global patterns in the image may occasionally be disturbed by the localised contrast enhancement provided by CLAHE. ViTs may find it more difficult to recognise and make use of these patterns as a result of this disruption. If we compare the results obtained on the original subsets with those obtained on augmented ones, original with Gamma correction and original with CLAHE, respectively, in all tested scenarios the results are better on the augmented subsets. If we compare the results obtained on Gamma-augmented subsets with those obtained on CLAHE-augmented subsets, we can notice that the accuracy and sensitivity are better for the Gamma case, while the specificity is better just in the case of MIAS and DDSM mass, both CC and MLO views subsets.

This study investigated how image pre-processing techniques and imaging perspectives impact the effectiveness of ViT for mammography-based breast cancer classification. We showed that merging the original and enhanced images can improve the classification performance.

Building on these findings, future research could explore the integration of additional pre-processing techniques, such as denoising or deep learning-based enhancement, to further improve model robustness. Investigating the generalisability of these approaches across diverse datasets, as well as incorporating clinical metadata, may also enhance the clinical applicability of ViT-based models in breast cancer detection.

Acknowledgements. This study was supported by: The MCID, as Intermediary Body for the Operational Programme Competitiveness 2014-2020 project code SMIS 2014+ 127725, contract no. 352/390028/23.09.2021, INSPIRE; The MIPE as Managing Authority for the Smart Growth, Digitalisation and Financial Instruments Programme 2021 - 2027 and the MCID as Intermediary Research Body, project code SMIS 2021+ 324771, contract MIPE no. G-2024-71962/23.10.2024 and contract MCID no.390005/23.10.2024, INSPIRE-II.

References

1. Abd Elaziz, M., Dahou, A., Aseeri, A.O., Ewees, A.A., Al-Qaness, M.A., Ibrahim, R.A.: Cross vision transformer with enhanced growth optimizer for breast cancer detection in iomt environment. Comput. Biol. Chem. **111**, 108110 (2024)
2. Abimouloud, M.L., Bensid, K., Elleuch, M., Ammar, M.B., Kherallah, M.: Advancing breast cancer diagnosis: token vision transformers for faster and accurate classification of histopathology images. Visual Comput. Ind. Biomed, Art **8**(1), 1 (2025)
3. Al-Dhabyani, W., Gomaa, M., Khaled, H., Fahmy, A.: Dataset of breast ultrasound images. Data Brief **28**, 104863 (2020). https://doi.org/10.1016/j.dib.2019.104863
4. Al-Hejri, A.M., Al-Tam, R.M., Fazea, M., Sable, A.H., Lee, S., Al-Antari, M.A.: Etecadx: ensemble self-attention transformer encoder for breast cancer diagnosis using full-field digital x-ray breast images. Diagnostics **13**(1), 89 (2022)
5. Al Mansour, A.G., Alshomrani, F., Alfahaid, A., Almutairi, A.T.: Mammovit: a custom vision transformer architecture for accurate birads classification in mammogram analysis. Diagnostics **15**(3), 285 (2025)
6. Al-Tam, R.M., Al-Hejri, A.M., Alshamrani, S.S., Al-antari, M.A., Narangale, S.M.: Multimodal breast cancer hybrid explainable computer-aided diagnosis using medical mammograms and ultrasound images. Biocybern. Biomed. Eng. **44**(3), 731–758 (2024)
7. Al-Tam, R.M., et al.: A hybrid workflow of residual convolutional transformer encoder for breast cancer classification using digital x-ray mammograms. Biomedicines **10**(11), 2971 (2022)
8. Alsolami, A.S., Shalash, W., Alsaggaf, W., Ashoor, S., Refaat, H., Elmogy, M.: King abdulaziz university breast cancer mammogram dataset (kau-bcmd). Data **6**(11), 111 (2021). https://doi.org/10.3390/data6110111
9. Chawla, N.V., Bowyer, K.W., Hall, L.O., Kegelmeyer, W.P.: Smote: synthetic minority over-sampling technique. J. Artif. Intell. Res. **16**, 321–357 (2002). https://doi.org/10.1613/jair.953. https://www.jair.org/index.php/jair/article/view/10302
10. Dosovitskiy, A., et al.: An image is worth 16×16 words: transformers for image recognition at scale. arXiv preprint arXiv:2010.11929 (2020)
11. Gheflati, B., Rivaz, H.: Vision transformers for classification of breast ultrasound images. In: 2022 44th Annual International Conference of the IEEE Engineering in Medicine & Biology Society (EMBC), pp. 480–483. IEEE (2022)
12. Gutierrez-Cardenas, J.: Breast cancer classification through transfer learning with vision transformer, pca, and machine learning models. Int. J. Adv. Comput. Sci. Appl. **15**(4) (2024)
13. Jahan, I., et al.: Deep learning and vision transformers-based framework for breast cancer and subtype identification. Neural Comput. Appl. 1–20 (2025)
14. Jolliffe, I.T.: Principal Component Analysis. Springer Series in Statistics, 2nd edn. Springer, Heidelberg (2002). https://doi.org/10.1007/b98835
15. Kingma, D.P., Ba, J.: Adam: a method for stochastic optimization. arXiv preprint arXiv:1412.6980 (2014)
16. Moreira, I.C., Amaral, I., Domingues, I., Cardoso, A., Cardoso, M.J., Cardoso, J.S.: Inbreast: toward a full-field digital mammographic database. Acad. Radiol. **19**(11), 1364–1370 (2012). https://doi.org/10.1016/j.acra.2012.06.003
17. Muthukrishnan, R., Balasubramaniam, A., Krishnasamy, V., Ravichandran, S.K.: An efficient lightweight multi head attention gannet convolutional neural network based mammograms classification. Int. J. Med. Rob. Comput. Assist. Surg. **21**(1), e70043 (2025)

18. Pedregosa, F., et al.: Scikit-learn: machine learning in python. J. Mach. Learn. Res. **12**, 2825–2830 (2011)
19. Sawyer-Lee, R., Gimenez, F., Hoogi, A., Rubin, D.: Curated breast imaging subset of digital database for screening mammography (cbis-ddsm). In: International Research and Innovation Summit (IRIS2017) (2017)
20. Sickles, E.A.: Acr bi-rads® atlas, breast imaging reporting and data system. Am. Coll. Radiol. 39 (2013)
21. Sokolova, M., Lapalme, G.: A systematic analysis of performance measures for classification tasks. Inf. Process. Manag. **45**(4), 427–437 (2009). https://doi.org/10.1016/j.ipm.2009.03.002
22. Suckling, J., et al.: The mammographic image analysis society digital mammogram database. Int. Congr. Ser. **1069**, 375–378 (1994)
23. Wang, W., Jiang, R., Cui, N., Li, Q., Yuan, F., Xiao, Z.: Semi-supervised vision transformer with adaptive token sampling for breast cancer classification. Front. Pharmacol. **13**, 929755 (2022)
24. Zuiderveld, K.: Contrast limited adaptive histogram equalization. Graph. Gems **4**, 474–485 (1994)

Hierarchical Multi-task Siamese Networks for Vehicle Model Classification

Alexandru Manole[(✉)] and Laura Dioşan

Department of Computer Science, Babes-Bolyai University, Mihai Kogalniceanu nr.1, Cluj-Napoca, Romania
`{alexandru.manole,laura.diosan}@ubbcluj.ro`

Abstract. This work investigates a hierarchical Siamese multi-task learning approach for vehicle model similarity, using the StanfordCars dataset as a benchmark. The main contribution is the combined learning objective, which aims to jointly learn to compute the similarity between two, or more hierarchic concepts (i.e., car make comparison and the fine-grained car model similarity), a novel technique, especially for vehicle model classification. Two multi-task architectures are presented and evaluated: parallel and cascaded, which incorporate auxiliary objectives such as car make and car type similarity to enhance the main task of model-level similarity estimation. The experiments span nine established classifiers and examine both single-task and multi-task configurations. The results demonstrate that multi-task learning improves performance by up to 8.8% over single-task baselines. Notably, four of the top five configurations leverage a multi-task setup, with the Parallel Efficient-Net model, and make similarity objective achieving the highest accuracy (86.9%). Incorporating car make as an auxiliary objective consistently boosts model similarity accuracy by 1âĂŞ2%, and parallel architectures outperform cascade in most scenarios. Given the complexity of vehicle make and model classification (VMMC), a domain with hundreds of thousands of class variants and rapid model turnover, the findings support multi-task learning as a scalable solution. This is particularly relevant for intelligent transportation systems, where accurate recognition of vehicle types is essential for applications in traffic monitoring, driver assistance, and smart city applications.

Keywords: Multi-Task Learning · Siamese Networks · Vehicle Model Similarity

1 Introduction

Vehicle make and model classification (**VMMC**) is a complex compound task where the number of classes is extremely large and dynamic, with numerous new models appearing on the market every year. This problem is an essential step in all future intelligent traffic applications and could be used in driver assistance (predicting the behavior of cars based on model, as larger models may need more time to brake), surveillance, planning, and smart parking systems [5,7].

© The Author(s), under exclusive license to Springer Nature Switzerland AG 2026
C. Chira et al. (Eds.): InnoComp 2025, CCIS 2793, pp. 37–50, 2026.
https://doi.org/10.1007/978-3-032-12478-4_3

Traditionally, classifiers based on a convolutional neural network (**CNN**), a recurrent neural network (**RNN**) or a transformer were used for the task of VMMC [7]. However, using this otherwise established approach has a number of limitations in this field. Although high-volume data can be relatively easily obtained for VMMC (online traffic footage), annotating enough samples for a robust classifier capable of distinguishing between almost all car models used in the present is a tremendous task. Public datasets offer images of a limited number of car models (196 models in StanfordCars [5] and 4446 in CompCars [25]) compared to the estimated number of existing models, which is more than 400.000 [19]. The estimated number of different car models can vary significantly depending on the source and differentiation criteria used. This variability arises from several ambiguous scenarios, such as distinguishing between model years of the same vehicle, market-specific variants, ~~mid-cycle "facelifts"~~ and models with identical exteriors but different powertrains (e.g., internal combustion vs. electric versions). In addition, the high turnover rate of vehicle models, with new variants frequently introduced, even on a monthly basis, poses a significant challenge for traditional classification models, as they would require frequent retraining to stay up-to-date and effective.

Siamese networks [1,12] offer a robust alternative to traditional classification approaches by reformulating the direct categorization task. In this paradigm, the model is required to learn a similarity metric between input pairs such that the similarity is high for same-class samples and low for inputs representing different classes. This approach is particularly useful for addressing the ambiguity inherent in VMMC, where subtle differences, such as the ones mentioned above, can make rigid class definitions problematic. Rather than requiring explicit labels for every possible variant, Siamese networks learn to distinguish vehicles based on relative similarity. Additionally, their ability to generalize to unseen classes without retraining makes them highly adaptable in contexts characterized by high model turnover. At runtime, this is achieved by comparing the investigated sample against a small set of reference examples, eliminating the need to frequently retrain the model as new data become available.

The qualities of Siamese networks made them an impressive solution with effective results in other tasks such as: face verification [17], sentence similarity [15], handwriting author identification [22], malware detection [20], drug discovery through molecule similarity [9], and many more.

Fine-grained classification tasks, including vehicle make and model classification, also benefited from the efficiency of Siamese networks [6,11,16]. Although these approaches obtain impressive results, which push the boundaries of performance, achieving new state-of-the-art accuracies for VMMC, they do not exploit the hierarchic relation between model and make. To this end, this work proposes a novel Multi-Task Siamese Network that computes model similarity and make similarity simultaneously with the aim of embedding a better understanding of this concept in the feature extractor to effectively handle the challenges posed by high intra-class variance and inter-class similarity. The proposed architecture exploits the hierarchical relationship between objectives as the common encoder is tasked with obtaining features that can dis tangle the difference between car makes (coarse objective) and car model (fine-grained objective). The hierarchic characteristic of the framework is even more present in the cascaded multi-task, where initially the model predicts the similarity between car makes and then uses the result as part of the car model prediction. This paper aims to answer the following research questions: **RQ1:** *How is the performance of vehicle model similarity estimation influenced by the inclusion of vehicle make similarity as an additional objective?* **RQ2:** *How does the choice of multi-task learning architecture impact performance on the VMMC task?*

The structure of the rest of this paper is the following: Sect. 2 describes the related methods used for solving this task or other similar problems, Sect. 3 presents the proposed approach for the vehicle model and make classification task and Sect. 4 describes the conducted experiments and the analysis of their results. The paper is concluded in Sect. 5.

2   Backgrounds and Related Work

In this section, the two core paradigms used in the proposed approach are presented: Siamese networks and multi-task learning (**MTL**). Furthermore, existing methods that integrate these paradigms to address various tasks and highlight prior Siamese network-based solutions applied to VMMC are reviewed.

2.1   Siamese Networks

Siamese networks [1,12] are a class of neural networks designed to learn the similarity between pairs of inputs, which can have matching or non-matching classes. They consist of two subnets with shared weights that process input samples in parallel. The network learns to project inputs into a common embedding space where similar pairs are close together and dissimilar pairs are far apart, usually trained using contrastive or triplet loss [17].

The contrastive formulation, which is the base of our approach, can also be seen as a single encoder that processes two input samples to obtain their respective embeddings. These two embeddings are then concatenated and fed into a feedforward neural network (**FFNN**) which outputs the similarity score.

2.2   Multi-Task Learning

Multi-Task Learning (**MTL**) is a paradigm introduced in [2,3] in which the performance of a main objective is enhanced through the addition of symbiotic objectives. This can have a regulatory effect on the learning process and increase the generalization capability. Since its initial introduction, this design philosophy was adapted in Deep Learning, being successfully used in numerous tasks from the Computer Vision and Natural Language Processing domains [26].

In an MTL network, a common feature encoder is used to obtain embeddings which are then used for the two (or more) task-specific decision processes.

Depending on the structure of these layers multi-task architectures can be categorized as: parallel, cascaded, or interacted [4,26].

The form in which MTL was introduced is the parallel structure, where the embeddings are fed into two or more task-specific paths, usually in the form of MLPs. The parallel approach is most similar to the interacted MTL, also known as cross-talk multi-task. The difference between the two is that in this second architecture type, information is shared between the two previously parallel and independent paths. Lastly, the cascaded approach starts by solving one task and continues by using the result as an additional feature that is concatenated to the common embeddings and fed as input to the remaining tasks.

2.3   Multi-task Siamese Networks

Multi-task Siamese networks have been investigated in diverse vision applications, usually by combining classification objectives with similarity learning.

For example, in medical imaging, Wang et al. [21] proposed a Siamese multi-task model for retinal analysis that jointly classifies the type of vessel (artery vs. vein) and computes the similarity between the vessel segments. By combining the classification head with a Siamese similarity comparison head, their network could disentangle complex structures of retinal vessels, leading to a more robust understanding of the types of vessels. Another medical application was introduced in [24] in which similarity is calculated between breast mammogram images from a dual view input (craniocaudal and mediolateral-oblique mammograms). This task is combined with two classification heads, each required to identify masses in one of the two views. The proposed method generalizes well and is capable of reducing false positives.

In object tracking, Siamese networks often integrate multi-task learning to perform target classification and localization concurrently. This object tracking formulation uses an anchor image that shows the target object as input to the Siamese architecture. It additionally employs an image that displays a scene in which the object's region of interest must be located.

In [23] a Siamese-based tracker is presented that jointly trains an embedding similarity function and a correlation filter-based detection module. The proposed architecture leverages both the deep feature matching and the attention mechanism to exploit the benefits of both to better recognize similar objects. Similarly, Huang et al. [8] developed an end-to-end Siamese tracking network with a residual hierarchical attention mechanism and a learnable correlation-filter layer. In this case, the classification between the target object and the background is combined with a bounding-box regression. Both task-specific sub-networks exploit the rich semantic feature map obtained by attention layers in order to obtain state-of-the-art performance for real-time object tracking.

Beyond traditional recognition tasks, multi-task Siamese architectures have been applied to domains like physiological signal estimation and medical change detection. Lee et al. [14] introduced a Multi-Task Siamese model using twin 3D-CNN branches in facial regions to simultaneously predict cardiac pulse (photoplethysmography) and respiratory waveforms from video. By learning heart rate and breathing signals together from the same face video (with ROIs for forehead and cheek fed as two separate Siamese inputs), their network achieved accurate vital sign estimation while drastically reducing parameters compared to separate models.

Besides all applications presented above, multi-task Siamese networks hybrids are most often used for the task of change detection in high-resolution remote sensing data. In this task, the Siamese framework is employed, as this task requires comparing the same entity at two moments in time and identifying the changes. The MTL paradigm is also well-suited to this problem, as it allows the intelligent system to perform change detection gradually by initially obtaining scene-understanding information. For example, the authors of [18] presented a feature-constrained Siamese network (**SFCCD**) for identifying change detection maps that adds semantic segmentation constraints to the change detection task. By learning to classify land cover in each input image along with predicting the binary change mask, their model improved the performance of detected change regions and yielded more regular boundaries. Zheng et al. [23] developed MDESNet, a multitask difference-enhanced Siamese network to build change detection. MDESNet uses a shared Siamese encoder to produce a change map as the main output, supplemented by two auxiliary segmentation outputs for the buildings in the pre- and post-event images. Continuing this trend, recent work has further refined multi-task Siamese models for change detection. In [27], EGMS-Net is introduced. This architecture combines coarse-to-fine multi-task learning with Siamese feature fusion and attention mechanisms to improve change localization.

Unlike other works in the literature that combine similarity learning with a heterogeneous task, in [10], two similarity objectives are jointly optimized. A Siamese deep metric learning framework is proposed for age estimation based on facial images. Instead of directly classifying samples, the feature encoder is trained to embed the features of pairs of faces so that the decision subnetwork can distinguish between individuals based on age (if the age is close enough, two persons are considered as part of the same class). MTL is used to augment the main objective with an auxiliary gender similarity learning task to guide the shared features, improving age prediction accuracy.

2.4 Siamese Networks for Vehicle Make and Model Classification

Although MTL Siamese networks are not yet employed for this task, the impressive results obtained by single-task Siamese networks in other domains made this design a suitable choice for the classification of vehicles by make and model.

In [6] the authors propose a pairwise confusion (PC) loss to mitigate overfitting in fine-grained visual classification tasks, including vehicle images from Stanford Cars. Their approach trains a Siamese network on image pairs and minimizes the distance between their predicted class distributions. By introducing confusion between visually similar classes, the network avoids over-discriminative features and better generalizes to unseen data, increasing the accuracy due to its regulatory effect. The authors test this loss on 4 CNNs, and obtain the best performance using ResNet50, reaching a 93.43% accuracy, 1.72% being attributed to the addition of the PC loss.

The paper [16] offers a reinterpretation of Siamese networks through the lens of Riemannian geometry. They show that the standard contrastive loss has hidden symmetries that can hinder training, and propose a solution by embedding samples on two manifolds to capture the structure of the data better. This manifold-aware formulation leads to improved convergence behavior and higher accuracy in fine-grained classification tasks. Their work provides both conceptual insight into the geometry of contrastive learning and practical improvements over traditional Euclidean-space optimization. They compare their stochastic gradient descent policy with other approaches from the literature on the InceptionNet architecture, and achieve a 73.2% recall, exceeding the previously established state-of-the-art method by almost 1%. To our knowledge, it remains to be tested whether or not this Riemannian-based approach would increase the accuracy presented in [6].

Ji et al. [11] address label scarcity in fine-grained settings using Siamese networks. They generate two complementary views of each image via attention-based semantic part selection and random erasing. Their method enforces the alignment between embeddings using a Siamese network. A center loss further enforces intra-instance consistency, resulting in view-invariant representations without relying on manual annotations. The method achieves competitive performance on several fine-grained benchmarks, demonstrating that Siamese self-supervised learning can effectively extract discriminative features even in the absence of labeled data. On the Stanford Cars dataset, using a ResNet101 backbone, they reach an accuracy of 95.5%, increasing the performance of their baseline by around 3%.

Unlike other fields in which multi-task Siamese networks are extensively used, as seen in Subsect. 2.3, in vehicle model classification the problem is solved in a single-task setting. The research is conducted on the Stanford dataset in all three cases and uses mostly only CNNs (DenseNet, InceptionNet and ResNet), with the exception of a basic attention module. They obtain impressive performance reaching an accuracy of 95.5%.

One novelty of this work comes from the investigation of Siamese networks based on Transformer backbone, which offer enhanced capacity for modeling complex dependencies. Furthermore, while existing studies for the VMC problem typically address a single task in isolation, a multi-task learning framework that enables concurrent optimization of multiple objectives is proposed, leading to improved performance and robustness.

3   Proposed MTL Framework for Siamese Networks

This work investigates vehicle model classification using numerous Siamese networks built upon two types of classifiers: CNNs (MobileNetV3 small, MovileNetV3 large, ResNet18, ResNet50, EfficientNet, ConvNext base) and

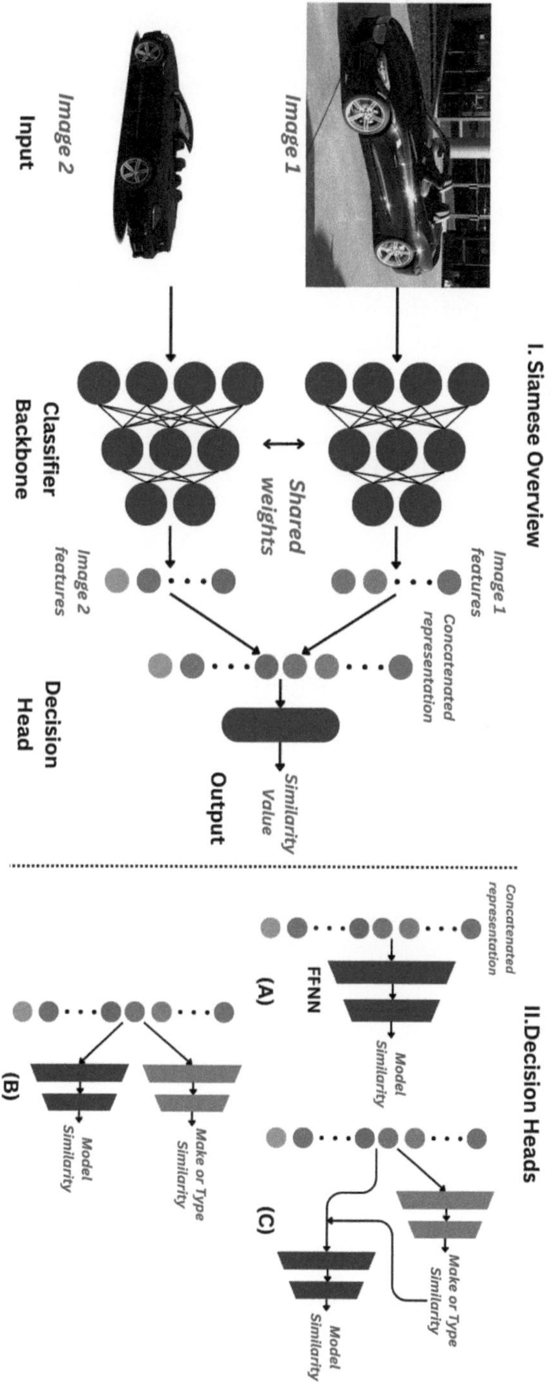

Fig. 1. (Left) Overview of the Siamese Network approach (Right) Decision heads for car model similarity, which in the MTL approach also contains car type similarity or car make similarity: (A) Single-task approach (B) Parallel approach with 2 objectives (C) Cascaded approach with 2 objectives.

Transformers (ViT, Swin). Furthermore, we experiment with the single-task architecture and two types of multi-task architectures, in the form of cascaded MT and parallel MT. The auxiliary tasks were chosen to enhance the performance of the car model similarity computation. The rationale behind the choice of car make and car type comes from our hypothesis which argues that understanding the general concept is a key factor in distinguishing between finer-grained aspects. Both the car make and car type label could easily be computed from the car model ground-truth allowing us to explore this idea without additional annotations.

Starting from one of the previously mentioned models pretrained on the ImageNet dataset, the last layer of the classifier is removed, which is specific to that dataset. Depending on the MT architecture, this MLP has different structures.

As seen in Fig. 1, each input image, I_1 and I_2, is first passed through the shared feature extractor $f(\cdot)$. This encoder produces dense embeddings for both images. To compare these two representations, they are concatenated to form a single input vector:

$$x_1 = f(I_1), \quad x_2 = f(I_2), \quad z = [x_1; x_2] \tag{1}$$

In the single-task architecture, the concatenated embedding z is passed through an FFNN to predict the similarity of the two images. The FFNN first reduces the combined representation to 512 features, which, after being passed through the dropout layer, is again distilled by a fully connected layer to a single value:

$$h = \mathrm{ReLU}(\mathrm{Linear}_1(z)) \quad \text{where } h \text{ has shape (batch\_size} \times 512) \tag{2}$$
$$h' = \mathrm{Dropout}(h) \tag{3}$$
$$s_{\mathrm{model}} = \mathrm{Sigmoid}(\mathrm{Linear}_2(h')) \tag{4}$$

In the parallel multi-task setup, the same concatenated embedding z is used by three separate branches to predict similarity for car model, car type, and car make, respectively. Depending on the experiment, we test all possible combinations: (car model, car make); (car model, car type), and (car model, car make, car type).

$$s_{\mathrm{model}} = \mathrm{Sigmoid}(\mathrm{Linear}_2^{\mathrm{model}}(\mathrm{Dropout}(\mathrm{ReLU}(\mathrm{Linear}_1^{\mathrm{model}}(z))))) \tag{5}$$
$$s_{\mathrm{type}} = \mathrm{Sigmoid}(\mathrm{Linear}_2^{\mathrm{type}}(\mathrm{Dropout}(\mathrm{ReLU}(\mathrm{Linear}_1^{\mathrm{type}}(z))))) \tag{6}$$
$$s_{\mathrm{make}} = \mathrm{Sigmoid}(\mathrm{Linear}_2^{\mathrm{make}}(\mathrm{Dropout}(\mathrm{ReLU}(\mathrm{Linear}_1^{\mathrm{make}}(z))))) \tag{7}$$

In the cascaded setup, similarity predictions for model and type are computed first using the base embedding z. These outputs are then concatenated to the feature vector. Similarly to the parallel scenario, we investigate all possible combinations. Below we present the complete one containing both secondary objectives:

$$z' = [x_1; x_2; s_{\text{model}}; s_{\text{type}}] \tag{8}$$

$$s_{\text{make}} = \text{Sigmoid}\big(\text{Linear}_2^{\text{make}}\big(\text{Dropout}\big(\text{ReLU}\big(\text{Linear}_1^{\text{make}}(z')\big)\big)\big)\big) \tag{9}$$

The model is trained using a combined loss, where each loss is the binary cross-entropy:

$$\mathcal{L}_{\text{total}} = \lambda_1 \mathcal{L}_{\text{model}} + \lambda_2 \mathcal{L}_{\text{type}} + \lambda_3 \mathcal{L}_{\text{make}} \tag{10}$$

These combined objectives are optimized using Adam, enhanced by the OneCycle scheduler with an initial learning rate of $3\text{e}-4$, which during the warm-up phase increases to $1\text{e}-3$, then slowly decreases. Throughout all experiments, the dropout value is 0.3, and the batch size is set to 32. Both λ values are set to 0.2. This value was fixed empirically during the early experimentation phase, without rigorously running all possible combinations of models and parameters; a task which remains as part of our future work.

4   Experiments

4.1   Dataset

The Stanford Cars dataset [13] contains 16,185 images of cars, divided into 8,144 training samples and 8,041 testing samples. Each image depicts a single car labeled with its make, model, and year. There are 196 model classes grouped under 49 car makes, with a balanced class distribution across both splits. Since the test set labels are not publicly available, only the training set is used and divided into training (70%), validation (20%), and test (10%) splits, as we were unable to find a public submission platform to evaluate on the original test set.

The images vary in resolution, typically around 360×240 pixels, and are captured in real-world settings with diverse angles and lighting, leading to high intra-class variation. Inter-class similarity is also challenging, especially among models from the same manufacturer. Despite these complexities, the dataset's hierarchical structure and in-the-wild characteristics make it well-suited for evaluating multi-task learning methods. All images are resized to 224×224.

4.2   Numerical Results

Table 1 presents a comparison of various backbone architectures evaluated across different combinations of auxiliary objectives (make and type). For each configuration, model accuracy is reported under three training setups: single-task (ST), parallel multi-task, and cascaded multi-task. The rows are grouped according to the presence of the auxiliary objectives, highlighting their effect on the accuracy at the model level. To facilitate the comparison, the single-task accuracy values are reported for all multi-task combinations (make objectives, type objective, and both make objective and type objective), although this metric is not influenced by the choice of the secondary objective (as the ST model does not use the secondary objective).

The first trend which can be observed is the superiority of the parallel MT architecture over the cascaded one in this experiment. In most cases, the parallel version outperforms the cascaded one with some notable exceptions. The cascaded SwinB achieves 74. 0%, while the parallel variant reaches 73.1%. The other main beneficiary is ConvNext Base which with the help of the cascaded make objectives increases the single task performance by 2% from 83.6% to 85.6%.

Next, almost all single-task models have an MT variant with better performance. One exception is Vit B 16, which does not appear suitable for this task on this data set, as even its single-task variant only achieves 60.3% accuracy in what is essentially a binary classification task. The other model, which drastically loses performance when the MT objective is attempted, is DenseNet121. SwinB, another transformer-based classifier, slightly benefits from the addition of the make secondary objective. In this case, the ST model achieves 72.5%, while parallel MT and cascaded MT yield accuracies of 73.1% and 74.0%. Whenever a secondary objective of car-type is present, either alone or in combination with the make, the Swin's performance suffers greatly. This is somehow similar to VitB suggesting that transformer-based architectures require better parameterization to function in a multi-task manner.

The MobileNet models are improved when the make objective is added in parallel to the main one, increasing the performance by around 1%. An interesting aspect is that the small variant of this architecture is the only one from this study which benefits from the addition of the car-type similarity objective. The other models become worse when car-type is added by itself or in combination with car-make. The best accuracy is achieved by parallel EfficientNet B0 which successfully uses the additional car make objective to reach an accuracy of 86.9%. Although this is the best performance, the model most improved by the MT approach is ResNet50, which has its accuracy increased from 66.8% to 75.6%.

These experiments offer answers to the research questions. For **RQ1**, it can be said that using two objectives is better than using all three objectives, especially when the secondary objective is the car make classification. Usually, this task yields better results when compared to car type classification (Parallel ResNet18, Cascaded MobileNet Large and Both MT DenseNet121 being the only exceptions). This is true when the secondary (and tertiary) objective weights are set to 0.2. It is possible that the car type or the combination of car type and car make could be more accurate if hyper-parameter optimization were performed on those variables.

In the case of **RQ2**, the results suggest that CNNs are better as the siamese backbone than Transformers. Moreover, they appear to react better to the addition of multi-task objectives. Overall, Parallel EfficientNet produces the best results, reaching 86.9% when the secondary make objective is added. This is a 2% increase over its single-task counterpart. ResNet50 is the most improved backbone with the addition of MTL, having its accuracy increased from 66.8% to 75.6%.

Table 1. Comparison of base models and auxiliary objectives on model test accuracy. Bold text represents the best overall performance for each type of model (ST, Parallel MT, Cascaded MT). Additionally, underlined results show the best results for each combination of tasks (model and make; model and type; make and type).

Base Model	Make Obj.	Type Obj.	ST Model Acc.	Parallel Model Acc.	Cascaded Model Acc.
MobileNet Small	Yes	No	79.7%	80.6%	79.7%
MobileNet Large			84.1%	85.00%	82.9%
ResNet18			79.4%	76.7%	76.8%
ResNet50			66.8%	75.6%	75.1%
DenseNet121			79.3%	51.6%	47.7%
EfficientNet B0			**84.8%**	**86.9%**	81.1%
SwinB			72.5%	73.1%	74.00%
Vit B 16			60.3%	50.1%	52.9%
ConvNext Base			83.6%	85.3%	**85.6%**
MobileNet Small	No	Yes	79.7%	74.8%	81.2%
MobileNet Large			84.1%	80.8%	81.1%
ResNet18			79.4%	79.6%	76.4%
ResNet50			66.8%	73.6%	49.9%
DenseNet121			79.3%	60.1%	59.6%
EfficientNet B0			84.8%	82.4%	81.4%
SwinB			72.5%	67.9%	57.1%
Vit B 16			60.3%	49.4%	51.0%
ConvNext Base			83.6%	82.2%	83.2%
MobileNet Small	Yes	Yes	79.7%	74.9%	51.2%
MobileNet Large			84.1%	81.2%	49.9%
ResNet18			79.4%	78.3%	50.0%
ResNet50			66.8%	54.5%	51.5%
DenseNet121			79.3%	48.8%	47.9%
EfficientNet B0			84.8%	83.7%	50.6%
SwinB			72.5%	49.1%	64.7%
Vit B 16			60.3%	53.1%	51.1%
ConvNext Base			83.6%	80.3%	74.9%

5   Conclusions and Future Work

This work investigated a hierarchical Siamese multi-task approach for image similarity. The analysis was focused on car similarity using the StanfordCars dataset. Two multi-task architectures: parallel and cascaded, and their effects on 9 established classifiers were investigated. Moreover, experiments were conducted with two possible additional objectives: car make similarity and car type similarity. The MT approach was evaluated with each objective individually and with both objectives together.

Experimental results show promising results, in which multitask learning improves the performance of single-task models by up to 8. 8%. Moreover, from the top five best overall results, 4 were obtained using an MT configuration: Parallel EfficientNet with make objective (86.9%), Cascaded ConvNextBase with make objective (85.6%), Parallel ConvNextBase with make objective (85.3%) and Parallel MobileNetV3 Large with make objective (85%).

The validity of the presented results could be strengthened in the future by addressing the following main limitations of our study: an investigation of the effects of the λ values and extension of each experiment with multiple runs using different splits to reduce the possible variance in the numerical results.

In our future work, we strive to further improve this approach and develop a more complex study by focusing on the following aspects: extend the study on datasets from other fields (i.e. medical); study cross-talk MT models; experiment more with the λ values or techniques which are able to autonomously obtain the best values for these weights; focus on the structure of the feedforward neural network, which separates the multi-task objectives: deeper FFNN to better understand task-specific features or allow the task-specific final layers to also share a part of the FFNN. Most importantly, we aim to apply the proposed MTL framework on state-of-the-art approaches to test whether this paradigm could be used to further improve the upper limits of performance in vehicle model classification.

Acknowledgement. This research is supported by the project "Romanian Hub for Artificial Intelligence - HRIA", Smart Growth, Digitization and Financial Instruments Program, 2021–2027, MySMIS no. 334906.

References

1. Bromley, J., Guyon, I., LeCun, Y., Säckinger, E., Shah, R.: Signature verification using a "siamese" time delay neural network. Adv. Neural Inf. Process. Syst. **6** (1993)
2. Caruana, R.: Multitask learning: a knowledge-based source of inductive bias1. In: Proceedings of the Tenth International Conference on Machine Learning, pp. 41–48. Citeseer (1993)
3. Caruana, R.: Multitask learning. Mach. Learn. **28**, 41–75 (1997)
4. Crawshaw, M.: Multi-task learning with deep neural networks: a survey. arXiv preprint arXiv:2009.09796 (2020)
5. Dehghan, A., Masood, S.Z., Shu, G., Ortiz, E., et al.: View independent vehicle make, model and color recognition using convolutional neural network. arXiv preprint arXiv:1702.01721 (2017)
6. Dubey, A., Gupta, O., Guo, P., Raskar, R., Farrell, R., Naik, N.: Pairwise confusion for fine-grained visual classification. In: Proceedings of the European Conference on Computer Vision (ECCV), pp. 70–86 (2018)

7. Gayen, S., Maity, S., Singh, P.K., Geem, Z.W., Sarkar, R.: Two decades of vehicle make and model recognition-survey, challenges and future directions. J. King Saud Univ.-Comput. Inf. Sci. **36**(1), 101885 (2024)

8. Huang, W., Gu, J., Ma, X., Li, Y.: End-to-end multitask Siamese network with residual hierarchical attention for real-time object tracking. Appl. Intell. **50**(6), 1908–1921 (2020)

9. Jeon, M., et al.: ReSimNet: drug response similarity prediction using Siamese neural networks. Bioinformatics **35**(24), 5249–5256 (2019)

10. Jeong, Y., Lee, S., Park, D., Park, K.H.: Accurate age estimation using multi-task Siamese network-based deep metric learning for frontal face images. Symmetry **10**(9), 385 (2018)

11. Ji, R., Li, J., Zhang, L.: Siamese self-supervised learning for fine-grained visual classification. Comput. Vis. Image Underst. **229**, 103658 (2023)

12. Koch, G., Zemel, R., Salakhutdinov, R., et al.: Siamese neural networks for one-shot image recognition. In: ICML Deep Learning Workshop, Lille, vol. 2, pp. 1–30 (2015)

13. Krause, J., Stark, M., Deng, J., Fei-Fei, L.: 3D object representations for fine-grained categorization. In: Proceedings of the IEEE International Conference on Computer Vision Workshops, pp. 554–561 (2013)

14. Lee, H., et al.: Multitask Siamese network for remote photoplethysmography and respiration estimation. Sensors **22**(14), 5101 (2022)

15. Mueller, J., Thyagarajan, A.: Siamese recurrent architectures for learning sentence similarity. In: Proceedings of the AAAI Conference on Artificial Intelligence, vol. 30 (2016)

16. Roy, S.K., Harandi, M., Nock, R., Hartley, R.: Siamese networks: the tale of two manifolds. In: Proceedings of the IEEE/CVF International Conference on Computer Vision, pp. 3046–3055 (2019)

17. Schroff, F., Kalenichenko, D., Philbin, J.: FaceNet: a unified embedding for face recognition and clustering. In: Proceedings of the IEEE Conference on Computer Vision and Pattern Recognition, pp. 815–823 (2015)

18. Shen, Q., Huang, J., Wang, M., Tao, S., Yang, R., Zhang, X.: Semantic feature-constrained multitask Siamese network for building change detection in high-spatial-resolution remote sensing imagery. ISPRS J. Photogramm. Remote. Sens. **189**, 78–94 (2022)

19. Tracy, D.: How many different models of car are there? (2024). https://www.theautopian.com/how-many-different-models-of-car-are-there/. Accessed 24 Mar 2025

20. Ullah, S., Oh, H.: BinDiff NN: learning distributed representation of assembly for robust binary diffing against semantic differences. IEEE Trans. Softw. Eng. **48**(9), 3442–3466 (2021)

21. Wang, Z., Jiang, X., Liu, J., Cheng, K.T., Yang, X.: Multi-task Siamese network for retinal artery/vein separation via deep convolution along vessel. IEEE Trans. Med. Imaging **39**(9), 2904–2919 (2020)

22. Xing, L., Qiao, Y.: DeepWriter: a multi-stream deep CNN for text-independent writer identification. In: 2016 15th International Conference on Frontiers in Handwriting Recognition (ICFHR), pp. 584–589. IEEE (2016)

23. Xuan, S., Li, S., Zhao, Z., Han, M.: The multi-task fully convolutional Siamese network with correlation filter layer for real-time visual tracking. In: Lin, Z., et al. (eds.) PRCV 2019. LNCS, vol. 11857, pp. 123–134. Springer, Cham (2019). https://doi.org/10.1007/978-3-030-31654-9_11

24. Yan, Y., Conze, P.H., Lamard, M., Quellec, G., Cochener, B., Coatrieux, G.: Towards improved breast mass detection using dual-view mammogram matching. Med. Image Anal. **71**, 102083 (2021)
25. Yang, L., Luo, P., Change Loy, C., Tang, X.: A large-scale car dataset for fine-grained categorization and verification. In: Proceedings of the IEEE Conference on Computer Vision and Pattern Recognition, pp. 3973–3981 (2015)
26. Zhang, Y., Yang, Q.: A survey on multi-task learning. IEEE Trans. Knowl. Data Eng. **34**(12), 5586–5609 (2021)
27. Zuo, X., et al.: Multi-task Siamese network guided by enhanced change information for semantic change detection in bi-temporal remote sensing images. IEEE J. Sel. Top. Appl. Earth Observ. Remote Sens. (2024)

Compact Real-Time Speech Enhancement for ONNX

Andrei Filip[1,2], Honorius Gâlmeanu[1,2(✉)], and Alexandru Drîmbărean[3]

[1] Department of Mathematics and Informatics, Transilvania University of Braşov,
Braşov, Romania
galmeanu@unitbv.ro
[2] FotoNation SRL, Braşov, Romania
[3] FotoNation LTD, Galway, Ireland
alexandru.drimbarean@fotonation.com

Abstract. Real-time speech enhancement is crucial for applications such as VoIP calls and hearing aids, which require low-latency noise suppression on resource-constrained devices. We present a causal speech enhancement model based on Facebook's Denoiser (a Demucs-inspired architecture), modified for deployment as an ONNX model. Key contributions include fixing the input to 2 s of audio at 16 kHz (32,000 samples), removing dynamic input length computation by padding to a constant length (32,085 samples), and trimming outputs accordingly during the training process. The model uses a convolutional encoder-decoder architecture with a gated LSTM bottleneck, trained on the CSTR VCTK Corpus+DEMAND (Valentini) dataset with data augmentation, and fine-tuned on the Microsoft Scalable Noisy Speech Dataset (MS-SNSD). The enhanced model achieves high speech quality, achieving results that match the state-of-the-art benchmarks for causal speech enhancement. The proposed model achieves PESQ 2.996 (vs. 3.07 for the original Denoiser), STOI 0.944 (vs. 0.95), CSIG 4.016, CBAK 3.453, and COVL 3.525 on the Valentini test set on par with state-of-the-art causal methods while remaining compatible with ONNX Opset 9. We also implemented a model based on a GRU bottleneck and observed a slight performance loss for offline processing, but better results during the online evaluation compared to the LSTM based model.

Keywords: Speech denoising · LSTM · GRU · U-Net · Deep CNN · ONNX

1 Introduction

Speech signals recorded in everyday environments often suffer from background noise that degrades intelligibility and overall quality. Speech enhancement aims to suppress noise and improve speech quality. In many applications (e.g., online meetings, telephony, hearing aids), an enhancement system must operate with low latency (causal, no future frames) and run efficiently on edge devices [11].

© The Author(s), under exclusive license to Springer Nature Switzerland AG 2026
C. Chira et al. (Eds.): InnoComp 2025, CCIS 2793, pp. 51–66, 2026.
https://doi.org/10.1007/978-3-032-12478-4_4

This requirement motivates the development of models that are both causal and easily deployable on a variety of platforms. An emerging standard for portable neural networks is ONNX (Open Neural Network Exchange), which allows models trained in frameworks like PyTorch to run on mobile or embedded hardware.

Facebook's Denoiser demonstrated that a U-Net convolutional architecture operating on raw waveforms can achieve a quality of speech enhancement comparable to state-of-the-art methods [1]. Their model, based on the Demucs source separation architecture, runs in real-time on CPU and attains excellent objective scores. However, the original implementation was not designed for ONNX export and had dynamic handling of input lengths. This work addresses these limitations by adapting the Denoiser architecture for fixed-size inputs and splitting the model to enable a specific ONNX Opset 9 compatibility.

We demonstrate that these changes allow for successful deployment in ONNX runtimes without a significant loss in quality. The model is trained on two standard datasets (Valentini [9] and MS-SNSD [7]) and evaluated with objective metrics (PESQ, STOI, CSIG, CBAK, COVL). Our results show that the model achieves competitive performance to other state-of-the-art enhancement systems while meeting real-time and deployment constraints.

Compared with existing Demucs architecture that we started from, the current work brings in the following original contributions:

- We modified the Denoiser architecture based on Demucs and created a lightweight causal model. In particular, we exchanged the original LSTM bottleneck with a Gate Recurrent Unit (GRU), which reduced the number of parameters by 12.5%, also reducing the computational load while maintaining the ability to learn temporal dependencies. The real-time lag was improved from 10.1 ms to 7.9 ms, while the RTF decreased by 20%;
- We developed a simpler two-step model training formula using 2 readily available datasets: i) the pre-training on the Valentini dataset, composed of recordings of 56 separate speakers (28 men and 28 women) with various background sounds added later, characteristic of everyday environments (living room, office, public transportation, cafe or public square), and ii) continuing training on the MS-SNSD dataset, composed of the same set of 56 separate speakers and configured with several types of background sound (air conditioning, household appliances, car noise, printer, door closing, chewing, babbling of several speakers, furniture creaking, traffic, keyboard sounds) and several SNR levels of the added noise;
- We trained this modified model using different training setups in order to observe and quantify the impact of the data augmentation strategy used, and to observe the performance trade-offs introduced by using a GRU bottleneck;
- For inference, we split the model into 3 modules (encoder, bottleneck, decoder) and exported each one using ONNX opset 9. Currently, the internal toolchain compatibility is with ONNX opset 9, so we were specifically constrained by that. We wanted thus to perform the optimization based on this particular opset.
- The entire implementation of this project can be found on GitHub[1].

---

[1] https://github.com/GitStroberi/demucs-onnx.

The rest of this paper is structured as follows. In Sect. 2 we present the most relevant works in waveform-domain speech enhancement domain. Section 3 details our model architecture, including the Demucs-inspired encoder, the causal LSTM bottleneck, and the model of the decoder with skip connections. It also contains our adaptations for ONNX Opset 9: the split encoder/LSTM/decoder design and flexible window-length handling. Section 4 outlines the experiments and the results, showing the training process of the pipeline, the datasets and the augmentations used, as well as the hyperparameters. We also introduce here the combined time-domain and multi-resolution STFT loss function, as well as the evaluation metrics and experimental protocol used. Finally, Sect. 5 concludes the article and offers some directions for future work.

2   Related Work

Early speech enhancement methods used signal processing techniques (spectral subtraction, Wiener filtering) to estimate noise and enhance speech, but struggled with non-stationary noises. Deep learning methods have since achieved superior performance on diverse noise types [1].

Notably, SEGAN [6] introduced a GAN-based enhancement model operating end-to-end on waveforms, using a CNN encoder-decoder to learn a mapping from noisy to clean speech. SEGAN achieved significant quality improvements (e.g. PESQ 2.16 for clean vs. 1.97 for noisy) over traditional methods by learning non-linear noise suppression.

RNNoise [10] took a different approach, combining a classic DSP pipeline with a small recurrent network. It uses an LSTM to estimate spectral gains in critical bands, yielding a lightweight noise suppressor that runs in real-time on CPUs with very low complexity. RNNoise proved that even a tiny RNN can reduce the noise effectively.

. Conv-TasNet [5] pioneered fully-convolutional time-domain speech separation. It uses a learned encoder and decoder with a temporal convolutional network (TCN) in between. Conv-TasNet was originally developed for speech separation, but its principles have been applied to enhancement. It achieves extremely low latency (frame sizes of a few milliseconds) and outperforms traditional time-frequency masking.

Facebook's Demucs [2] is a U-Net convolutional architecture initially designed for music source separation. Denoiser is an adaptation of Demucs for speech enhancement [1]. It features an encoder/decoder convolutional stack with skip connections and a bidirectional LSTM as the bottleneck. The model is trained on the waveform domain with a combination of time and frequency losses. This approach sets a new state-of-the-art for speech enhancement, matching or exceeding prior methods like SEGAN and even non-causal models in objective metrics.

3   Proposed Model

3.1   Model Architecture

The model follows an encoder - LSTM - decoder architecture with skip connections, inspired by the Demucs U-Net. The input is a 1-channel noisy waveform, and the output is a 1-channel enhanced waveform of the same length. The entire model is causal: it does not incorporate future context at any layer. We present the detailed model in Table 1.

Table 1. Layer-by-layer summary of the causal Denoiser model.

Layer	Input Shape (C × L)	Output Shape (C × L)	#Params	Kernel Shape (C_O × C_I × K)
Denoiser	1 × 32,085	1 × 32,085	18,867,937	
ModuleList			4,709,616	
Sequential			5,136	
Conv1d	1 × 128,340	48 × 32,084	432	48 × 1 × 8
ReLU	48 × 32,084	48 × 32,084	0	
Conv1d	48 × 32,084	96 × 32,084	4,704	96 × 48 × 1
ModuleList			4,708,849	
Sequential			5,089	
GLU ($skip_1 \rightarrow$)	96 × 32,084	48 × 32,084	0	
ModuleList			4,709,616	
Sequential			55,584	
Conv1d	48 × 32,084	96 × 8,020	36,960	96 × 48 × 8
ReLU	96 × 8,020	96 × 8,020	0	
Conv1d	96 × 8,020	192 × 8,020	18,624	192 × 96 × 1
ModuleList			4,708,849	
Sequential			5,089	
GLU ($skip_2 \rightarrow$)	192 × 8,020	96 × 8,020	0	
ModuleList			4,709,616	
Sequential			221,760	
Conv1d	96 × 8,020	192 × 2,004	147,648	192 × 96 × 8
ReLU	192 × 2,004	192 × 2,004	0	
Conv1d	192 × 2,004	384 × 2,004	74,112	384 × 192 × 1
ModuleList			4,708,849	
Sequential			5,089	
GLU ($skip_3 \rightarrow$)	384 × 2,004	192 × 2,004	0	
ModuleList			4,709,616	
Sequential			885,888	
Conv1d	192 × 2,004	384 × 500	590,208	384 × 192 × 8

(*continued*)

Table 1. (*continued*)

Layer	Input Shape (C × L)	Output Shape (C × L)	#Params	Kernel Shape (C_O × C_I × K)
ReLU	384 × 500	384 × 500	0	
Conv1d	384 × 500	768 × 500	295,680	768 × 384 × 1
ModuleList			4,708,849	
Sequential			5,089	
GLU ($skip_4 \rightarrow$)	768 × 500	384 × 500	0	
ModuleList			4,709,616	
Sequential			3,541,248	
Conv1d	384 × 500	768 × 124	2,360,064	768 × 384 × 8
ReLU	768 × 124	768 × 124	0	
Conv1d	768 × 124	1,536 × 124	1,181,184	1,536 × 768 × 1
ModuleList			4,708,849	
Sequential			5,089	
GLU	1,536 × 124	768 × 124	0	
LSTM	768 x 124	768 x 124	9,449,472	
ModuleList			4,708,849	
Sequential			3,540,864	
Conv1d	768 × 124	1,536 × 124	1,181,184	1,536 × 768 × 1
Sequential			5,089	
GLU	1,536 × 124	768 × 124	0	
Sequential			3,540,864	
ConvTranspose1d	768 × 124	384 × 500	2,359,680	768 × 384 × 8
ReLU	384 × 500	384 × 500	0	
Sequential			885,696	
Conv1d	384 × 500	768 × 500	295,680	768 × 384 × 1
Sequential			5,089	
GLU	768 × 500	384 × 500	0	
Sequential			885,696	
Add ($skip_4 \leftarrow$)	384 × 500	384 × 500	0	
ConvTranspose1d	384 × 500	192 × 2,004	590,016	384 × 192 ×8
ReLU	192 × 2,004	192 × 2,004	0	
Sequential			221,664	
Conv1d	192 × 2,004	384 2,004	74,112	384 × 192 × 1
Sequential			5,089	
GLU	384 × 2,004	192 × 2,004	0	
Sequential			221,664	
Add ($skip_3 \leftarrow$)	192 × 2,004	192 × 2,004	0	

(continued)

Table 1. (continued)

Layer	Input Shape (C × L)	Output Shape (C × L)	#Params	Kernel Shape (C_O × C_I × K)
ConvTranspose1d	192 × 2,004	96 × 8,020	147,552	192 × 96 × 8
ReLU	96 × 8,020	96 × 8,020	0	
Sequential			55,536	
Conv1d	96 × 8,020	192 × 8,020	18,624	192 × 96 × 1
Sequential			5,089	
GLU	192 × 8,020	96 × 8,020	0	
Sequential			55,536	
Add ($skip_2$ ←)	96 × 8,020	96 × 8,020	0	
ConvTranspose1d	96 × 8,020	48 × 32,084	36,912	96 × 48 × 8
ReLU	48 × 32,084	48 × 32,084	0	
Sequential			5,089	
Conv1d	48 × 32,084	96 × 32,084	4,704	96 × 48 × 1
GLU	96 × 32,084	48 × 32,084	0	
Add ($skip_1$ ←)	48 × 32,084	48 × 32,084	0	
ConvTranspose1d	48 × 32,084	1 × 128,340	385	48 × 1 × 8

The encoder consists of $L = 5$ convolutional blocks. Each one employs a 1-D convolution that downsamples the time resolution and increases the feature channels. The first encoder layer takes the input waveform (16 kHz) and upsamples it by a factor of 4 using a sinc interpolation filter to 64 kHz. This high-resolution signal is then passed through a series of convolutional blocks. Each block i performs:

- Conv1d(ch_in, ch_out, kernel_size=8, stride=4, padding=0), followed by ReLU
- 1×1 Conv1d expanding to $2 \times$ channels,
- Gated Linear Unit (GLU) activation which halves the number of channels.

The initial number of channels is $H = 48$, and it grows by a factor of 2 at each layer (i.e., 48, 96, 192, 384, 768). Because of the $stride = 4$ in each Conv1d, the time length is reduced by 4× per layer. After 5 layers, the temporal dimension is greatly compressed, and the feature dimension is high (768 channels). Each encoder block's output (after GLU) is stored as a skip connection to be used in the decoder.

The output of the final encoder layer is a sequence with 768 features. This is fed into a two-layer LSTM (long short-term memory recurrent network) with hidden size 768. The LSTM processes the sequence causally (no bidirectional context) and outputs a sequence of the same length. The LSTM enables the model to capture long-range dependencies beyond the convolutional receptive fields. Notably, this LSTM is the only part of the model that carries state across time; all other layers are feed-forward on the fixed input segment.

The decoder mirrors the encoder, using transpose convolutions to upsample back to the original sampling rate and length. It takes the LSTM output (768 channels) as initial input. Then for each of the 5 decoder layers (from deepest to shallowest):

- A 1×1 convolution reduces the feature dimension (inverse of the encoder's 1×1 expansion) with GLU activation.
- A ConvTranspose1d (transposed convolution) with kernel size 8 and stride 4 upscales the time resolution by $4\times$.
- The stored skip output from the corresponding encoder layer is added (element-wise) to the decoder's feature map before the transpose convolution (a U-Net skip connection). This helps retain fine details from earlier high-resolution features.
- A ReLU is applied after each decoder except the last layer.

Because the model upsamples by $4\times$ at the input and downsamples by $4\times$ at output (to revert 64 kHz back to 16 kHz), we ensure that the final output length matches the original input length. By design, when the input length is a specific valid length (32,085 samples, see next section), the decoder output exactly equals the input length. Finally, the enhanced waveform at 16 kHz is obtained.

Except for GLUs in each block and the final output (which is linear), ReLU activations are used throughout the modules. No explicit sigmoid or tanh is needed on output since the model directly predicts the time-domain signal.

3.2 ONNX Opset 9 Adaptation

Unlike conventional ONNX Pytorch-exported models requiring fixed input lenghts, our model is designed to handle varying input durations dynamically, accommodating practical trade-offs between latency and quality. Initially, we configured the model to process a fixed window of 2 s (32,000 samples at 16 kHz), which yielded optimal speech enhancement performance but introduced a relatively high latency. However, by splitting the network into three components, those being the encoder, LSTM bottleneck and decoder, we achieve a flexible window size at deployment time.

Specifically, the encoder and decoder modules exported to ONNX opset 9 support arbitrary lengths, constrained only by ensuring symmetry between down-sampling and upsampling operations. We calculate an appropriate input length for each desired window duration using a helper function, ensuring that internal convolution operations align correctly. This approach allows the selection of short processing windows (following the original Denoiser's latency target for real time) [1] to significantly reduce latency. Short windows provide faster, more responsive real-time inference but may slightly impact enhancement quality due to reduced context.

At inference, each audio chunk is encoded separately and the LSTM bottleneck state is maintained and carried forward between successive chunks. This ensures temporal continuity and a causal processing. Consequently, shorter windows permit lower algorithmic latency with minimal loss in performance, enabling deployments tailored to the latency requirements of different applications.

4 Experiments and Performance

4.1 Loss Function

We optimize a weighted combination of time-domain and frequency-domain losses, similar to the original implementation:

$$\mathcal{L}_{\ell_1}(\hat{y}, y) = \frac{1}{T} \sum_{t=1}^{T} |\hat{y}_t - y_t| \tag{1}$$

$$\mathcal{L}_{\text{SC}}^{(r)}(\hat{X}^{(r)}, X^{(r)}) = \frac{\|\hat{X}^{(r)} - X^{(r)}\|_F}{\|X^{(r)}\|_F} \tag{2}$$

$$\mathcal{L}_{\text{Mag}}^{(r)}(\hat{X}^{(r)}, X^{(r)}) = \frac{1}{F_r T_r} \sum_{f=1}^{F_r} \sum_{t=1}^{T_r} |\log \hat{X}_{f,t}^{(r)} - \log X_{f,t}^{(r)}| \tag{3}$$

$$\mathcal{L}_{\text{MRSTFT}}(\hat{y}, y) = \alpha \frac{1}{R} \sum_{r=1}^{R} \mathcal{L}_{\text{SC}}^{(r)} + \beta \frac{1}{R} \sum_{r=1}^{R} \mathcal{L}_{\text{Mag}}^{(r)} \tag{4}$$

$$\mathcal{L}_{\text{combined}}(\hat{y}, y) = \mathcal{L}_{\ell_1}(\hat{y}, y) + \mathcal{L}_{\text{MRSTFT}}(\hat{y}, y) \tag{5}$$

where:

- T is the number of time-domain samples,
- $\hat{y}, y \in \mathbb{R}^T$ are the predicted and the true waveforms,
- for each resolution $r = 1, \ldots, R$, $X^{(r)}, \hat{X}^{(r)} \in \mathbb{R}^{F_r \times T_r}$ are the target and predicted STFT magnitude matrices,
- $\| \cdot \|_F$ is the Frobenius norm,
- F_r, T_r the frequency and time-frame counts at resolution r,
- α and β are the SC/magnitude weighting factors.

The following loss functions were used:

- *Waveform L1 Loss*: $\mathcal{L}_{\ell_1}(\hat{y}, y)$ is the mean absolute error between the enhanced waveform and the clean target. L1 loss in waveform domain encourages the enhanced signal to closely match the clean signal sample-by-sample, which correlates with reducing overall distortion;
- *Multi-Resolution STFT Loss*: $\mathcal{L}_{\text{MRSTFT}}(\hat{y}, y)$ measures the difference between the enhanced $\hat{X}_{f,t}^{(r)}$ and the clean audio signal $X_{f,t}^{(r)}$ in the time-frequency domain. We compute the STFT (Short-Time Fourier Transform) magnitudes at multiple resolutions using different FFT sizes and hop lengths.

The total loss $\mathcal{L}_{\text{combined}}(\hat{y}, y)$ is a sum of the L1 waveform loss $\mathcal{L}_{\ell_1}(\hat{y}, y)$ and all STFT losses $\mathcal{L}_{\text{MRSTFT}}(\hat{y}, y)$. The L1 waveform loss directly minimizes per-sample amplitude errors, promoting accurate waveform reconstruction and robust convergence, and the multi-resolution STFT loss enforces coherence of both fine temporal transients and broad spectral structure, and prevents overfitting to a single timeâĂŞfrequency representation. [12].

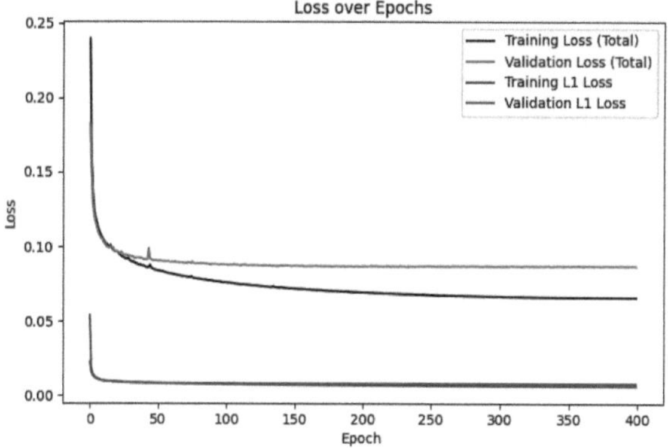

(a) Training on the Valentini dataset without augmentation.

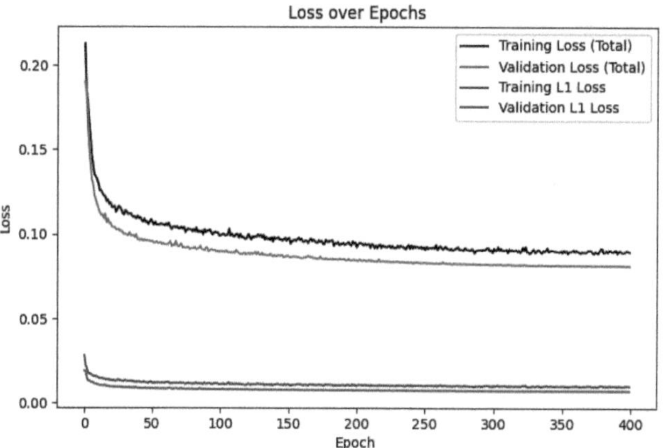

(b) Training on the Valentini dataset with augmentation.

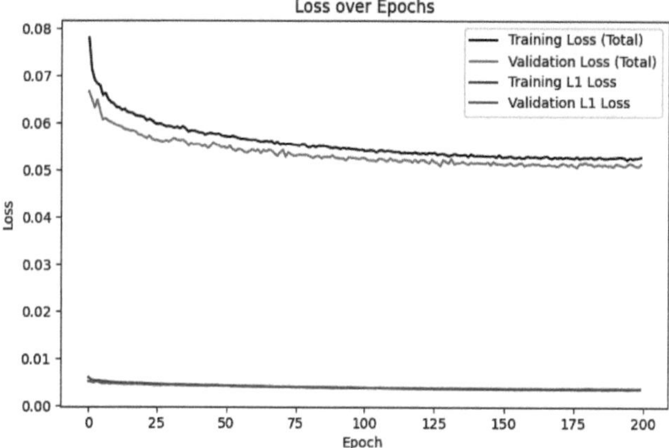

(c) Fine-tuning training on the MS-SNSD dataset continuing training 1b.

Fig. 1. Plot of the L1 and combined loss functions during training.

4.2   Training Pipeline

We train the model in two stages. First, we use the CSTR VCTK Corpus + DEMAND (Valentini-Botinhao dataset) which is a standard dataset for supervised speech enhancement. We then fine-tune the model on the Microsoft Scalable Noisy Speech Dataset (MS-SNSD). Fine-tuning on MS-SNSD helps the model generalize to more diverse real-world noises beyond the Valentini set.

All audio is resampled to 16 kHz and segmented into 2 s chunks. For the Valentini dataset, we pad the given segments with zeroes if they do not match the input size exactly, otherwise if the segments are too long, we use an identic randomly cut 2 s window from the noisy and clean samples. For MS-SNSD, we generated clean audio clips of at least 2 s in length, each being paired with 5 noisy samples with various added noise conditions at SNR levels of 0, 10, 20, 30 and 40 dB.

We applied data augmentation to the training data to improve robustness, following techniques similar to those in the original Denoiser implementation. The augmentations we used consist of:

- *Reverberation*: this augmentation technique adds a series of decaying echoes to the noisy sample in order to simulate different room environments that would potentially introduce echoing noise. This type of noise is particularly hard to isolate because it can contain the formant characteristics of the human voice;
- *BandMask*: we apply random frequency band dropouts on the noisy input, zeroing out random frequency bands. This augmentation would prevent the model from being dependant on certain frequency bands, and would use the whole spectrum to reconstruct the processed audio;
- *Time shift*: a slight random time shift was applied to each pair of noise and clean sound, essentially by cutting a random starting offset for the noise relative to the clean sound or by shifting one relative to the other by a small amount. This ensures that the model does not depend too much on the alignment of the noise relative to the clean recording;
- *Remix*: we randomly mix the noise within a batch in order to create new mixtures. This would further prevent the model from associating certain noises with particular clean samples.

We train the base model on the Valentini dataset for 400 epochs. A high epoch count was used since the dataset is relatively small; we observed the loss plateauing well before 400 epochs (around 300) but continued to ensure full convergence, as seen in Figs. 1a and 1b. We used the Adam optimizer with an initial learning rate of $1e-3$. After the Valentini dataset training, we fine-tuned on MS-SNSD for 200 epochs as can be seen in Fig. 1c, which allowed the model to adapt to new noise without overfitting the smaller Valentini dataset corpus characteristics.

After fine-tuning on MS-SNSD, we obtained a final model. We evaluated 4 model variants, trained in different scenarios: the model trained on the Valentini dataset with and without augmentations (to measure augmentation effect), and then fine-tuned on MS-SNSD with and without augmentations, starting from the model trained on Valentini with augmentations. For each variant, the checkpoint with lowest validation loss was chosen for evaluation.

4.3   Evaluation Metrics and Setup

We evaluated the enhanced speech using standard objective metrics widely reported in the speech enhancement literature. In order to make a proper performance assessment of the proposed model, we computed the same metrics as in [1], and used the reports in the original article as a reference point. These are as follows:

- *PESQ (Perceptual Evaluation of Speech Quality)*: this is an objective score (ITU-T P.862) [4] that predicts the mean opinion score (MOS) of speech quality as heard by a human. PESQ ranges from -0.5 to 4.5 (higher is better, with 4.5 corresponding to clean reference quality). It is sensitive to speech distortions and correlates reasonably with human judgments of quality;
- *STOI (Short-Time Objective Intelligibility)*: STOI is a measure of speech intelligibility (Taal et al., 2011) [8]. It is computed by comparing temporal envelope patterns of clean and enhanced speech in short overlapping segments. STOI ranges from 0 to 1 (higher is better), and values above 0.95 indicate near-transparent intelligibility. STOI is known to correlate with the ability of listeners to understand speech, especially in noise;
- *CSIG, CBAK, COVL (composite measures)*: these are the composite MOS measures proposed by Hu and Loizou [3] to approximate human ratings for signal distortion, background intrusiveness, and overall quality, respectively. They are computed as weighted combinations of basic objective criteria (like segmental SNR, PESQ, LLR). CSIG (signal distortion MOS) indicates the perceived cleanliness/naturalness of the target speech (5.0 is best, corresponding to no distortion). CBAK (background noise MOS) reflects how intrusive the remaining background noise is (5.0 means no noise). COVL (overall quality MOS) is an overall quality predictor.

The evaluation was performed in the following way: we used the test set from the Valentini dataset by running each clip through the model. The output is trimmed to the original length and aligned with the clean reference. We compute PESQ and STOI for each utterance by comparing it with the clean reference, then average across all test utterances to get the final score. CSIG, CBAK, COVL are computed using the composite formula on per-utterance values and then averaged. We used the implementation from the PySEPM toolkit[2].

4.4   Offline Configuration Evaluation Results

This experimental setup will use the model with a fixed input window size of 30,085 samples at 16 kHz, in order to evaluate the performance on windows

---

[2] https://github.com/schmiph2/pysepm.

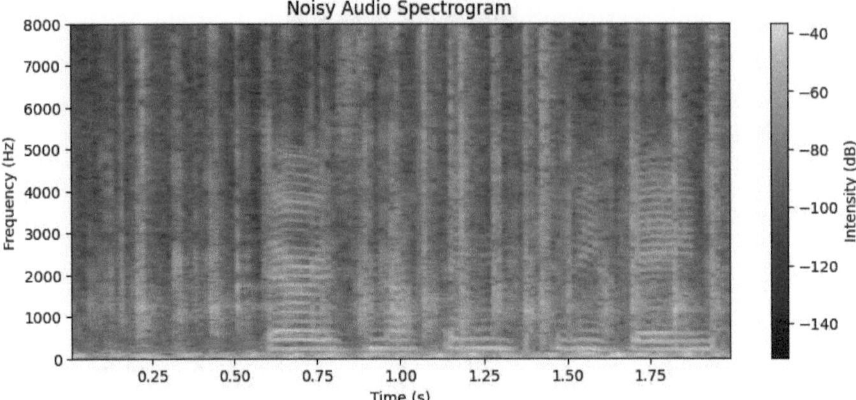

(a) **Noisy input:** clean speech mixed with keyboard clicks. The broadband transients of each keystroke appear as sharp, vertical high-energy bursts that disrupt the underlying harmonic bands.

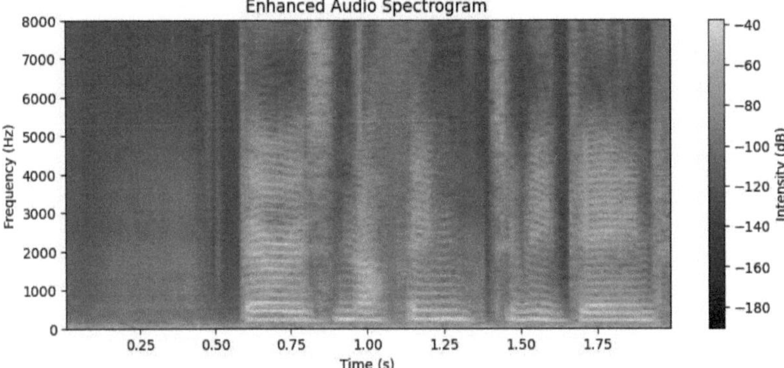

(b) **Enhanced output:** vertical bursts are substantially attenuated, while the formant trajectories associated with human speech remain clearly defined.

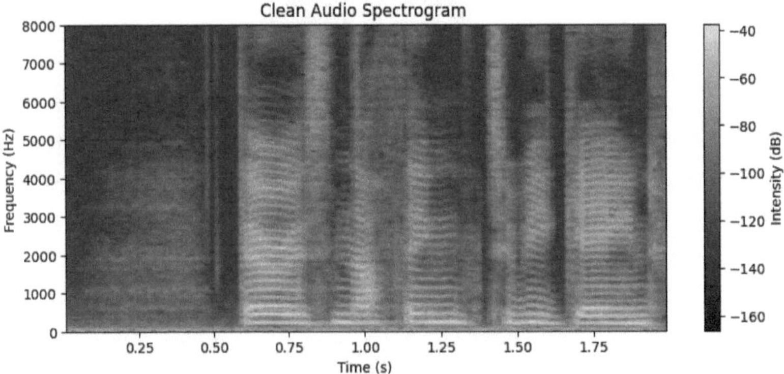

(c) **Clean reference:** the original speech signal without noise, showing continuous harmonic bands, and serving as the ground truth.

Fig. 2. Spectrograms for a representative utterance from the test set.

of the same audio clip length as the training data. The proposed variant used in these evaluations was initialized with the following model hyperparameters (H: initial number of hidden channels, S: stride, U: resampling factor) (H, S, U) = (48, 4, 4), while the original Denoiser model for offline processing used (H, S, U) = (64, 2, 2). This was done in order to analyze if the more compact architecture meant for causal inference can provide comparable results with the original setup.

We report the average metrics for four model variants in Table 2:

(A) *Valentini (no aug)*: trained on the Valentini dataset without augmentations;
(B) *Valentini (aug)*: trained on the Valentini dataset with augmentations;
(C) *MS-SNSD (no aug)*: model (B) further trained on MS-SNSD without augmentations;
(D) *MS-SNSD (aug)*: model (B) further trained on MS-SNSD with augmentations.

Table 2. Objective evaluation results. Higher scores indicate better speech quality/intelligibility. "Facebook Denoiser" refers to the original causal model used in [1].

Model	PESQ	STOI	CSIG	CBAK	COVL
Proposed (Valentini no aug)	2.828	0.940	3.912	3.333	3.377
Proposed (Valentini aug)	2.919	0.943	4.106	3.404	3.532
Proposed (MS-SNSD no aug)	2.906	0.941	3.918	3.381	3.430
Proposed (MS-SNSD aug)	2.996	0.944	4.016	**3.453**	**3.525**
Facebook Denoiser	**3.07**	**0.95**	**4.31**	3.40	3.63

In addition to the original architecture, we explored the option of replacing the LSTM bottleneck with a GRU bottleneck with the same configuration, in order to observe how a simpler bottleneck could affect the objective metric results. We present the results of this experiment in Table 3.

Table 3. Objective evaluation results for the Denoiser model with a GRU bottleneck.

Model	PESQ	STOI	CSIG	CBAK	COVL
Proposed GRU (Valentini aug)	2.911	0.943	3.863	3.403	3.377
Proposed GRU (MS-SNSD no aug)	2.920	0.941	3.954	3.393	3.456
Proposed GRU (MS-SNSD aug)	2.978	0.944	4.013	**3.455**	3.515
Proposed LSTM (MS-SNSD aug)	2.996	0.944	4.016	3.453	3.525
Facebook Denoiser	**3.07**	**0.95**	**4.31**	3.40	**3.63**

We can observe that for offline inference on 2 s windows, the GRU bottleneck maintains similar objective metrics performance with the LSTM based models.

4.5   Online Configuration Evaluation Results

This experimental setup will use both GRU and LSTM based models with a fixed input window size of 597 samples at 16 kHz, in order to evaluate the potential performance in a streaming scenario. Here, we only used the models that were trained using both datasets and used the data augmentation pipeline, since these models produced the best overall results in the offline evaluation.

We report the average metrics for the two model variants in Table 4:

Table 4. Objective evaluation results for the Denoiser model with different bottlenecks in a streaming scenario.

Model	PESQ	STOI	CSIG	CBAK	COVL
LSTM MS-SNSD (aug)	2.736	0.934	3.793	3.279	3.264
GRU MS-SNSD (aug)	**2.748**	**0.935**	**3.810**	**3.316**	**3.282**

For the streaming configuration, we can observe that the objective evaluation scores are lower compared to the offline setup. The GRU model proves to be slightly better than the LSTM model.

4.6   Real-Time Evaluation

In this scenario, we run our current implementation on a laptop CPU (AMD Ryzen 7 8845HS) and compute the Real-Time Factor (RTF) for both LSTM and GRU causal models, while also including the total disk size of the model (Table 5).

Table 5. Real-Time performance of both LSTM and GRU models

Model	Size	Stride	Frame time	RTF
LSTM	72 MB	16 ms	10.1 ms	0.63
GRU	**63 MB**	16 ms	**7.9 ms**	**0.50**

This experiment shows that the GRU bottleneck gives slightly faster real-time performance while being 12.5% smaller in disk size.

To qualitatively illustrate the impact of the proposed model, we select a representative utterance from the test set and compute its spectrogram (Fig. 2). The three panels that follow show, in order: (Fig. 2a) the noisy input signal formed by mixing clean speech with keyboard clicks, (Fig. 2b) the corresponding denoised output produced by proposed model, and (Fig. 2c) the original clean speech reference. By inspecting these spectrograms side-by-side, we can clearly see how the model suppresses the broadband transients caused by keystrokes while preserving the underlying harmonic structure of the human voice.

5 Conclusion

We have successfully adapted the Denoiser model, making it deployable via ONNX for real-time noise suppression. By leveraging a split-model approach, we allowed the ONNX encoder and decoder modules to accept arbitrary input lengths, facilitating latency-quality trade-offs tailored to application-specific requirements. Performance evaluations indicate minimal quality impact when shorter inference windows are used by employing causal streaming strategies for the original Denoiser.

In future research we intend to explore different quantizing strategies in order to reduce the model's footprint and also integrating and using this particular architecture in different runtimes and on different hardware, such as FPGAs or dedicated hardware for accelerating model inference, and evaluating the real-time performance, objective and subjective quality metrics.

Acknowledgements. The ISOLDE project, nr. 101112274 is supported by the Chips Joint Undertaking and its members Austria, Czechia, France, Germany, Italy, Romania, Spain, Sweden, Switzerland.

This work was supported by grant of the Ministry of Research, Innovation and Digitization, CNCS/CCCDI - UEFISCDI, project number PN-IV-P8-8.1-PME-2024-0024 within PNCDI IV.

References

1. Dèfossez, A., et al.: Real time speech enhancement in the waveform domain. In: Interspeech 2020 (2020). https://doi.org/10.48550/arXiv.2006.12847
2. Dèfossez, A., et al.: Music source separation in the waveform domain. arXiv preprint arXiv:1911.13254 (2020). https://doi.org/10.48550/arXiv.1911.13254
3. Hu, Y., Loizou, P.C.: Evaluation of objective quality measures for speech enhancement. IEEE Trans. Audio Speech Lang. Process. **16**(1), 229–238 (2008). https://ieeexplore.ieee.org/document/4389058
4. ITU-T Recommendation P.862. Perceptual Evaluation of Speech Quality (PESQ): An objective method for end-to-end speech quality assessment of narrow-band telephone networks and speech codecs. In: IEEE International Conference on Acoustics, Speech, and Signal Processing (ICASSP) (2001). https://doi.org/10.1109/ICASSP.2001.941023
5. Luo, Y., Mesgarani, N.: Conv-TasNet: surpassing ideal time-frequency magnitude masking for speech separation. IEEE/ACM Trans. Audio Speech Lang. Process. **27**(8), 1256–1266 (2019). https://ieeexplore.ieee.org/document/8707065
6. Pascual, S., Bonafonte, A., Serr, J.: SEGAN: speech enhancement generative adversarial network. In: Interspeech (2017). https://doi.org/10.48550/arXiv.1703.09452
7. Reddy, C.K.A., et al.: A scalable noisy speech dataset and online subjective test framework. In: Interspeech (2019). https://doi.org/10.48550/arXiv.1909.08050
8. Taal, C.H., Hendriks, R.C., Heusdens, R., Jensen, J.: An algorithm for intelligibility prediction of time-frequency weighted noisy speech. IEEE Trans. Audio Speech Lang. Process. **19**(7), 2125–2136 (2011). https://doi.org/10.1109/TASL.2011.2114881

9. Valentini-Botinhao, C.: Noisy speech database for training speech enhancement algorithms and TTS models. University of Edinburgh, Centre for Speech Technology Research (CSTR) (2016). https://doi.org/10.7488/ds/2117
10. Valin, J.M.: A hybrid DSP/deep learning approach to real-time full-band speech enhancement. arXiv preprint arXiv:1709.08243 (2017). https://doi.org/10.48550/arXiv.1709.08243
11. Wu, H., Braun, S., et al.: Ultra-low latency speech enhancement - a comprehensive study. arXiv preprint arXiv:2409.10358 (2024). https://doi.org/10.48550/arXiv.2409.10358
12. Yamamoto, R., Song, E., Kim, J.M.: Parallel WaveGAN: a fast waveform generation model based on generative adversarial networks with multi-resolution spectrogram. arXiv preprint arXiv:1910.11480 (2020). https://doi.org/10.48550/arXiv.1910.11480

Efficient Face Recognition with ResNet18

Ioan Chiţu[1,2], Honorius Gâlmeanu[1,2(✉)], and Alexandru Drîmbărean[3]

[1] Department of Mathematics and Informatics, Transilvania University of Braşov,
Braşov, Romania
galmeanu@unitbv.ro
[2] FotoNation SRL, Braşov, Romania
[3] FotoNation LTD, Galway, Ireland
alexandru.drimbarean@fotonation.com

Abstract. The domain of face recognition has shown a dramatic development in the recent years. Most impressive results were encountered using deep convolutional networks. Driven by reasons of efficiency, we approached the problem using a simple architecture, namely a ResNet18 network with 11 million parameters. By using the Additive Margin Softmax loss function we show how the performance of a plain architecture with no changes can be improved close to the one of state-of-the-art models with an order of magnitude more parameters. We obtain a train accuracy of 96% and a validation accuracy of 87% and a ROC area close to 0.98 on the LFW dataset. We show that even such a simple architecture is appropriate for face recognition.

Keywords: Deep CNN · image processing · face recognition · ResNet18

1 Introduction

The domain of face recognition has shown a dramatic development in recent years, with most impressive results encountered using deep convolutional networks. While state-of-the-art models achieve high accuracy, they often do so with deep, complex architectures that are computationally demanding, rendering them impractical for resource-constrained environments. This work, in contrast, focuses on achieving a compelling balance between accuracy and efficiency.

Our primary contribution is the demonstration that a simple architecture, a modified ResNet18 [5] with approximately 11 million parameters and paired with the Additive Margin Softmax (AM-Softmax) [10] loss function, can achieve performance close to that of much larger models or with smaller ones with complex architecture. This paper distinguishes itself from other recent works by explicitly prioritizing efficiency. For instance, while other research has enhanced *ResNet18* with complex additions like the *hybrid domain attention mechanism* [7], our approach deliberately avoids such modules to maintain a lower parameter count and faster inference speed. The novelty of our method lies not in the

C. Chira et al. (Eds.): InnoComp 2025, CCIS 2793, pp. 67–82, 2026.
https://doi.org/10.1007/978-3-032-12478-4_5

invention of new architectural components or a lengthy hyperparameter search, but in the validation and application of a simple model in tandem with a robust, pre-configured loss function. We demonstrate that by adopting the recommended hyperparameters from the original AM-Softmax paper—including learning rate, margin, and data preprocessing—our compact ResNet18 model can learn highly discriminative features. Our work provides a crucial methodological insight: it confirms that the AM-Softmax is highly transferable and that its established parameters are effective even when applied to a different, more lightweight architecture. This illustrates a practical and direct path to developing robust, efficient, and accurate face recognition systems suitable for real-world deployment, significantly reducing the engineering overhead typically associated with hyperparameter optimization. The complete implementation is available on GitHub[1].

The article is structured as follows. Section 2 provides a review of key deep learning face recognition models and establises a baseline for performance comparison. Section 3 details our methodology, outlining the ResNet18-based architecture and the configuration of the AM-Softmax loss function. The datasets and training procedures are described in Sect. 4. Section 5 presents our experimental results and a detailed performance analysis. Finally, Sect. 6 concludes the paper, summarizing our contributions.

2   Related Work

This section discusses several important state-of-the-art models for face recognition, highlighting their key characteristics and contributions.

Considered one of the most important models for face recognition, the Deep-Face [9] network is an early model achieving near-human accuracy (97.35%) on LFW [6]. The 9-layer CNN, with 137.8 million parameters, used 3D alignment and end-to-end learning. The large model size and computational demands made real-time performance difficult.

In DeepID2 [8], the authors improved upon DeepFace by introducing a joint identification-verification loss, a model with a smaller size which maintained the same amount of accuracy shown by the DeepFace model.

The VGG-Face [1] adapted the VGG-16 architecture to face recognition and was the state of the art of its time, but the size made its utilization very difficult. Following this, the Deep Convolutional Maxout Network (DCMN) [4] was designed for fine-grained face verification, which was an improvement from an architecture with small convolution filters.

---

[1] https://github.com/ionut200328/ResNet18-for-Face-Recognition.

Table 1. Comparison of face recognition models.

Model	Params	LFW Acc.	Key Innovation
DeepFace [9]	137.8 M	97.35%	3D align, end-to-end
DeepID2 [8]	10–50 M	99.15%	Joint ID-verify
VGG-Face [1]	145 M	98.90%	VGG adaptation
DCMN [4]	262.35 M	98.03%	Deep CNN+maxout
VGG Noisy Softmax [2]	145 M	99.18%	Softmax regularization

Another break-through model was the Noisy Softmax [2]. Using a VGG-based model, it introduced a training strategy using added noise (technique that could be applied to any CNN), improving the accuracy while maintaining a reduced computational usage.

Table 1 summarizes the key characteristics of these models.

2.1 Comparison with a Contemporary Lightweight Architecture

To accurately position the performance of our ResNet18-based model, it is essential to compare it against a relevant contemporary architecture trained under similar conditions. While many state-of-the-art models leverage massive, private datasets, the work by Zhou et al. [13] provides an ideal point of comparison. Their study, like ours, uses the public CASIA-WebFace dataset for training and evaluates performance on the LFW benchmark. Their approach involves taking the standard MobileNet V1 architecture and enhancing it with a custom weighted pooling layer designed to capture more discriminative global features. This makes the comparison particularly insightful: our work investigates the impact of an advanced loss function (AM-Softmax) on a standard architecture, while their work investigates the impact of a custom architectural module on a standard loss function. Table 2 provides a direct comparison of the two approaches.

2.2 Baseline Comparison

The comprehensive comparison in Table 2, made possible by the benchmarks provided in the study by Zhou et al., yields several important insights:

1. **A Clear Performance Hierarchy:** The results establish a clear performance spectrum among these lightweight models when trained on the same dataset. Our ResNet18 model, at 96.2% accuracy, significantly outperforms both the standard ShuffleNet (95.13%) and SqueezeNet (93.57%) architectures. It is only marginally outperformed by the "Improved" MobileNet, which benefits from a custom-designed pooling layer.

2. **The Trade-off Between Parameters and Performance:** While our ResNet18 model has the highest parameter count, the results demonstrate its efficiency in leveraging those parameters for higher accuracy. SqueezeNet

Table 2. Comparative Analysis of Lightweight Face Recognition Models on CASIA-WebFace

Model	Parameters (M)	Training Dataset	Key Methodological Difference	LFW Accuracy (%)
Improved MobileNet[1]	4.3	CASIA-WebFace	Custom Architectural Module (Weighted Pooling)	96.72%
Our ResNet18	**11.24**	**CASIA-WebFace**	**Advanced Loss Function (AM-Softmax)**	**96.2 ± 0.8%**
ShuffleNet[1]	4.0	CASIA-WebFace	Architecture (Channel Shuffle)	95.13%
SqueezeNet[1]	4.8	CASIA-WebFace	Architecture (Fire Modules)	93.57%

[1] Metrics for Improved MobileNet, ShuffleNet, and SqueezeNet are all sourced from the comparative study by Zhou et al. [13], ensuring a direct comparison as they use the same CASIA-WebFace training set. Note their different training parameters (e.g., lower initial learning rate and iteration-based schedule).

and ShuffleNet, despite being parameter-efficient, do not achieve the same level of performance under these training conditions. This suggests that for face recognition, the architectural choices of ResNet provide a stronger foundation for feature extraction, which is then capitalized on by the advanced loss function used.

3. **Validating the Power of the Loss Function:** This is the most critical insight. The fact that our standard, unmodified ResNet18 with a better loss function can outperform other standard lightweight architectures is a powerful insight to our methodology. It proves that focusing on the loss function is a highly effective strategy for performance enhancement. Furthermore, it achieves an accuracy that is highly competitive with the improved MobileNet, but without requiring any custom architectural engineering. This positions our approach as a practical and accessible framework for achieving high performance by focusing on the training objective rather than complex layer design.

3 Model Architecture and Loss Function

In the following we present the system's architecture, detailing the Multi-Tasked Cascading Neural Network (MTCNN) for face detection and a modified ResNet18 for extracting 128-dimensional L2-normalized embeddings. We also explain the Additive Margin Softmax loss function employed for training, including its configuration and rationale.

Our face recognition system consists of two main stages: the face detection and the feature extraction modules.

To achieve Face Detection, the MTCNN algorithm [12], a robust and widely-used face detection system, is applied to the input image. It employs a cascaded architecture with three stages of convolutional neural networks (P-Net, R-Net, and O-Net) to progressively refine face detections and simultaneously perform facial landmark localization. The algorithm efficiently identifies faces of varying sizes and orientations, outputting the bounding box coordinates for each face along with detecting its key facial landmarks.

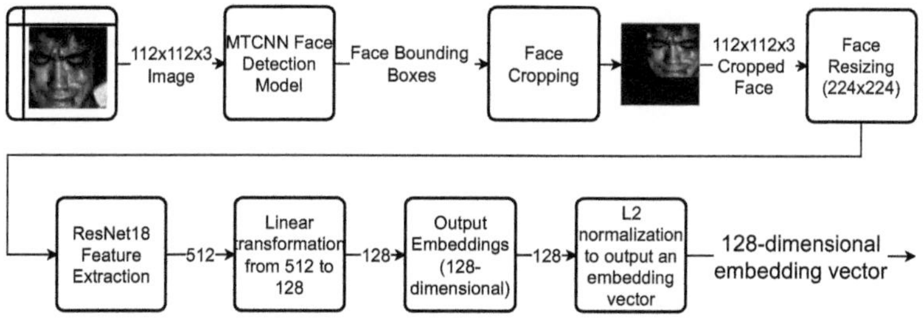

Fig. 1. Data Flow Diagram of the Model Architecture.

Once identified, the faces are then cropped and resized to 224×224 pixels, used by our modified ResNet18 model as input. A data flow diagram of the overall system is shown in Fig. 1.

The complete customized ResNet18 architecture is detailed in Table 3. Our implementation deviates from the standard architecture only in the final fully connected layer. Instead of the original 1000 outputs for ImageNet classification, we use a fully connected layer that produces 128 outputs. These outputs are then L2-normalized to generate a 128-dimensional embedding suitable for our recognition task.

To enhance the discriminative power of the learned face embeddings, we employ the Additive Margin Softmax (AM-Softmax) loss function. The AM-Softmax, as proposed in [10], encourages larger margins between different classes, leading to better separation of the learned feature representations.

We also experimented with Triplet Loss; however, the long training times associated with triplet generation and the significantly slower convergence compared to AM-Softmax proved to be unsuitable for our application.

The standard Softmax loss function is defined as:

$$L_{Softmax} = -\frac{1}{N} \sum_{i=1}^{N} \log \frac{e^{W_{y_i}^T x_i}}{\sum_{j=1}^{C} e^{W_j^T x_i}} \tag{1}$$

where:

- x_i is the feature vector of the i-th sample;
- W_j is the weight vector of the j-th class;
- y_i is the true label of the i-th sample;
- N is the number of samples in the batch;
- C is the total number of classes.

AM-Softmax modifies this by introducing a margin m and a scaling factor s:

$$L_{AM-Softmax} = -\frac{1}{N} \sum_{i=1}^{N} \log \frac{e^{s(W_{y_i}^T x_i - m)}}{e^{s(W_{y_i}^T x_i - m)} + \sum_{j=1,j\neq y_i}^{C} e^{sW_j^T x_i}} \tag{2}$$

where:

- s scales the logits. A higher s concentrates the probability distribution, making the decision boundary sharper;
- m is the additive margin penalty that is applied to the logit of the correct class. Increasing m forces the model to learn more discriminative features with larger inter-class separation.

A complete implementation of the AM-Softmax loss, including a numerically stable LogSumExp trick, is available on GitHub[2].

The AM-Softmax loss modifies the standard softmax by introducing a margin within the cosine of the angle between the feature vector and the weight vector

---

[2] https://github.com/dalisson/am_softmax.

Table 3. ResNet18 Architecture Details

Type	Subtype	Output Size	Kernel	Stride	Padding	No. Params.	MACs
Conv2d		$64 \times 112 \times 112$	7×7	2×2	3×3	9.41 k	118.01 MMac
BatchNorm2d		$64 \times 112 \times 112$				128	1.61 MMac
ReLU		$64 \times 112 \times 112$					802.82 KMac
MaxPool2d		$64 \times 56 \times 56$	3×3	2×2	1×1		802.82 KMac
layer1 - BasicBlock0						73.98 k	232.42 MMac
	Conv2d	$64 \times 56 \times 56$	3×3	1×1	1×1	36.86 k	115.61 MMac
	BatchNorm2d	$64 \times 56 \times 56$				128	401.41 KMac
	ReLU	$64 \times 56 \times 56$					401.41 KMac
	Conv2d	$64 \times 56 \times 56$	3×3	1×1	1×1	36.86 k	115.61 MMac
	BatchNorm2d	$64 \times 56 \times 56$				128	401.41 KMac
	ReLU	$64 \times 56 \times 56$					ac
layer1 - BasicBlock1						73.98 k	232.42 MMac
	Conv2d	$64 \times 56 \times 56$	3×3	1×1	1×1	36.86 k	115.61 MMac
	BatchNorm2d	$64 \times 56 \times 56$				128	401.41 KMac
	ReLU	$64 \times 56 \times 56$					401.41 KMac
	Conv2d	$64 \times 56 \times 56$	3×3	1×1	1×1	36.86 k	115.61 MMac
	BatchNorm2d	$64 \times 56 \times 56$				128	401.41 KMac
	ReLU	$64 \times 56 \times 56$					
layer2 - BasicBlock0						230.14 k	180.63 MMac
	Conv2d	$128 \times 28 \times 28$	3×3	2×2	1×1	73.73 k	57.8 MMac
	BatchNorm2d	$128 \times 28 \times 28$				256	200.7 KMac
	ReLU	$128 \times 28 \times 28$					200.7 KMac
	Conv2d	$128 \times 28 \times 28$	3×3	1×1	1×1	147.46 k	115.61 MMac
	BatchNorm2d	$128 \times 28 \times 28$				256	200.7 KMac
	Conv2d	$128 \times 28 \times 28$	3×3	2×2	1×1	8.19 k	6.42 MMac
	BatchNorm2d	$128 \times 28 \times 28$				256	200.7 KMac
	ReLU	$128 \times 28 \times 28$					
layer2 - BasicBlock1						295.42 k	231.81 MMac
	Conv2d	$128 \times 28 \times 28$	3×3	1×1	1×1	147.46 k	115.61 MMac
	BatchNorm2d	$128 \times 28 \times 28$				256	200.7 KMac
	ReLU	$128 \times 28 \times 28$					200.7 KMac
	Conv2d	$128 \times 28 \times 28$	3×3	1×1	1×1	147.46 k	115.61 MMac
	BatchNorm2d	$128 \times 28 \times 28$				256	200.7 KMac
	ReLU	$128 \times 28 \times 28$					
layer3 - BasicBlock0						919.04 k	180.23 MMac
	Conv2d	$256 \times 14 \times 14$	3×3	2×2	1×1	294.91 k	57.8 MMac
	BatchNorm2d	$256 \times 14 \times 14$				512	100.35 KMac
	ReLU	$256 \times 14 \times 14$					100.35 KMac
	Conv2d	$256 \times 14 \times 14$	3×3	1×1	1×1	589.82 k	115.61 MMac
	BatchNorm2d	$256 \times 14 \times 14$				512	100.35 KMac
	Conv2d	$256 \times 14 \times 14$	3×3	2×2	1×1	32.77 k	6.42 MMac
	BatchNorm2d	$256 \times 14 \times 14$				512	100.35 KMac
	ReLU	$256 \times 14 \times 14$					
layer3 - BasicBlock1						1.18 M	231.51 MMac
	Conv2d	$256 \times 14 \times 14$	3×3	1×1	1×1	589.82 k	115.61 MMac
	BatchNorm2d	$256 \times 14 \times 14$				512	200.7 KMac
	ReLU	$256 \times 14 \times 14$					100.35 KMac
	Conv2d	$256 \times 14 \times 14$	3×3	1×1	1×1	589.82 k	115.61 MMac
	BatchNorm2d	$256 \times 14 \times 14$				512	100.35 KMac
	ReLU	$256 \times 14 \times 14$					
layer4 - BasicBlock0						3.67 M	180.03 MMac
	Conv2d	$512 \times 7 \times 7$	3×3	2×2	1×1	1.18 M	57.8 MMac
	BatchNorm2d	$512 \times 7 \times 7$				1024	50.18 KMac
	ReLU	$512 \times 7 \times 7$					50.18 KMac
	Conv2d	$512 \times 7 \times 7$	3×3	1×1	1×1	2.36 M	115.61 MMac
	BatchNorm2d	$512 \times 7 \times 7$				1024	50.18 KMac
	Conv2d	$512 \times 7 \times 7$	3×3	2×2	1×1	131.07 k	6.42 MMac
	BatchNorm2d	$512 \times 7 \times 7$				1024	50.18 KMac
	ReLU	$512 \times 7 \times 7$					
layer4 - BasicBlock1						4.72 M	231.36 MMac
	Conv2d	$512 \times 7 \times 7$	3×3	1×1	1×1	2.36 M	115.61 MMac
	BatchNorm2d	$512 \times 7 \times 7$				1024	50.18 KMac
	ReLU	$512 \times 7 \times 7$					50.18 KMac
	Conv2d	$512 \times 7 \times 7$	3×3	1×1	1×1	2.36 M	115.61 MMac
	BatchNorm2d	$512 \times 7 \times 7$				1024	50.18 KMac
	ReLU	$512 \times 7 \times 7$					
AdaptiveAvgPool2d		$512 \times 1 \times 1$					25.09 KMac
Dropout with p=0.4		512					
Linear		128				65.66 k	65.66 KMac
L2 normalization		128					
Total						11.24 M	1.82 GMac

associated with the correct class. This means that the cosine of the angle between a feature vector and its correct class weight vector must be greater than the cosine of the angle with any other class weight vector plus the margin. This enforces good separation in feature space promotes better generalization and robustness.

Specifically, following the above mentioned implementation of the AM-Softmax, we L2-normalize the output of the last layer of our ResNet18 before applying the AM Softmax Loss, where the L2 normalization is defined as $\hat{\mathbf{x}} = \frac{\mathbf{x}}{||\mathbf{x}||_2}$, ensuring that all feature vectors lie on the unit hypersphere.

The key parameters of the AM-Softmax loss function in our experiments are:

- *Scaling factor $s = 30$*: the scaling factor s controls the concentration of the probability distribution. Higher values of s lead to a sharper decision boundary. We set $s = 30$ following the recommendations in [10], which found that this value provides a good performance.
- *Margin parameter $m = 0.4$*: the margin parameter m imposes a penalty on the logit of the correct class, encouraging the model to learn more discriminative features with larger inter-class separation. We set $m = 0.4$, because in the mentioned paper it was taken between 0.35 and 0.4; we chose 0.4 because we discovered it provides lower variance, a slightly better balance between training accuracy and generalization performance in our experiments. An ablation study for this parameter is presented in Sect. 5.

4   Datasets and Training Procedure

We trained our ResNet18 model on the CASIA-WebFace dataset [11], a large-scale face recognition dataset containing 10,572 identities. For testing and evaluation, we used the Labeled Faces in the Wild (LFW) dataset [6]. LFW is a standard benchmark for face verification, containing pairs of face images with the task of determining whether the two images belong to the same individual. To ensure a fair evaluation, we removed any overlapping identities between the CASIA-WebFace training set and the LFW test set.

We also used the Similar Looking Labeled Faces in the Wild (SLLFW) dataset [3], which is a more challenging subset of the LFW dataset, specifically designed to evaluate the fine-grained discrimination capabilities of face recognition systems.

The training set was split randomly into 80% training images and 20% validation images. During training, we applied the following data augmentation techniques to improve the model's generalization ability:

- *Normalization*: the input images were normalized by subtracting 128 from each pixel value and then dividing by 128, thus scaling the pixel values to the [-1, 1] range. This normalization scheme was chosen to follow the recommendations in existing face recognition literature, such as the AM-Softmax paper [10], and was found to provide a stable training;

- *Horizontal Mirroring*: random horizontal mirroring (flipping) was applied to the images. Horizontal mirroring is also widely used in face recognition to augment the training data without introducing unrealistic distortions, as it preserves the overall structure and identity of the face.

We experimented with other data augmentation techniques, including vertical mirroring, random tilting, random cropping, and color jittering. However, these techniques consistently led to a decrease in the model's validation performance. Therefore, we chose to limit our data augmentation to normalization and horizontal mirroring, as these techniques provided the best balance between data diversity and preservation of essential facial features.

The Stochastic Gradient Descent (SGD) optimizer was used for training with a momentum of 0.9 and a weight decay of 5e-4. The initial learning rate was set to 0.1, and it was divided by 10 at epochs 16, 24, and 28. The model was trained for a total of 30 epochs with a batch size of 256.

The entire training process, including both training and validation sets comprising approximately 0.5 million images, took approximately 5 h on a single Nvidia RTX A4000 GPU.

Table 4. Performance Comparison on LFW and SLLFW Datasets as reported by [3].

Method	Training Images	LFW	SLLFW
DeepFace [9]	0.5 M	92.87%	78.78%
DeepID2 [8]	0.5 M	95.00%	78.25%
VGG-Face [1]	2.6 M	96.70%	85.78%
DCMN [4]	0.5 M	98.03%	91.00%
Noisy Softmax [2]	0.5 M	99.18%	94.50%
Human	n/a	99.85%	92%
Our ResNet18 (AM-Softmax)	0.5 M	96.2 ± 0.8%	91.8 ± 1.1%

5 Experimental Results and Performance Comparison

We compared our model with DeepFace [9], DeepID2 [8], VGG-Face [1], and Noisy Softmax [2]. All models are all trained with CASIA-WebFace, a robust dataset for training images (see Table 4).

Our ResNet18 model, trained on CASIA-WebFace using AM-Softmax loss, was evaluated on both the LFW and SLLFW datasets using 10-fold cross-validation via scikit-learn's KFold method. Each fold comprised 6,000 face pairs, balanced with 3,000 matched and 3,000 mismatched pairs to ensure robust evaluation. The model achieved an accuracy of 96.2 ± 0.8% on LFW and 91.8 ± 1.1% on SLLFW. We showed the performance comparison with our model in

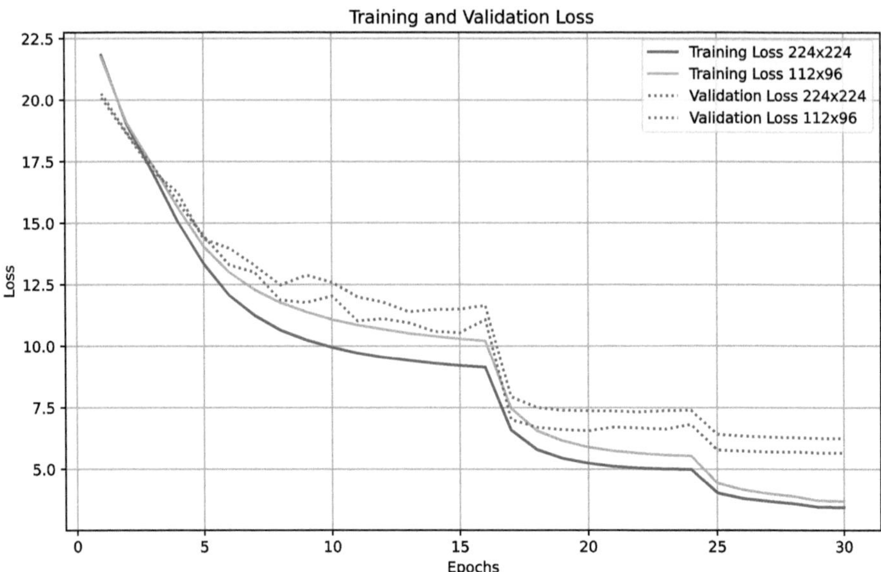

(a) Train and Validation Loss.

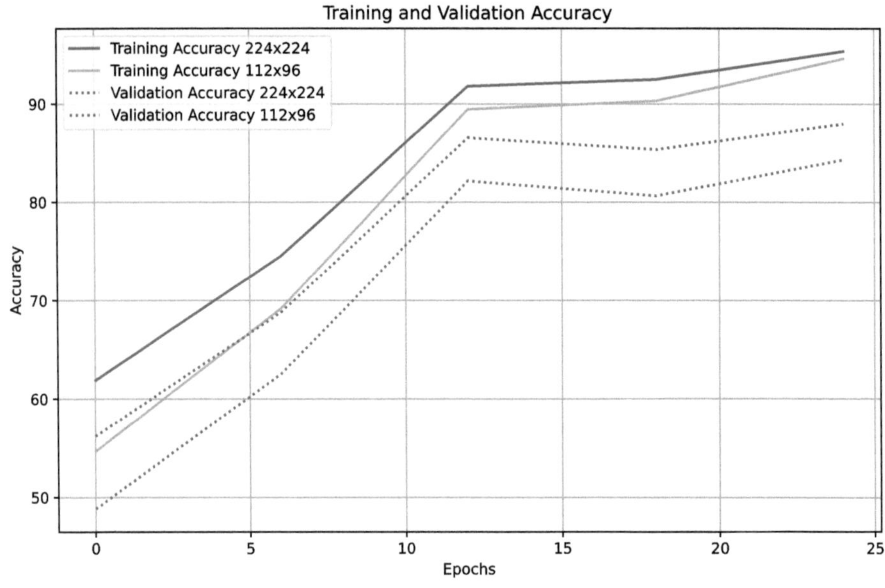

(b) Train and validation accuracy.

Fig. 2. Train and validation curves for ResNet18 with 224×224 and 112×96 input sizes.

Table 4. We computed the validation rates at a False Accept Rate (FAR) of 0.001. The validation rate for LFW was 87.83 ± 2.26%, while the validation rate for SLLFW was significantly lower, 60.87 ± 3.14%.

Figures 2a and 2b show the loss and training and validation accuracy curves for the ResNet18 model trained with both 224×224 and 112×96 input sizes on CASIA-WebFace. One can observe that a significant drop in the training and validation loss occurs at epochs 16, 24, and 28 (as seen in Fig. 2a). This corresponds to the scheduled learning rate decay, where the learning rate was reduced by a factor of 10. These drops are expected, as a smaller learning rate allows the model to fine-tune its weights more precisely and reduce the error.

Both configurations converge to similar high accuracy (around 95%) by the end of training. However, the validation accuracy for the 224×224 input size consistently outperforms that of the 112×96 input across all epochs, resulting in a final validation accuracy of approximately 88% for the former and 84% for the latter. This persistent gap underscores the significantly better generalization performance achieved with the larger input size.

Table 5. Face Validation Performance on LFW

Input Size	Accuracy	Validation Rate @ FAR=0.001
224×224	0.962 ± 0.008	0.87833 ± 0.02257
112×96	0.944 ± 0.008	0.77167 ± 0.02790

Table 5 presents the face validation performance of our ResNet18 model with both 224×224 and 112×96 input sizes on the LFW dataset.

The Receiver Operating Characteristic (ROC) curves (Figs. 3a and 3b) provide further insight into the model's discriminative ability on the LFW dataset. These ROC curves were generated using a model trained with the settings described so far: SGD optimizer, initial learning rate of 0.1, batch size of 256, data normalization, and horizontal mirroring.

The ROC curve for the 224×224 input size achieves an Area Under the Curve (AUC) of 0.977 (Fig. 3a), indicating excellent discriminative ability for general face recognition. The Fine-Grained ROC curve generated from the SLLFW data shows a lower AUC of 0.952, confirming the greater challenge of fine-grained face recognition, as it is better to differentiate.

The ROC curve for the 112×96 input size achieves an AUC of 0.972 and a Fine Grained 0.92, indicating a somewhat lower discriminative ability (Fig. 3b). The model with 224×224 input size achieves a better performance, as expected.

Table 6. Fine-Grained Performance on SLLFW

Input Size	Accuracy	Validation Rate @ FAR=0.001
224×224	0.918 ± 0.011	0.60867 ± 0.03138
112×96	0.870 ± 0.014	0.45900 ± 0.04110

Table 6 presents the fine-grained performance results on the SLLFW dataset. While the accuracy on the SLLFW dataset decreases by approximately 5% when using the 112×96 input size, the validation rate experiences a more substantial drop. This suggests that distinguishing fine-grained differences between similar-looking faces becomes more difficult.

5.1   Ablation Study on Margin Parameter

To determine the optimal margin parameter for the AM-Softmax loss function within our framework, we conducted an ablation study. We trained our ResNet18 model with five different margin values, from $m = 0.2$ to $m = 0.6$, keeping all other parameters (224×224 input, learning rate schedule, etc.) constant. The performance was evaluated on both the LFW and the more challenging SLLFW datasets.

Table 7. Ablation Study of Margin Parameter 'm' on LFW and SLLFW Datasets

Margin	LFW Performance		SLLFW Performance	
(m)	Accuracy (%)	Val. Rate (%)	Accuracy (%)	Val. Rate (%)
0.6	95.5 ± 0.8	85.9 ± 1.9	89.2 ± 1.2	39.0 ± 4.8
0.5	96.1 ± 0.8	80.2 ± 2.6	90.8 ± 0.8	49.3 ± 2.7
0.4	$\mathbf{96.2 \pm 0.8}$	$\mathbf{87.8 \pm 2.2}$	$\mathbf{91.6 \pm 1.1}$	$\mathbf{60.8 \pm 3.1}$
0.3	96.2 ± 0.9	87.8 ± 2.6	90.8 ± 1.2	39.0 ± 3.5
0.2	96.2 ± 0.8	85.3 ± 2.1	90.6 ± 0.9	42.3 ± 3.7

The results, presented in Table 7, show a clear trade-off between standard and fine-grained recognition performance. While multiple margin values achieve the same peak accuracy of 96.2% on LFW, their ability to discriminate between similar-looking faces varies significantly. A margin of $m = 0.4$ provides the best overall performance, achieving not only high LFW accuracy but also the highest accuracy (91.6%) and validation rate (60.8%) on the difficult SLLFW dataset. Margins that are too low (e.g., $m = 0.2$) or too high (e.g., $m = 0.6$) fail to generalize as effectively on the fine-grained task. Therefore, we selected $m = 0.4$ as the optimal value for our final model, as it demonstrates the most robust feature learning for both general and fine-grained face verification.

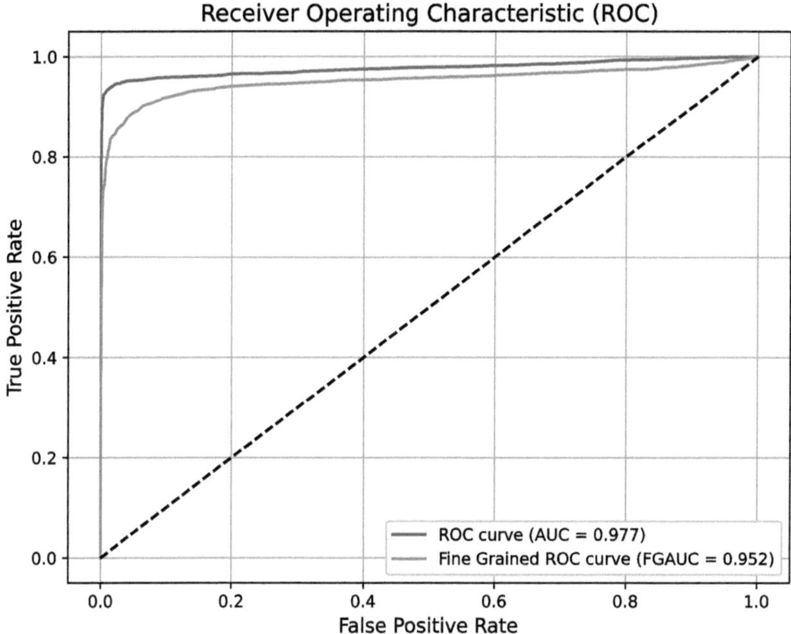

(a) ROC curve for 224x224 input size on LFW

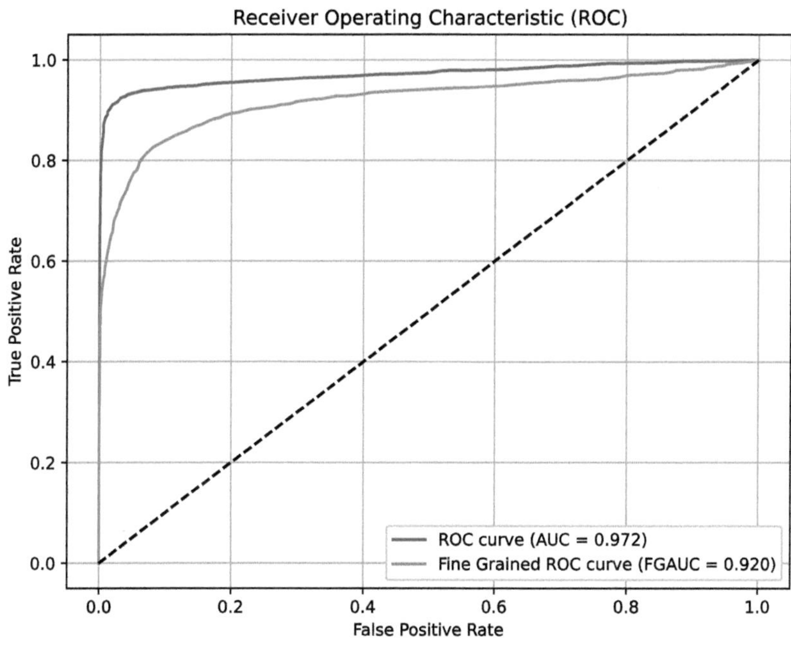

(b) ROC curve for 112x96 input size on LFW

Fig. 3. ResNet18 classification performance ROC curves on LFW for a 224x224 input size (a) and for a 112x96 input size (b).

5.2 Ablation Study on Input Image Size

The resolution of the input image is a critical hyperparameter that directly influences both model accuracy and computational demand. To quantify this trade-off, we conducted an ablation study by training our model on four different input sizes, ranging from 64×64 to 300×300 pixels. All other training conditions, including the margin parameter $m = 0.4$, were held constant.

Table 8. Ablation Study of Input Image Size on LFW and SLLFW Datasets

Input Size (pixels)	LFW Performance		SLLFW Performance	
	Accuracy (%)	Val. Rate (%)	Accuracy (%)	Val. Rate (%)
64×64	93.6 ± 1.2	63.0 ± 3.1	82.4 ± 1.7	25.3 ± 4.2
112×96	94.4 ± 0.8	77.2 ± 2.8	87 ± 1.4	45.9 ± 4.1
112×112	95.1 ± 0.9	79.5 ± 2.2	87.4 ± 1.1	40.0 ± 3.8
$\mathbf{224 \times 224}$	$\mathbf{96.2 \pm 0.8}$	$\mathbf{87.8 \pm 2.3}$	$\mathbf{91.8 \pm 1.1}$	$\mathbf{60.9 \pm 3.1}$
300×300	96.3 ± 0.5	89.6 ± 2.1	92.0 ± 0.9	59.7 ± 2.6

The results, detailed in Table 8, reveal a clear and expected trend: model performance improves consistently as input resolution increases. The additional detail available in larger images allows the network to learn more discriminative features, which is particularly evident in the significant performance jump on the SLLFW dataset when moving from 112×112 to 224×224.

However, the analysis also highlights a point of diminishing returns. While the 300×300 input size provides a marginal accuracy increase over 224×224 (e.g., 96.3% vs. 96.2% on LFW), this small gain comes at a substantial cost in terms of computational complexity and memory usage during training and inference. The 224×224 resolution, therefore, represents the optimal balance in our study. It delivers near-peak performance and a significant improvement over smaller sizes, without the excessive overhead associated with the 300×300 input. For this reason, the 224×224 input size was selected for our final model configuration.

6 Conclusions

This work demonstrates the effectiveness of combining a lightweight ResNet18 architecture with the AM-Softmax loss function for efficient and accurate face recognition. Our contributions center on creating an efficient and practical solution for face verification.

Our implementation successfully utilized the AM-Softmax loss function by applying the hyperparameters recommended in the original work. This decision underscores a key finding: the AM-Softmax framework is robust enough to be applied effectively to a suitable simple architecture like ResNet18, bypassing a time-intensive hyperparameter search. This differentiates our work from studies focused on novel architectural modifications or extensive parameter optimization. Instead, we validate that a judicious combination of an efficient model (ResNet18) and a powerful, established loss function (AM-Softmax) offers an optimal balance of accuracy and efficiency.

The key methodological insight is that high-performing, resource-aware face recognition systems can be built upon the solid foundation of existing research, emphasizing replicability and practical deployment. Our results, which are competitive with much larger models, confirm that this streamlined approach is not only viable but highly effective for real-world applications.

Acknowledgements. The ISOLDE project, nr. 101112274 is supported by the Chips Joint Undertaking and its members Austria, Czechia, France, Germany, Italy, Romania, Spain, Sweden, Switzerland.

This work was supported by grant of the Ministry of Research, Innovation and Digitization, CNCS/CCCDI - UEFISCDI, project number PN-IV-P8-8.1-PME-2024-0024 within PNCDI IV.

References

1. Cao, Q., Shen, L., Xie, W., Parkhi, O.M., Zisserman, A.: VGGFace2: a dataset for recognising faces across pose and age. In: 2018 13th IEEE International Conference on Automatic Face & Gesture Recognition (FG 2018), pp. 67–74 (2018)
2. Chen, B., Deng, W., Du, J.: Noisy softmax: improving the generalization ability of DCNN via postponing the early softmax saturation. In: 2017 IEEE Conference on Computer Vision and Pattern Recognition (CVPR), pp. 4021–4030 (2017)
3. Deng, W., Jiani, H., Zhang, N., Chen, B., Guo, J.: Fine-grained face verification: FGLFW database, baselines, and human-DCMN partnership. Pattern Recogn. **66**, 63–73 (2017)
4. Feng, Y., et al.: DCMN: double core memory network for patient outcome prediction with multimodal data. In: 2019 IEEE International Conference on Data Mining (ICDM), pp. 200–209 (2019)
5. He, K., Zhang, X., Ren, S., Sun, J.: Deep residual learning for image recognition (2015)
6. Huang, G.B., Mattar, M., Berg, T., Learned-Miller, E.: Labeled faces in the wild: a database for studying face recognition in unconstrained environments. Technical Report Technical Report 07-49, University of Massachusetts, Amherst (2008)
7. Mei, Y.: ResNet18 facial feature extraction algorithm improved based on hybrid domain attention mechanism. PLoS ONE **20**(3), e0319921 (2025)
8. Sun, Y., Wang, X., Tang, X.: Deeply learned face representations are sparse, selective, and robust. In: 2015 IEEE Conference on Computer Vision and Pattern Recognition (CVPR), pp. 2892–2900 (2015)
9. Taigman, Y., Yang, M., Ranzato, M.A., Wolf, L.: DeepFace: closing the gap to human-level performance in face verification. In: 2014 IEEE Conference on Computer Vision and Pattern Recognition, pp. 1701–1708 (2014)
10. Wang, F., Cheng, J., Liu, W., Liu, H.: Additive margin softmax for face verification. IEEE Signal Process. Lett. **25**(7), 926–930 (2018)

11. Yi, D., Lei, Z., Liao, S., Li, S.Z.: Learning face representation from scratch. CoRR, abs/1411.7923 (2014)
12. Zhang, L., Wang, H., Chen, Z.: A multi-task cascaded algorithm with optimized convolution neural network for face detection. In: 2021 Asia-Pacific Conference on Communications Technology and Computer Science (ACCTCS), pp. 242–245 (2021)
13. Zhou, Y., Liu, Y., Han, G., Fu, Y.: Face recognition based on the improved MobileNet. In: 2019 IEEE Symposium Series on Computational Intelligence (SSCI), pp. 2776–2781 (2019)

Automated Model Interpretation, Pattern Analysis and Insights in Explainable AI

David-Ioan Stoica[1], Iulian Popa[2(✉)], and Alexandra Băicoianu[2]

[1] Faculty of Mathematics and Informatics, Transilvania University of Braşov,
Braşov, Romania
`david.stoica@student.unitbv.ro`
[2] Department of Mathematics and Informatics, Transilvania University of Braşov,
Braşov, Romania
`{iulian.popa,a.baicoianu}@unitbv.ro`

Abstract. Deep learning has revolutionized artificial intelligence by enabling highly accurate models across diverse domains such as computer vision and autonomous driving. However, the inherent complexity of deep neural networks often results in *black-box* models, whose decision-making processes lack transparency and interpretability. Explainable Artificial Intelligence addresses this challenge by developing techniques that make model predictions more understandable to humans. Among these, Class Activation Mapping (CAM) and its variants: Grad-CAM, Grad-CAM++ and Score-CAM, have become pivotal in providing visual explanations for convolutional neural networks. This paper presents a systematic comparison of several prominent CAM-based explanation methods applied to the AlexNet architecture on a standardized Dogs vs. Cats image dataset. We evaluate these methods using quantitative metrics including Intersection over Union, Dice coefficient and Deletion and Insertion tests, aiming to reveal their respective strengths, limitations and practical utility. These findings offer valuable insights to guide practitioners in selecting appropriate explanation techniques. Specifically, we show that combining CAM-based techniques enables a deeper understanding of pixel importance by identifying critical regions composed of both object and background pixels and by guiding data preprocessing promotes more stable training and improves the model's focus. Furthermore, we demonstrate the potential for constructing robust feature attribution strategies grounded in layer-wise analysis and assess the reliability of various XAI algorithms by evaluating their behavior across multiple threshold levels.

Keywords: Explainable AI · CAM techniques · Saliency maps

1 Introduction

Deep learning, a subset of machine learning, has become a cornerstone of artificial intelligence (AI). Inspired by the human brain's neural networks, deep learning has found applications across a wide range of fields, including computer vision, natural language processing, healthcare and autonomous driving. However, as AI

C. Chira et al. (Eds.): InnoComp 2025, CCIS 2793, pp. 83–95, 2026.
https://doi.org/10.1007/978-3-032-12478-4_6

systems increase in complexity and capability, a critical challenge has surfaced: the lack of transparency and interpretability in the decision-making processes of deep neural networks [1,2]. Often referred to as "black-box" models, these systems can be remarkably accurate while remaining opaque, making it difficult to understand why a particular prediction was made.

To address this challenge, Explainable AI (XAI) [10,11] has become a key area in machine learning, aiming to make model decisions more transparent and interpretable. In the context of Convolutional Neural Networks (CNNs), visual explanation methods such as Class Activation Mapping (CAM) [12], along with its variants like Grad-CAM [13], Grad-CAM++ [14] or Score-CAM [15], have emerged as foundational tools in decisions understanding and interpreting.

These algorithms provide visual explanations by highlighting input regions that most influence a model's decision. Grad-CAM, for instance, uses class-specific gradients from the final convolutional layer to generate coarse heatmaps that reveal model reasoning. Later variants improved spatial resolution, class sensitivity, and robustness across architectures. However, no single method is universally optimal, prompting comparative evaluations of CAM-based techniques [4,5]. Each variant involves trade-offs in fidelity, localization accuracy, computational cost, and resilience to input changes. Their effectiveness also depends on the model architecture, dataset type, and interpretability metrics. Choosing the best CAM variant for a specific task remains challenging, motivating further comparative analysis under controlled conditions.

Recent advances in Explainable Artificial Intelligence (XAI) aim to improve trust, accountability, and transparency in black-box models. Modern XAI methods are structured around the types of questions users might ask, such as: "Why was this decision made?", "Why not another outcome?" or "How can I achieve a different prediction?" [3] [10]. Studies demonstrate that regions highlighted by activation maps can effectively be used to suppress background noise or emphasize salient regions, thereby improving model focus and training stability [7].

A diverse set of XAI techniques contributes to model interpretability at different levels. Instance-level methods such as LIME and SHAP provide localized insights by identifying key input features that drive predictions [2,22], while counterfactual approaches like CEM and DiCE explain how minimal input changes could alter model output [23]. Global techniques, including feature importance scores and surrogate models, offer a broader understanding of the model's decision logic. To more rigorously assess the quality of saliency-based explanations, insertion and deletion tests have been introduced [8], which measure the impact of progressively removing or enhancing salient regions on model confidence. Additionally, layer-wise attribution strategies—such as aggregating Class Activation Maps (CAMs) from multiple convolutional layers using weighted averages, max-pooling, or PCA fusion—have been shown to produce richer and more reliable heatmaps [7].

The purpose of this article is to make a systematic comparison of several prominent CAM-based explanation algorithms with a focus on understanding their strengths, limitations and practical utility. By applying each method to a standardized image classification task, we seek to evaluate them using concrete and, in some cases, already well-established metrics, that we find more suitable than visual coherence and human intuition, such as Intersection over Union (IoU) [25], Dice [26], Deletion and Insertion Tests [24].

The main innovation of our work lies in combining systematic benchmarking of CAM-based methods, layer-wise visual analysis, and the practical use of XAI techniques to guide data preprocessing and improve model training-all evaluated using standardized quantitative metrics. While CAM, Grad-CAM, Grad-CAM++, and Score-CAM are established methods, our systematic and architecture-specific evaluation on AlexNet offers new practical insights. Furthermore, demonstrating that CAM-based explainability tools can support preprocessing decisions and enhance training stability represents a practical and relatively underexplored application. We show that visualizations can help optimize both input selection and model behavior, which we believe is a meaningful contribution to the field.

2   Materials and Methods

2.1   Class Activation Mapping Techniques

Gradient-weighted Class Activation Mapping (Grad-CAM) [13] generates visual explanations for CNN-based models by computing the gradients of the target class score with respect to convolutional feature maps. These are globally averaged to obtain weights, which produce a heatmap highlighting important input regions. Grad-CAM++ [14], an extension of Grad-CAM, improves object feature localization by generating more precise, interpretable heatmaps. It uses a weighted sum of positive partial derivatives from the last convolutional layer with respect to the target class score.

A gradient-free approach is exemplified by Score-CAM [15], which determines the importance of each activation map by forward-passing masked inputs through the model and measuring the change in the target class score. This method often produces even more reliable and visually appealing explanations, as it is less susceptible to issues such as gradient saturation.

2.2   Attribution Methods: Neuron, Layer and Feature Attribution

Neuron attribution [18] methods provide the most in-depth perspective by assigning relevance to individual neurons. They aim to identify which neurons are most active for a given output and how their activation patterns impact the prediction. Although this level of detail provides deep insight into the model's inner workings, it often presents challenges in terms of intuitive human interpretation.

Focusing on the contributions of intermediate layers, Layer attribution [18] aims to explain how different layers of the network influence the model's output. Instead of analyzing input features directly, layer attribution reveals how information is transformed and propagated through the network's hierarchical structure.

CAM techniques naturally lend themselves to layer attribution. Grad-CAM output offers a novel semi-temporal insight into how representations evolve throughout the convolutional layers of the network. Although CNNs are not inherently sequential models in time, viewing activations as a sequence reveals patterns of increasing abstraction and spatial refinement as one moves deeper into the network. This methodology allows an intuitive understanding of how progressively deeper layers refine and abstract the input information.

Feature attribution [17] methods aim to assign importance scores to individual input features (such as pixels in an image) relative to a model's prediction. These techniques attempt to identify which parts of the input were most influential for this decision. Prominent examples include gradient-based saliency maps, integrated gradients or LIME [6].

Complementing the layer attribution analysis, several feature attribution strategies can be adopted to gain a more detailed understanding of how individual input features contribute to a model's prediction.

Simple averaging the Grad-CAM outputs across all considered layers to produce a composite heatmap, summarizing feature importance throughout the network. In contrast, a weighted average assigns linearly increasing weights to deeper layers, reflecting the intuition that deeper features tend to be more abstract and semantically meaningful. The weights increase linearly with the layer depth. Max Pooling Across Layers selects the pixel-wise maximum from all individual Grad-CAM maps, highlighting the most salient signals regardless of the originating layer.

Principal Component Fusion is a widely used, model-agnostic, and user-friendly interpretability technique that produces easily understandable outputs without requiring additional training or retraining. This approach stacks all Grad-CAM maps into a 3D tensor with shape $[L, H, W]$, where L denotes the number of layers and H, W are the spatial dimensions of the input. Principal Component Analysis (PCA) is then applied to extract the dominant attribution patterns across layers, effectively summarizing the most informative features.

2.3   Model Description

The AlexNet model [16] marked a significant breakthrough in deep learning, especially in computer vision. Its relatively shallow depth and simple architecture make it well-suited for early explorations of XAI methods, enabling visualization techniques like Grad-CAM to clearly reveal the progressive feature transformations across convolutional layers. This model was chosen for its simplicity and widespread recognition in the literature. In our study, AlexNet serves solely as a tool for conducting a comparative analysis of different XAI approaches.

To simplify the task and enable focused analysis, a binary classification problem was used: distinguishing cats from dogs. This approach employs the well-known Dogs vs. Cats dataset [9], which includes approximately 25,000 labeled training images and 12,154 labeled test images, with a balanced 50/50 class split.

The goal was not to achieve state-of-the-art accuracy, but to obtain a sufficiently trained model for generating and analyzing attribution maps.

3   Explainable AI Approach

XAI methods were employed to enhance model interpretability and guide data preprocessing decisions. Specifically, the presented CAM techniques were used to generate saliency heatmaps that highlight the most relevant regions in the input images. These techniques build upon prior work that combines CAM-based analysis with data augmentation strategies presented in Table 1.

Table 1. Overview of preprocessing strategies used in the experiments.

Heading level	Example Description
Original dataset	Standard training on the original dataset, used as a reference baseline.
Blurred background	Background regions (outside the animal) are blurred with a Gaussian filter, encouraging the model to focus on the object of interest.
Blacked-out background	Background pixels are replaced with black, isolating the subject more clearly.
Cropped background	Images are cropped tightly around the foreground object (the cat or dog) to eliminate contextual background.
CAM-based blurring	Gaussian blur with a 21×21 kernel is applied to regions where the CAM heatmap value is below a threshold (0.4, 0.3, or 0.2).
CAM deletion	Salient pixels are progressively removed based on their importance and the model's confidence. The resulting confidence degradation is summarized by the Area Under the Curve (AUC). A low AUC (below 0.50) suggests those regions can be discarded.
Augmented dataset	CAM-guided crops retain only the most discriminative regions of each image. Images tend to appear bluer due to the reduced presence of warm-toned background areas.

The flexibility of augmentation and preprocessing methods offers a wide range of training strategies to explore model behavior and improve result quality, while gaining a deeper understanding of the data's specific characteristics.

Each training condition was quantitatively evaluated using a dedicated evaluation pipeline. In addition to classification metrics, a separate benchmarking procedure was applied to measure model efficiency under each training condition. To this end, each trained model was executed multiple times over the same test dataset to record average inference time and standard deviation.

4   Results and Interpretation

The performance of each described technique is evaluated using different metrics and from various perspectives. Implementations of the following methods are available in a public GitHub repository [19].

4.1   Heatmap Visualization

Figure 1 presents heatmap visualizations for each layer of the pretrained AlexNet model [7], based on a randomly selected sample from the test dataset. Each of the five layers shown includes heatmaps generated using the CAM-based methods: Grad-CAM, Grad-CAM++ and Score-CAM, presented in this order.

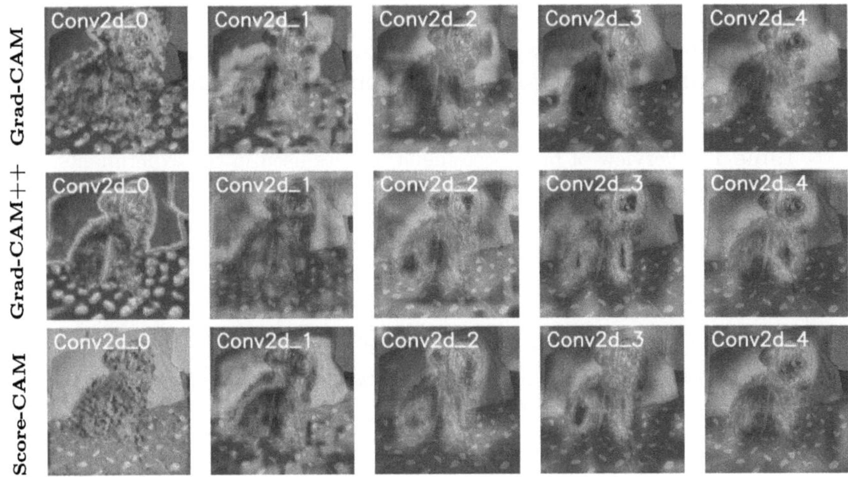

Fig. 1. Layer attribution results.

The intensity of the colors in the heatmap reflects how much each pixel contributes to the model's decision. Red and yellow areas indicate high importance, red typically representing attribution values around 0.77 and about 0.63 for yellow. Meanwhile, green and light blue areas signify lower importance, usually values of about 0.50. In contrast, dark blue and purple regions are typically disregarded by the model, often corresponding to background or irrelevant areas, with attribution values around 0.16. These color-coded activations are obtained by applying a ReLU operation to the weighted sum of the final convolutional layer's feature maps, highlighting only the positive contributions to the predicted class. Based on the frames in Fig. 1, we observe how each pixel's importance varies across layers, revealing the regions most relevant to the model's prediction. Each CAM method highlights different areas throughout the network, particularly around the object of interest—the dog. However, at the final convolutional layer, all three methods converge on the dog's face, indicating it as the key region driving the classification.

For a deeper understanding and more robust explainability, it is important to observe how the heatmaps across different layers converge to highlight consistent regions of interest in the image. This convergence supports meaningful conclusions about which parts of the input are most relevant to the model's prediction. A visual representation of this convergence is provided in Fig. 2.

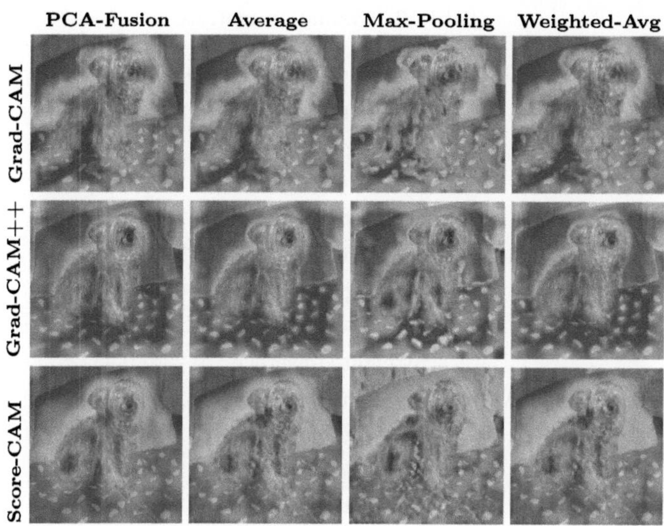

Fig. 2. Feature attribution maps.

Analyzing the generated heatmaps allows us to assess the fidelity of CAM-based explanations and the alignment between saliency regions and the actual semantic object in the image. It becomes evident that the region containing the dog is crucial for the model's decision, especially the defining features of the animal. Additionally, the visualizations reveal how the model handles background areas and transitional pixels, indicating a relevant influence from those regions.

4.2   Overlap with Segmentation Ground Truth

To support a stronger and more precise conclusion regarding the importance of the region of interest compared to the background, an evaluation was conducted using a binary segmentation mask. This ground truth mask assigns a value of 1 to pixels belonging to the animal and 0 to those corresponding to the background.

To objectively assess the quality of the generated attribution maps, we adopt several established evaluation metrics. By applying a threshold of 0.5 to the CAM maps, binary importance masks were produced, used to compute overlap metrics. These values are computed using two standard metrics, as shown in Eq. 1: the IoU, which measures the ratio between the overlap and the total area covered by both masks, and the Dice Coefficient, which represents the harmonic mean of precision and recall at the pixel level.

$$IoU(A, B) = \frac{|A \cap B|}{|A \cup B|}, \quad Dice(A, B) = \frac{2\,A \cap B|}{|A| + |B|} \tag{1}$$

The evaluation results presented in Table 2 show that all scores remain below 0.45, indicating that less than half of the object of interest is consistently iden-

Table 2. Evaluation metrics

CAM Algorithm	Feature Attribution	IoU	Dice
GradCAM	PCA-fusion	0.2111	0.3487
	Average	0.2150	0.3540
	Max-pooling	0.2199	0.3606
	Weighted-average	0.2150	0.3540
GradCAM++	PCA-fusion	0.2912	0.4511
	Average	0.2757	0.4322
	Max-pooling	0.2295	0.3734
	Weighted-average	0.2757	0.4322
ScoreCAM	PCA-fusion	0.2909	0.4507
	Average	0.2546	0.4059
	Max-pooling	0.2135	0.3519
	Weighted-average	0.2546	0.4059

tified as relevant by the model. This suggests limited overlap between the attribution maps and the ground-truth segmentation.

The results also support that advanced CAM variants, such as Score-CAM, especially when combined with richer feature aggregation methods like PCA fusion, improve the fidelity of attribution. These combinations produce heatmaps that better align with the true object mask, enhancing interpretability.

4.3   Insertion/Deletion Tests

To evaluate whether the regions highlighted by an attribution method are truly informative, we analyzed how the model performs when only parts of the image are used for prediction. Figure 3 presents a graph showing the CAM scores obtained by progressively revealing the most important regions, helping estimate the number of pixels required for the model to make an accurate prediction.

The Insertion test [8] evaluates how quickly the model's confidence increases as important features are progressively introduced into a blank or blurred baseline. A higher area under the curve (AUC) indicates more faithful attribution, as key features contribute significantly to the prediction when reintroduced. The Deletion test [8] measures how quickly the model's confidence in the predicted class decreases as the most important features, as identified by the attribution map, are progressively removed from the input. A lower AUC reflects greater attribution quality, suggesting that the identified regions play a critical role in the model's decision-making process.

The Deletion metric showed that removing just 5%–10% of the most important pixels (according to the heatmaps) caused a large drop in prediction confidence. The Insertion metric, rose quickly as the first 10% of pixels were added, indicating that these pixels carry most of the class-relevant information. Good

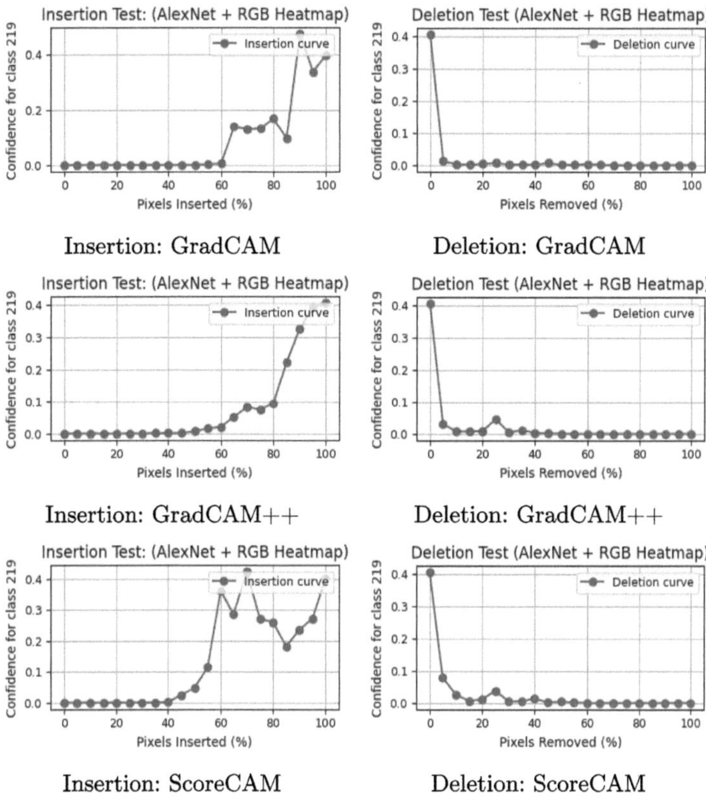

Insertion: GradCAM   Deletion: GradCAM

Insertion: GradCAM++   Deletion: GradCAM++

Insertion: ScoreCAM   Deletion: ScoreCAM

Fig. 3. Insertion and Deletion curves.

attribution map will identify both features that raise the model's confidence when inserted, as well as features that cause the confidence to drop when removed.

4.4   CAM-Guided Preprocessing

Based on the methodology from Table 1, another way to utilize CAM maps is by combining CAM techniques with standard preprocessing steps during training. Table 3 presents the evaluation metrics, including Accuracy, F1 score, Precision and Recall. The results were computed on the testing dataset after training the model with consistent parameters but under different conditions, to evaluate the impact of modifying or removing pixels considered unimportant.

Removing irrelevant regions through CAM-based deletion and background alterations of blacking or blurring the background often provided slight performance gains for pretrained models, while cropping reduced context and tended to lower recall. Targeted blurring using CAM heatmaps showed threshold-dependent effects. For example, using Grad-CAM with a 0.4 threshold yielded

Table 3. Performance metrics under each data preprocessing strategy.

Condition	Accuracy	F1Score	Precision	Recall
Full original dataset	0.96955	0.97012	0.97091	0.96934
Cropped background	0.96502	0.96638	0.96341	0.96931
Blurred background	0.96857	0.96918	0.96918	0.96918
Blacked-out background	0.96083	0.96191	0.95412	0.96982
CAM Augmented images	0.96207	0.96331	0.95038	**0.97660**
CAM deletion	0.96651	0.96693	0.97381	0.96014
GradCAM (threshold 0.4)	**0.97031**	**0.97089**	**0.97320**	0.96860
GradCAM (threshold 0.3)	0.96842	0.96944	0.97015	0.96874
GradCAM (threshold 0.2)	0.96715	0.96872	0.96930	0.96814
GradCAM++ (threshold 0.4)	0.96911	0.96970	0.97104	0.96837
GradCAM++ (threshold 0.3)	0.96784	0.96890	0.96945	0.96835
GradCAM++ (threshold 0.2)	0.96670	0.96822	0.96741	0.96904
ScoreCAM (threshold 0.4)	**0.97085**	**0.97133**	**0.97380**	0.96912
ScoreCAM (threshold 0.3)	0.96945	**0.97001**	0.97151	0.96853
ScoreCAM (threshold 0.2)	0.96770	0.96820	0.96972	0.96669

a higher F1-score and accuracy than the full-dataset baseline, indicating that focusing on the most important 40% of pixels improved learning.

In this study, we start empirically with a threshold of 0.2 (only the top 20% most important pixels were retained) and progressively increased it by 0.1. This process continued as long as the classification accuracy remained lower than or equal to the baseline accuracy obtained on the original dataset.

Applying a CAM-based preprocessing methodology before inference affects processing time. Benchmarked on an AMD Ryzen 5 7600X CPU, Score-CAM introduced an average delay of 7.8 milliseconds, compared to 5.9 milliseconds for Grad-CAM and 6.1 milliseconds for Grad-CAM++. These delays are negligible when using a more powerful GPU.

5   Conclusions

Our experiments highlight the potential of CAM-based explanations to meaningfully guide data preprocessing in ways that improve model interpretability and performance. By leveraging relevance maps to focus the model on salient image regions, we observed consistent improvements in classification accuracy, particularly when using Grad-CAM++ and Score-CAM with medium thresholds (e.g., 0.2–0.4). These settings provided a optimal balance between model focus and contextual information retention. While background blurring or deletion showed minor yet positive effects, aggressive cropping often led to performance degradation, notably in recall. This indicates that seemingly irrelevant background elements can still provide useful context for the model's decisions.

Among the CAM variants, Score-CAM achieved the best spatial alignment with ground truth segmentation masks, as indicated by the highest IoU and Dice scores. However, this came at a substantial computational cost, significantly increasing inference time. In contrast, Grad-CAM++ provided a more favorable trade-off, offering notable performance gains with much lower latency.

Importantly, simpler preprocessing approaches such as CAM-based deletion or targeted blurring incurred minimal computational overhead while still enhancing interpretability. Interestingly, inference time analysis revealed that to the baseline AlexNet performed fastest, due to its lack of learned activations— empha- sizing the computational load imposed by trained filters. Finally, perturbation-based evaluation metrics such as Insertion and Deletion proved more faithful in assessing explanation quality than segmentation-overlap metrics (IoU, Dice), aligning with prior findings [20,21]. Notably, deletion experiments showed that removing as little as 5% of the most salient pixels (according to heatmap scores) caused a sharp drop in model confidence, confirming the model's strong dependence on a small set of critical features.

Together, these findings offer actionable insights for leveraging explanation techniques not only for model transparency, but also as tools for performance optimization and input refinement.

ACKNOWLEDGMENT. This paper was supported by the project "NETWORK OF EXCELLENCE IN DIGITAL TECHNOLOGIES AND AI SOLUTIONS FOR ELECTROMECHANICAL AND POWER SYSTEMS APPLICATIONS – DITARTIS", funded from the European Union's Horizon Europe research and innovation programme, HORIZON-WIDERA-2021-ACCESS-03-01, under the grant agreement No 101079242. It reflects only the author's view; the EU Commission is not responsible for any use that may be made of the information it contains.

References

1. Zeiler, M.D., Fergus, R.: Visualizing and understanding convolutional networks. In: Computer Vision – ECCV 2014. Springer International Publishing, pp. 818–33. https://doi.org/10.1007/978-3-319-10590-1_53
2. Ribeiro, M.T., Singh, S., Guestrin, C.:'Why Should I Trust You?': explaining the predictions of any classifier. In: Proceedings of the 22nd ACM SIGKDD International Conference on Knowledge Discovery and Data Mining, pp. 1135–44. ACM (2016). https://doi.org/10.1145/2939672.2939778
3. Miller, T.: explanation in artificial intelligence: insights from the social sciences. Artif. Intell. **267**, 1–38 (2019)
4. Adebayo, J., Gilmer, J., Muelly, M., Goodfellow, I., Hardt, M., Kim, B.: Sanity checks for saliency maps. arXiv preprint: arXiv:1810.03292 (2018)
5. Szczepankiewicz, K., et al.: Ground truth based comparison of saliency maps algorithms. Sci. Rep. **13**(1), 16887 (2023). https://doi.org/10.1038/s41598-023-42946-w

6. Garreau, D., von Luxburg, U.: Explaining the explainer: a first theoretical analysis of LIME. arXiv preprint: arXiv:2001.03447 (2020)
7. Krizhevsky, A.: One weird trick for parallelizing convolutional neural networks. arXiv preprint: arXiv:1404.5997 (2014)
8. Hama, N., Mase, M., Owen, A.B.: Deletion and insertion tests in regression models. arXiv preprint: arXiv:2205.12423 (2022)
9. Kaggle: Dogs vs. Cats dataset. https://www.kaggle.com/c/dogs-vs-cats/data
10. Doshi-Velez, F., Kim, B.: Towards a rigorous science of interpretable machine learning. arXiv preprint: arXiv:1702.08608 (2017)
11. Adadi, A., Berrada, M.: Peeking inside the black-box: a survey on explainable artificial intelligence (XAI). IEEE Access **6**, 52138–52160 (2018). https://doi.org/10.1109/ACCESS.2018.2870052
12. Zhou, B., Khosla, A., Lapedriza, A., Oliva, A., Torralba, A.: Learning deep features for discriminative localization. In: IEEE Conference on Computer Vision and Pattern Recognition (CVPR), Las Vegas, NV, USA, pp. 2921–2929 (2016). https://doi.org/10.1109/CVPR.2016.319
13. Selvaraju, R.R., Cogswell, M., Das, A., Vedantam, R., Parikh, D., Batra, D.: Grad-CAM: visual explanations from deep networks via gradient-based localization. In IEEE International Conference on Computer Vision (ICCV), Venice, Italy, pp. 618–626 (2017). https://doi.org/10.1109/ICCV.2017.74
14. Chattopadhay, A., Sarkar, A., Howlader, P., Balasubramanian, V.N.: Grad-CAM++: generalized gradient-based visual explanations for deep convolutional networks. In: IEEE Winter Conference on Applications of Computer Vision (WACV), Lake Tahoe, NV, USA, pp. 839–847 (2018). https://doi.org/10.1109/WACV.2018.00097
15. Wang, H., et al.: Score-CAM: score-weighted visual explanations for convolutional neural networks. In: IEEE/CVF Conference on Computer Vision and Pattern Recognition Workshops (CVPRW), Seattle, WA, USA, pp. 111–119 (2020). https://doi.org/10.1109/CVPRW50498.2020.00020
16. Krizhevsky, A., Sutskever, I., Hinton, G.E.: ImageNet classification with deep convolutional neural networks. In: Advances in Neural Information Processing Systems (NeurIPS), vol. 25, pp. 1097–1105 (2012)
17. Wang, Y., Zhang, T., Guo, X., Shen, Z.: Gradient based feature attribution in explainable AI: a technical review. arXiv preprint: arXiv:2403.10415 (2024)
18. PyTorch Tutorials: Interpretability with Captum. https://docs.pytorch.org/tutorials/beginner/introyt/captumyt.html. Accessed 30 May 2025
19. GitHub repository: https://github.com/Stoica-David/XAI-Experiments
20. Fong, R.C., Vedaldi, A.: Interpretable explanations of black boxes by meaningful perturbation. In: IEEE International Conference on Computer Vision (ICCV), Venice, Italy, pp. 3449–3457 (2017). https://doi.org/10.1109/ICCV.2017.371
21. Zhang, et al.: Segmentation and evaluation metrics in medical imaging. Med. Image Anal. (2020)
22. Lundberg, S.M., Lee, S.I.: A unified approach to interpreting model predictions. In: Advances in Neural Information Processing Systems, vol. 30 (2017). arXiv preprint arXiv:arXiv:1705.07874 (2024)
23. Amit, D., et al: Explanations based on the missing: towards contrastive explanations with pertinent negatives. In: Advances in Neural Information Processing Systems (2018). https://doi.org/10.48550/arXiv.1802.07623
24. Petsiuk, V., Das, A., Saenko, K.: RISE: randomized input sampling for explanation of black-box models. arXiv preprint: arXiv:1806.07421 (2018)

25. Zhang, J., et al: Top-down neural attention by excitation backprop. Int. J. Comput. Vis. **126**, 1084–1102 (2018). https://doi.org/10.1007/s11263-017-1059-x
26. Guindon, B., Zhang, Y.: Application of the dice coefficient to accuracy assessment of object-based image classification. Can J. Remote Sens. **43**(1), 48–61 (2016). https://doi.org/10.1080/07038992.2017.1259557

Pairwise 3D Fragment Matching
Classification with Graph Neural Networks

Raluca-Diana Chiş[(✉)] [ID] and Camelia Chira [ID]

Babeş-Bolyai University, Cluj-Napoca, Romania
`raluca.chis@ubbcluj.ro`

Abstract. Automatic object reassembly is an important task in archaeology, robotics and medicine. The creation of a tool for finding fragments that complement each other and assembling them into a more or less complete object would facilitate the work of restorers. Recent works focus on predicting the pose of a set of fragments relative to each other or on estimating the matching of different surfaces with mathematical approaches. With this article, we propose a deep-learning method for pairwise matching classification of point clouds. The research methodology is based on the use of Siamese neural networks combined with graph convolutions. Pairs of matching and non-matching point clouds have been created using the fragmented objects provided in the Breaking Bad dataset. These objects are represented as point clouds and are divided into multiple pieces. Our best experiment reveals an accuracy of 82.7% and an F1 score of 81.09%, which demonstrates the power of the implemented method.

Keywords: Point cloud matching · Graph networks · Breaking Bad

1 Introduction

The reassembly of an object revolves around finding fragments of broken objects that match, similar to solving a jigsaw puzzle. The problem can be approached in two or three dimensions. For flat fragments, such as frescoes, papyrus pieces, torn papers or actual puzzle pieces, the general approach is based on identifying the edges and the curvature of each piece, then finding the matching fragments by comparing the found features and finally merging all pairs. However, in real life, the majority of situations require a 3D solution, where each piece is a tridimensional digital representation of an object. The 3D approach differs by the fact that extraction and matching are done for surfaces, not edges. Moreover, the input of the models is no longer represented by images, but by point clouds. The motivation behind investigating this problem comes from the applications of the problem in different real-life scenarios. In archaeology, restoring an object is prone to error, not scalable, and the fragments are fragile. Furthermore, the

---

https://www.cs.ubbcluj.ro/.

C. Chira et al. (Eds.): InnoComp 2025, CCIS 2793, pp. 96–110, 2026.
https://doi.org/10.1007/978-3-032-12478-4_7

reassembly of an object has applications in the correct setting of bone fragments following a comminuted fracture, as presented in [21]. Industrial robots frequently perform assembly tasks and could highly benefit from a robust system, that quickly adapts to new challenges, as described in [6].

The challenging task of pairwise matching of fragments represents an important initial step in the reassembly process. This objective involves identifying pieces of objects that were once connected. In the existing literature, few research works focus solely on pairwise matching ([1,12,26]), and most of the recent ones estimate the positions of a set of pieces in the reassembled object ([11,14,15,27,30]). The pairwise matching methods mostly use geometrical approaches ([1,12,24]), which have flaws. Son et al. [24] mention the lack of usability of their method on shell objects, and Wang et al. [27] indicate the use of deep learning technologies for improving the efficiency and the robustness of their model.

This study focuses on filling this research gap by proposing a Siamese neural network with graph convolutions. We started with an idea inspired by the MatchMakerNet model [26] and set out to investigate multiple architectures in an attempt to find out if performance changes and if a simpler model could meet this challenge. The benefits of a simpler architecture model are closely linked to shorter training times and reduced inference times. The architectures we propose are based on the use of graph convolutions (GCNConv [8]) layers. We also investigate the usefulness of using edges that point from a node to itself, as well as DynamicEdgeConv [29] layers, in which edges are regenerated at each step.

The novelty of the models we propose comes from regenerating the edges of the two point clouds after concatenating them, so that their proximity can be analyzed in the best possible way. What sets us apart from MatchMakerNet is obtaining, using a much simpler model, simlar results, close to those mentioned by these authors. Our best accuracy for the binary matching classification of two fragments is 82.7%. In addition, the transparency in terms of research methodology makes the model proposed by us easy to reproduce and further use to develop methods for object reassembly. For our experiments, we have generated pairs of matching and non-matching fragments, extracted from the Breaking-Bad dataset [23].

This paper is structured in 6 sections as follows. This initial section introduced the problem of 3D object reassembly. Section 2 details the research done in this field, along with the open challenges. The third section presents our approach to the problem, followed by a description of the research methodology in Sect. 4. Experiments and results are described in 5, and lastly, the conclusion and future work perspectives will be discussed in Sect. 6.

2   Related Work

In this section, a review of the most recent research works on object reassembly is provided, focusing on the pairwise or multi-part fragment matching, along with the datasets used. Based on our observation, the works related to object reassembly follow two main paths: some of the authors approach the problem as a pose estimation one, while others consider analysing the surface features of the fragments as a better solution.

3D Pose Prediction. Recent research works concentrate on the anticipatory modelling of fragment poses. The input components are acknowledged as constituents of an object, with the models forecasting in most cases, their 6-DoF alignment. Despite the good results, this specificity makes the models challenging to extrapolate and adapt for broader applications, such as discerning corresponding pairs within a collection of objects. For enhanced results, some papers propose an initial feature extraction with PointNet [22] [4, 11, 13, 15, 18, 28, 30]).

The models proposed for this task fall into several categories. One of them is represented by the GNNs (graph neural networks). Huang et al. [4] propose a dynamic graph-learning method, that is based on an iterative graph neural network. Li et al. [13] implemented a method for fragment faces segmentation with a conditional U-Net and PointNet [22], followed by a pose prediction model, that integrates two ResNet networks, pre-trained on ImageNet, and a graph message-passing algorithm. Narayan et al. [18] designed a recurrent graph learning network, inspired by [4] and [13] and based on progressive assembly. The created framework is based on a message-passing graph backbone. Cheng et al. [7] developed a Graph Neural Network for modelling a score function, which employs an embedding layer, composed of a Gaussian Fourier Projection. MatchMakerNet, proposed by Villegas et al. [26], is a simplified version of the Dynamic Graph CNN from [29].

Other techniques include the use of transformer networks [25], such as Neural Shape Mating (NSM) [2], FiT proposed by Wang et al. [28] (a framework for finding missing parts of a 3D object), *Jigsaw* (Joint Learning of Segmentation and Alignment Framework) [15], Geometric Point Attention Transformer (GPAT) [11], Proxy Match TransformeR (PMTR) [10], PuzzleFusion++ [30]. Moreover, non-deep learning strategies exist. Alagrami et al. [1] present a pose prediction method, which uses breaking curves extraction, a region-growing algorithm, and the Iterative Closest Point (ICP) algorithm for alignment.

Most of these approaches ([4, 7, 11, 13, 14, 18, 28]) use in their experiments the PartNet dataset [17] or Breaking Bad dataset [23] ([1, 2, 10, 11, 14, 15, 26, 30]). The main difference between these sets is that for Breaking Bad, the fragmentation was not performed semantically, but based on the natural tendency of objects to break, which makes it more robust.

Surface Feature Matching. Some studies propose analysing the surfaces of the fragments, and in this case, the pipeline comprises the segmentation of the surfaces of the fragments, the analysis of their geometrical structure and the identification of surfaces that complement each other. In this regard, some authors have proposed region-growing methods, combined with a Hierarchical Agglomerative Clustering, for identifying the surfaces of a fragment ([19, 27]), while others have analyzed the curvature of each surface and depending on the level of unevenness, assigned concave, convex and other labels to keypoints ([12, 24, 31]). Many of the authors ([12, 19, 31]) have approached coarse matching to eliminate obviously mismatched pairs, followed by fine matching. Multiple ICP algorithm variants have been applied for the final reassembly or for reducing the search space ([12,19,21,31]). Moreover, search graphs were implemented in [19] and [27] for optimal reassemblies. Many of these research works ([1,12,24,26,27,31]) apply the Puzzle3D dataset [5], containing scans of 6 objects (cake, brick, Venus, sculpture, head, gargoyle statue) and of 20 pieces from the Forma Urbis Romae set [9], pieced into a total of 101 fragments.

The literature review shows that the researchers have been focused in the last years on pose-prediction solutions. This can be explained by the rapid progress of the deep-learning models, which represent the base for most of these solutions. On the other side, the articles proposing surface-matching methods require detailed scans of real objects, fragmented before scanning, to be effective and to our knowledge, very few such datasets exist. Our work positions among the pairwise matching approaches, for which few authors propose solutions (namely [1,12,26,28,30]). This direction is worth studying, as this use case is highly required in real-life scenarios. In most of the works studied, it is assumed that the fragments used as input form an object and the proposed models indicate the position in the reassembled object. Alagrami et al. [1] mention the need for detecting non-matching fragments across many broken object pairs, and this paper aims to fill this gap.

3   Proposed Approach

To tackle the task of pairwise matching classification, we propose several models suitable for working with 3D data represented as point clouds. In what follows, we will present 3 models for the binary classification of the match state between two given point clouds, using graph convolutional networks (GCN) and dynamic graph convolutions.

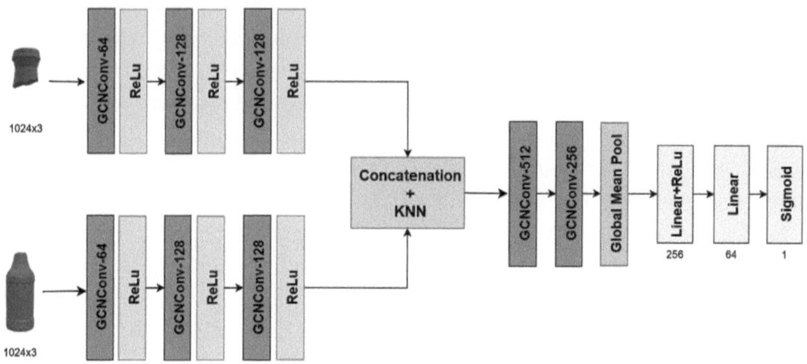

Fig. 1. Siamese model architecture with 3 GCN layers (*3GCNNet*).

3GCN. Graph Convolutional Networks represent a well-known technique for managing graph-structured data. At the time of their introduction [8], they outperformed the existing methods for graph classification. The implicit method works similarly with CNNs, but unlike image pixels, a node has varying numbers of neighbours, making them being unsuitable for the classical convolutions (unless a padding is performed, which is memory-consuming). PyTorch provides a graph convolutional layer (GCNConv), implemented based on the research results of Kipf et al. [8]. To process graphs, GCNConv aggregates information from the neighbours of a node and updates its representation. The propagation rule is the following:

$$H^{(l+1)} = \sigma \left(\tilde{D}^{-1/2} \tilde{A} \tilde{D}^{-1/2} H^{(l)} W^{(l)} \right),$$ (1)

where $H^{(l)}$ is the feature matrix at layer l, $W^{(l)}$ is the trainable weight matrix, $\tilde{A} = A + I$ is the adjacency matrix with self-loops added, $\tilde{D}$ is the diagonal degree matrix of $\tilde{A}$, σ is a non-linear activation function, ReLu in our case.

Considering this, we have developed a Siamese network architecture, which makes use of the aforementioned GCNConv layer, as displayed in Fig. 1. For each branch of the model, the two point clouds were processed through 3 convolutional blocks, each block consisting of one GCN layer (with output channel sizes of 64, 128 and 128) and the rectified linear unit (ReLu) activation. The results of the two branches were concatenated, and for the new node information, new edges were created with help from the k-nearest neighbour algorithm (KNN). KNN uses the Euclidean distance metric to compute the distances between nodes. For each node, it assigns edges between it and the k nearest neighbours. This step represents a novelty for the match finding task and is essential for the model's success, as the new concatenated graph requires new edges to analyse the proximity within the feature space. Without it, the concatenation would be less beneficial, since the graphs resulted from the 2 branches would be completely independent from each other. The new graph was passed through two graph convolutions with an output channel size of 256. Then, the node features were aggregated into one value per node with the *global_mean_pool* method from PyTorch Geometric [3] and passed through a multilayer perceptron.

In an initial phase of the development, the model was not using the edges directed from a node to itself (we will reference this model as *3GCN*). We later concluded that there is a strong benefit from using the self loops: if a node is isolated (which is possible to happen after the sampling process), the self-loop could help avoid losing its features during the message passing step, ensuring at least that the isolated node retains its own feature information. Consequently, we have experimented with the *3GCN* model with the existing self-loops (*3GCN-partial-self-loops*). Then, in order to make sure that all isolated nodes have a self-loop, we have added all possible self-loops (*3GCN-all-self-loops*).

4GCN. Starting from *3GCN*, we wanted to investigate the need for a more complex architecture, and we have added one more convolutional block with output size 256 for both branches. The following sections will refer to this architecture as *4GCN*. This time, the increase in model complexity did not bring great improvements in results, so it was decided that we would experiment with different methods for generating the edges of the input graphs. Inspired by the dynamic edge convolutions [29], we have generated all initial edges with the KNN algorithm (*4GCN-KNN*).

DynamicNet. Inspired by the PointNet model [22], a state-of-the-art model in point cloud classification that is highly flexible and invariant to permutations of the input points, Wang et al. proposed a dynamic graph CNN model [29], in which the graph is re-constructed at each layer, by computing the edges with the k-nearest neighbour algorithm. Although computationally expensive, this enables adaptive neighbourhood learning, capturing better local structures over layers.

This approach inspired us to develop a model based on the idea of dynamically generating edges for our point clouds. Similarly to the other two models, this architecture is a Siamese. In this case, the point clouds are processed with 2 Dynamic Edge Convolution layers (DynamicEdgeConv in PyTorch Geometric). Each of these DynamicEdgeConv layers starts with the generation of new edges with KNN. Following this, an analysis with an edge convolution function is performed, in which each node's features are concatenated with the neighbouring node's features, then passed through a multi-layer perceptron (MLP), as depicted in Fig. 2. The MLP comprises 2 Linear layers, with output sizes of 64 and 128 for the first DynamicEdgeConv block and 256, 128 for the second one, and 2 ReLu functions. The resulting features are aggregated with the *global_mean_pool* method. This process is described by the following formula:

$$h_i^{(l+1)} = mean_{j \in \mathcal{N}(i)} \mathrm{ReLU} \left(\theta \cdot \left(h_j^{(l)} - h_i^{(l)} \right) + \phi \cdot h_i^{(l)} \right), \tag{2}$$

where $h_i^{(l)}$ is the feature of node i at layer l, $h_j^{(l)}$ is the feature of neighbor node j, $\mathcal{N}(i)$ is the set of neighbors of node i, θ and ϕ are the learnable weight matrices.

The output of the branches is concatenated, new edges are created with KNN and the rest of the model is exactly as the ones described before. This model will be referenced as *DynamicNet*.

4   Methodology

In this section, the dataset with which we conducted experiments will be described, along with the preprocessing steps implied. In addition to this, the criteria used for evaluation will be discussed.

4.1   Dataset

To perform the proposed experiments, one of the latest fragmented object datasets has been used, named Breaking Bad [23]. Its main advantage, compared to the other datasets comprising fragmented objects, is that the fragmentation is done considering the natural tendency of an object to break, replicating

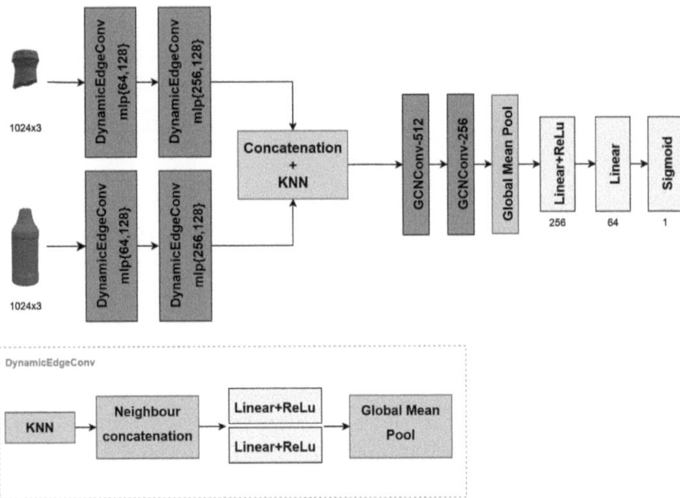

Fig. 2. Dynamic Graph Convolutional Network Architecture for point cloud match classification.

real-life situations. We used two of the three object categories provided, namely *Artifact* and *Everyday*. The *Artifact* category is intended to be used for archaeological reconstruction and contains the meshes from Thingi10k [32], which were labelled as *sculpture* or *scan*. This category provides a total of 204 objects, with 100 fragmentation modes, each divided into several pieces ranging from 2 to 99. The *everyday* category comprises 20 object categories from the PartNet dataset [17], representing common items (e.g. bottle, cup, toy). Each object category has 4 to 106 items, summing up to 542 objects.

Considering the objects provided in the Breaking Bad dataset [23], multiple sets of matching and non-matching pairs have been established. Each object divided into two parts has been considered a matching pair, and the non-matching pairs were created by combining the pieces of two consecutive fragmentation modes of the same object. This procedure has been applied to both categories, *Artifact* and *Everyday*, and the first one was used for training and the second one for testing. Since the obtained datasets were unbalanced, the number of non-matching pairs was sampled to equal the number of matching pairs. For *Artifact*, 12448 were obtained and for *Everyday* 27625.

The point clouds consist of an enormous number of vertices and faces. The faces represent sets of 3 vertices that are all connected 2 by 2, from which we extracted the edges necessary for the graphs inputed to our models. The fragments with which we have worked comprise between 6 and 75195 vertices and faces ranging from 8 to 150392.

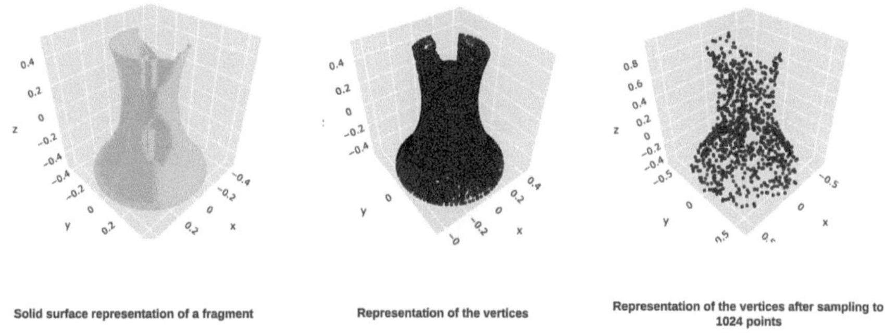

Fig. 3. Example of point cloud object, before and after sampling.

4.2   Data Preprocessing

Due to the massive computational costs for these point clouds, our research
has been directed to finding a suitable downsampling method. For this reason,
Farthest Point Sampling (FPS) has been applied. This algorithm starts from a
random point and manages to find the points that are farthest away from each
other, thus preserving the contour of the point cloud. Downsampling the vertices
leads to several useless faces, which were filtered out. Following this, a process of
either downsampling or upsampling was applied. The downsampling was done
by randomly selecting the faces, with higher chances for those with bigger areas.
In the opposite situation, where the FPS of the vertices led to having very few
faces, Delaunay triangulation was applied. This method generates faces with the
condition that no vertex lies in the circumcircle of any face in the set. Different
values for faces and vertices were experimented with, as we further describe in
the next section. Figure 3 depicts an object before and after the preprocessing
phase.

The FPS method was chosen after thoroughly analysing the available downsampling methods. We have noticed that many recent papers in this field describe
in their research methodology the use of FPS with a downsampling value of 1000
([4,13,18,28]). Others, such as [2] and [31], propose uniform sampling, where
points are selected randomly with the condition that they are evenly distributed
across the point cloud. This algorithm is faster than FPS, but the Farthest Point
Sampling method provides a better spread of the points, and for our task, this is
very important, as the points on the outskirts of a point cloud provide valuable
information for finding matching pairs. For this reason, we have chosen to work
with FPS, and we downsampled our point clouds to 1024 points initially and
then to 2048, which helped most of our models to improve their results, as we
further describe in the next section.

4.3   Evaluation Metrics

In order to evaluate our experiments, several metrics have been utilised. We
computes the accuracy, F1 score, Binary Cross-Entropy (BCE) loss and we have

also observed the time required for training the model for the first epoch, as
this restricts the number of experiments that can be done in a given time. The
accuracy is defined as the percentage of pairs correctly classified. The F1 score is
the harmonic mean between precision and recall, with values ranging from 0 to 1,
where 1 indicates perfect precision and recall. The BCE loss is a common function
for binary classification tasks, that penalizes the model heavily for predictions
that are confidently wrong and is expressed as

$$l(x, y) = -[y \cdot \log(x) + (1 - y) \cdot \log(1 - x)], \tag{3}$$

where y is the true label (0 or 1) and x is the predicted label. A low value is
desirable.

5   Computational Experiments

In this section, the implementation details of our approach will be presented. In
addition to this, the performance of our models will be discussed and compared
to similar approaches.

5.1   Experiment Setup

The implementation was done with the framework PyTorch [20], the optimiser
used for all experiments was Adam, with a learning rate of 0.001, which has been
established experimentally. The experiments were done on Google Colaboratory,
using GPU Nvidia T4, with 16 GB GDDR6 memory. Due to the computer's
limited processing capacity and RAM storage, it was decided to create a subset
of 1000 pairs for training and 1000 for testing. Sampling was done using the
Pandas library [16], preserving the data distribution. Regarding the number of
points included in each point cloud, experiments were done with sampling to
1024 and 2048 points.

5.2   Results

Multiple experiments have been conducted for the 3 architectures presented.
Table 1 exhibits the results of our main experiments evaluated on the testing
data. Fluctuating the number of neighbours found (k) when applying the KNN
algorithm has proven to be very beneficial, as we discovered that a value of 10 for
k is optimal for our models. By far, our best model is the one with k=10 neighbours, graphs sampled to 2048 points and 4096 faces (4096*3=12,288 edges)
and the *3GCNN* architecture, obtaining an accuracy of 82.7%. To reach this
conclusion, hyperparameter tuning was achieved for each architecture. Figure 4
presents a few examples of pair classifications with our best model (id 3), for
both matching and non-matching situations. A more complex model is not necessarily a better one and this is proven by experiments 6 and 7, where the
architecture with 4 convolutions was used, *4GCNNet*, along with an increased

number of points, 2048, and of faces, 4096. Despite this, the accuracy was only 77.30% for the model with ID 11. The use of self-loops did not bring any advantages (*3GCN-partial-self-loops* and *3GCN-all-self-loops*), so that the accuracies obtained were neither good, nor bad, even though the lowest loss of all experiments was retrieved by the *3GCN-all-self-loops* architecture.

The results of the experiments related to the DynamicEdgeConv layer are displayed in Table 2, and they are approaching the best ones achieved with the *3GCN* architecture. The highest accuracy was retrieved for the experiment with ID 9, 79.10%. The idea of generating all initial edges with KNN for the *4GCN-KNN* architecture did not yield great results, and what persuaded us to stop experimenting with it was the increased time needed for training. The longest training process was needed by this model, 119 min for the first epoch alone, which happened due to the edge generation process.

Table 1. Experimental results on the test set for the *GCNNet* model variations (k is the number of neighbours considered for KNN and for the time we considered only the approximate time needed to train the first epoch in minutes)

Id	Model	Pairs	Points	Faces	Batch size	k	Acc. % ↑	F1 % ↑	Loss ↓
1	*3GCN*	1000	1024	2048	64	10	81.60	79.82	0.2390
2	*3GCN*	1000	2048	4096	64	5	82.70	80.88	0.2140
3	*3GCN*	1000	2048	4096	64	**10**	**82.70**	**81.09**	0.2073
4	*3GCN-partial-self-loops*	1000	2048	4096	32	10	74.30	71.66	0.2190
5	*3GCN-all-self-loops*	1000	2048	4096	32	10	74.80	71.23	**0.1982**
6	*4GCN*	1000	2048	2048	32	40	71.30	60.08	0.6270
7	*4GCN*	1000	2048	4096	32	40	77.30	71.94	0.3809

Table 2. Experimental results on the test set for the *DynamicNet* model variations (n is the number of neighbours considered for the KNN applied in DynamicEdgeConv, k is the number of neighbours considered for KNN after the concatenation)

Id	Model	Pairs	Points	Batch size	n	k	Acc. % ↑	F1 % ↑	Loss ↓
8	*DynamicNet*	1000	1024	16	5	10	78.80	78.59	0.4130
9	*DynamicNet*	1000	1024	16	10	10	**79.10**	**77.45**	**0.4116**
10	*DynamicNet*	1000	2048	16	10	10	78.30	76.44	0.4168
11	*4GCN-KNN*	1000	1024	16	5	10	64.80	63.00	0.6018

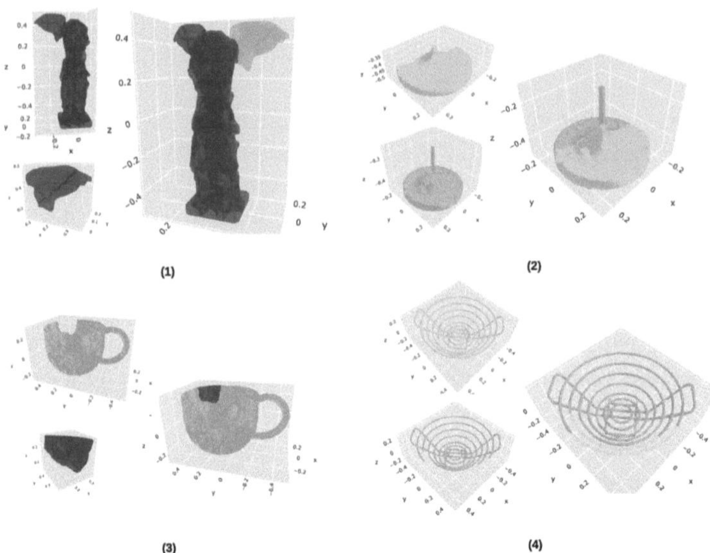

Fig. 4. Examples of predictions made with our best model: (1)&(3) matching pairs, (2)&(4) not-matching pairs; (1)&(2) incorrect predictions, (3)&(4) correct predictions.

5.3   Discussion

The models following the *3GCN* architecture have obtained the overall best results, with a maximum accuracy of 82.7% and an F1 score of 81.09%. Figure 5 shows on the left the confusion matrix for this model. It can be noticed from this figure that the model manages to predict a non-matching pair with a probability of 87.69%, while for the matching case, the probability drops to 71.34%, suggesting that this task is more difficult. This observation is reinforced by another one. On the experiments performed, in most cases, even though the accuracy is raised, the F1 score is low.

On the other hand, the model *DynamicNet* with id 9 has performed incredibly well and even though the accuracy is lower than for the *3GCN* models (78.80%) and the F1 score too (78.59%), the architecture using dynamic edge convolutions delivered the highest prediction probability for the matching case (recall), 81.04%. Figure 5 depicts on the right the confusion matrix of the discussed model and points out that the model is more prone to classify a pair as matching, including more true and false positives than for *3GCN*.

Our work was inspired by the MatchMakerNet model [26], where the authors also propose a pair classification model with remarkable results. The MatchMakerNet was also trained on pairs from *Artifact* [23] and tested on objects from Everyday [23]. The authors mention that for the process of pair creation, they considered all possible pairs between parts of the same fragmentation mode, computed the minimum distance between them, and if the distance was below a given threshold the pair was labelled as matching. However, the threshold is

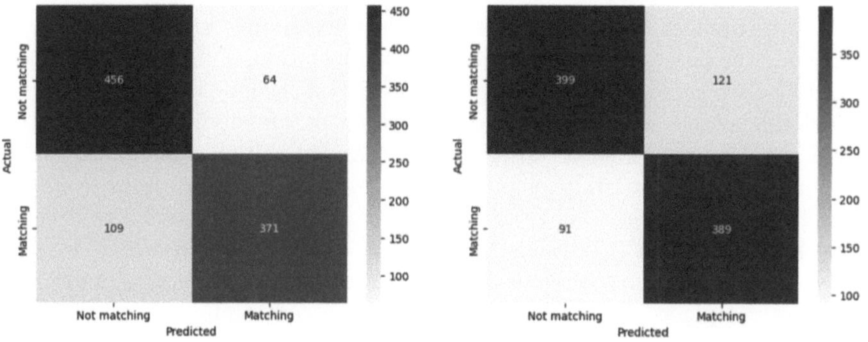

Fig. 5. (Left) Confusion matrix of our best model with an accuracy of 82.7% (*3GCN*). (Right) Confusion matrix of the *DynamicNet* model with id 16 (highest recall of all models, 81.04%).

not mentioned, and for the pairs generated from *Artifact* it is specified that the obtained set was balanced, but it is not mentioned how exactly, which makes the exact reproduction of their dataset a bit difficult. In comparison with [26], our results are similar. Their mean accuracy on *Artifact* is equal to 87.31% and on *Everyday* they reach an accuracy of 86.93%.

Another model with a task similar to ours is described in [30]. The verifier module does a binary classification of fragment pairs, starting from their predicted alignments, and obtains a very good accuracy, especially when using a threshold of 0.9, 90.77%, and 87.99% for a 0.5 threshold. However, the recall for both cases is very low, 45.95% for threshold 0.9 and 73.51% for threshold 0.5, which suggests many false negative predictions. This could be happening due to the fact that the module is applied on an unbalanced set of fragment pairs. In an initial phase of object reassembly, the denoiser module outputs a lot more alignments of fragments that do not match with the anchor (the starting point of the reassembly), thus leading to unbalance. The authors fail to specify exactly how many pairs are used for training the model. In our case, the dataset is balanced, with approximately half of the pairs forming a match, thus indicating more relevant results and a better recall (81.04%).

In comparison with the FiT framework [28], proposed by Wang et al., our model performes a bit worse. In this work, the authors present results on the completion of an object with a piece from a given set of candidate parts (toolkit with size 30). They provide an accuracy of matching equal to 92.93% on the PartNet dataset [17], but it should be mentioned that even though this value is with 10.23% higher than ours, the PartNet provides a semantic fragmentation of the objects, which significantly eases the task. Wang et al. also provide results on adapted versions of the DGL model [4] and on the model by Li et al. [13]. For DGL, the accuracy is also high, 91.05%, and for the mother model, the value is lower than ours, 81.75%.

As presented in Sect. 2, most of the recent techniques in this field focus on estimating the pose, in terms of rotation and translation, of fragments of objects, followed by a complete object reassembly. While on the long term this is our goal too, in this article we have specifically focused on classifying whether 2 point clouds could be matching or not. Our input is very valuable thanks to the fact that, in a real-life situation, a program should reassemble objects from a set of unclustered fragments, and in the studied papers, it is assumed that certain pieces form an object. It is difficult to compare our results with others. However, we can consider that a low value for part accuracy (PA) indicates many pieces that were not placed were they belong, which we could interpret as pieces that were matched with the wrong parts. With this idea, we can say that our model exceedes the models presented in Sect. 2, the highest values for PA being 73.1%, 71.6% and 70.56% (these values are obtained for complete object assembly).

6 Conclusions and Future Work

This article proposes several models for pairwise 3D object matching classification, as this is an important step in the reassembly process of a broken object. With applications in cultural heritage, robotics and even medicine, the reassembly problem presents several challenges, including degraded fragments, irregular shapes, and missing components. Additionally, another obstacle arises from the absence of a comprehensive dataset featuring genuinely damaged objects. One of the most useful point cloud datasets is Breaking Bad [23], which closely mirrors real-world scenarios by simulating the natural fragmentation tendencies of objects during generation.

We introduce multiple architectures that use a parallel processing of two point clouds with graph convolutions, followed by concatenation, edge regeneration, followed by more processing with graph convolutions and fully connected layers. We experimented on the dataset from Breaking Bad, from where pairs of point clouds were selected and the number of points for each of them was downsampled to 1024 and 2048 with Furthest Point Sampling. Out of all our experiments, the best accuracy obtained was 82.7%, for a model with three graph convolutions (GCNConv [8]) before concatenation, followed by two more after concatenation.

In the future, we would like to explore the use of graph convolutional attention networks for the matching classification task and explore more time-efficient methods for point cloud preprocessing step. Furthermore, to reach our goal of complete object reassembly, we will begin developing models for estimating the precise pose of matching point clouds. The current literature review revealed an impressive strength of the transformer networks for this task thus, we consider experimenting with a model of such matter. In addition to this, we are eager to develop a hybrid model that will first find possible matches in a set with the current classification model, and then find the optimal position for each fragment. From our knowledge, there are a few models addressing this specific research question, and we would like to investigate it deeper.

Acknowledgment. This research is supported by the project "Romanian Hub for Artificial Intelligence - HRIA", Smart Growth, Digitization and Financial Instruments Program, 2021-2027, MySMIS no. 334906.

References

1. Alagrami, A., Palmieri, L., Aslan, S., Pelillo, M., Vascon, S.: Reassembling broken objects using breaking curves (2023)
2. Chen, Y.C., Li, H., Turpin, D., Jacobson, A., Garg, A.: Neural shape mating: self-supervised object assembly with adversarial shape priors. In: Proceedings of the IEEE/CVF Conference on Computer Vision and Pattern Recognition (CVPR), pp. 12724–12733 (2022)
3. Fey, M., Lenssen, J.E.: Fast graph representation learning with PyTorch geometric (2019). https://arxiv.org/abs/1903.02428
4. Huang, J., et al.: Generative 3D part assembly via dynamic graph learning. In: Larochelle, H., Ranzato, M., Hadsell, R., Balcan, M., Lin, H. (eds.) Advances in Neural Information Processing Systems, vol. 33, pp. 6315–6326. Curran Associates, Inc. (2020)
5. Huang, Q.X., Flöry, S., Gelfand, N., Hofer, M., Pottmann, H.: Reassembling fractured objects by geometric matching. ACM Trans. Graphics **25**(3), 569–578 (2006)
6. Hughes, J., Gilday, K., Scimeca, L., Garg, S., Iida, F.: Flexible, adaptive industrial assembly: driving innovation through competition. Intel. Serv. Robot. **13**(1), 169–178 (2019). https://doi.org/10.1007/s11370-019-00292-9
7. Junfeng, C., Mingdong, W., Ruiyuan, Z., Guanqi, Z., Chao, W., Hao, D.: Score-PA: score-based 3D part assembly (2023)
8. Kipf, T.N., Welling, M.: Semi-supervised classification with graph convolutional networks (2017). https://arxiv.org/abs/1609.02907
9. Koller, D., Trimble, J., Najbjerg, T., Gelfand, N., Levoy, M.: Fragments of the city: stanford's digital forma Urbis Romae project. In: Proceedings of the Third Williams Symposium on Classical Architecture, vol. 61, pp. 237–252 (2006)
10. Lee, N., et al.: 3D geometric shape assembly via efficient point cloud matching. In: Salakhutdinov, R., Kolter, Z., Heller, K., Weller, A., Oliver, N., Scarlett, J., Berkenkamp, F. (eds.) Proceedings of the 41st International Conference on Machine Learning. Proceedings of Machine Learning Research, vol. 235, pp. 26856–26873. PMLR (2024). https://proceedings.mlr.press/v235/lee24s.html
11. Li, J., Cheng, C., Ma, J., Liu, G.: Geometric point attention transformer for 3D shape reassembly (2024). https://arxiv.org/abs/2411.17788
12. Li, Q., Geng, G., Zhou, M.: Pairwise matching for 3D fragment reassembly based on boundary curves and concave-convex patches. IEEE Access **8**, 6153–6161 (2020). https://doi.org/10.1109/ACCESS.2019.2961391
13. Li, Y., Mo, K., Shao, L., Sung, M., Guibas, L.: Learning 3D part assembly from a single image (2020)
14. Lu, J., Hua, G., Huang, Q.: Jigsaw++: imagining complete shape priors for object reassembly (2024). https://doi.org/10.48550/arXiv.2410.11816
15. Lu, J., Sun, Y., Huang, Q.: Jigsaw: learning to assemble multiple fractured objects (2023). https://arxiv.org/abs/2305.17975
16. McKinney, W., et al.: Data structures for statistical computing in python. In: Proceedings of the 9th Python in Science Conference, Austin, TX, vol. 445, pp. 51–56 (2010)

17. Mo, K., et al.: PartNet: a large-scale benchmark for fine-grained and hierarchical part-level 3D object understanding. In: Proceedings of the IEEE/CVF Conference on Computer Vision and Pattern Recognition (CVPR) (2019)
18. Narayan, A., Nagar, R., Raman, S.: RGL-NET: a recurrent graph learning framework for progressive part assembly. In: Proceedings of the IEEE/CVF Winter Conference on Applications of Computer Vision (WACV), pp. 78–87 (2022)
19. Papaioannou, G., et al.: From reassembly to object completion: a complete systems pipeline. ACM J. Comput. Cult. Heritage 10, 1–22 (2017). https://api.semanticscholar.org/CorpusID:17170033
20. Paszke, A., et al.: PyTorch: an imperative style, high-performance deep learning library (2019)
21. Paulano-Godino, F., Jimenez-Delgado, J.J.: Identification of fracture zones and its application in automatic bone fracture reduction. Comput. Methods Programs Biomed. 141, 93–104 (2017). https://doi.org/10.1016/j.cmpb.2016.12.014
22. Qi, C.R., Su, H., Mo, K., Guibas, L.J.: PointNet: deep learning on point sets for 3D classification and segmentation. In: Proceedings of the IEEE Conference on Computer Vision and Pattern Recognition (CVPR) (2017)
23. Sellan, S., Chen, Y.C., Wu, Z., Garg, A., Jacobson, A.: Breaking bad: a dataset for geometric fracture and reassembly (2022)
24. Son, T., Lee, J., Lim, J., Lee, K.: Reassembly of fractured objects using surface signature. Vis. Comput. 34(10), 1371–1381 (2017). https://doi.org/10.1007/s00371-017-1419-0
25. Vaswani, A., et al.: Attention is all you need. In: Guyon, I., Luxburg, U.V., et al. (eds.) Advances in Neural Information Processing Systems, vol. 30. Curran Associates, Inc. (2017)
26. Villegas-Suarez, A.M., Lopez, C., Sipiran, I.: MatchMakerNet: enabling fragment matching for cultural heritage analysis. In: Proceedings of the IEEE/CVF International Conference on Computer Vision (ICCV) Workshops, pp. 1632–1641 (2023)
27. Wang, H., Zang, Y., Liang, F., Dong, Z., Fan, H., Yang, B.: A probabilistic method for fractured cultural relics automatic reassembly. J. Comput. Cult. Herit. 14(1) (2021). https://doi.org/10.1145/3417711
28. Wang, W., Zhang, R., You, M., Zhou, H., He, B.: 3D assembly completion. Proc. AAAI Conf. Artif. Intell. 37(3), 2663–2671 (2023). https://doi.org/10.1609/aaai.v37i3.25365
29. Wang, Y., Sun, Y., Liu, Z., Sarma, S.E., Bronstein, M.M., Solomon, J.M.: Dynamic graph CNN for learning on point clouds. ACM Trans. Graph. 38(5) (2019). https://doi.org/10.1145/3326362
30. Wang, Z., Chen, J., Furukawa, Y.: PuzzleFusion++: auto-agglomerative 3D fracture assembly by denoise and verify (2025). https://arxiv.org/abs/2406.00259
31. Zhao, F., Zhou, M., Geng, G., Zhu, L.: Rigid blocks matching method based on contour curves and feature regions. IET Comput. Vis. 12(1), 76–85 (2018). https://doi.org/10.1049/iet-cvi.2016.0392
32. Zhou, Q., Jacobson, A.: Thingi10K: a dataset of 10,000 3D-printing models (2016)

A Comparison of Ranking Methods for Feature Selection

Grace Villacrés[1]([✉]) [ID], Marius Marinescu[1] [ID], Luca Martino[2] [ID], and Óscar Barquero[1] [ID]

[1] Universidad Rey Juan Carlos, Fuenlabrada, Spain
grace.villacres@urjc.es
[2] Universitá di Catania, Catania, Italy

Abstract. Feature selection, also known as variable selection, plays a critical role in statistics, signal processing, and machine learning by enhancing model interpretability, identifying relevant variables for specific tasks, and enabling effective dimensionality reduction. Central to this process is the accurate ranking of variables. In this study, we conduct a systematic and fair comparison of several wrapper-based feature selection methods to evaluate their performance. To ensure a controlled and insightful analysis, we generate synthetic datasets that incorporate variable correlations, different distributions, and interdependencies. Our findings reveal that two specific wrapper methods consistently outperform others, offering practical guidance for both researchers and practitioners. Notably, the widely discussed leave-one-covariate-out (LOCO) method-closely related to Shapley value-based approaches-performs the worst in our evaluations. In contrast, the backward elimination method, which constructs a variable ranking by iteratively removing the least important features, achieves the highest performance. These results underscore the importance of method selection in feature ranking and provide actionable insights for future applications.

Keywords: Feature selection · Ranking method · Wrapper methods

1 Introduction

Feature selection, also known as variable selection, is a relevant problem in statistics, data analysis, and machine learning. As datasets increase in scale and complexity, selecting the most informative features is critical for ensuring model accuracy and interpretability. Effective feature selection simplifies models by mitigating overfitting, which is achieved by eliminating non-relevant variables, thereby reducing noise and minimizing the effect of bias in predictions. A feature selection process can also be employed as a dimension reduction technique.

Since feature selection is essential in data analysis, it is currently the object of several studies; see, e.g., [1–5,14,16,19,20,23] to name a few. Different approaches to feature selection have been proposed in the literature,

© The Author(s), under exclusive license to Springer Nature Switzerland AG 2026
C. Chira et al. (Eds.): InnoComp 2025, CCIS 2793, pp. 111–122, 2026.
https://doi.org/10.1007/978-3-032-12478-4_8

which can broadly classified into four classes based on their evaluation strategies [3,18,20]: (a) wrapper methods, which iteratively evaluate subsets of variables by optimizing a cost function to assess their contribution to model performance, (b) filter-based methods, which rank variables based on statistical criteria independently of the model, (c) embedded methods, which integrate the selection process into model training, and (d) hybrid methods, which combine the strengths of multiple approaches, such as using filters to reduce the feature space followed by a wrapper or embedded method for a refined selection. Moreover, more recently, the concepts of the importance of a variable have been studied and used [1,10,13,22,23].

Theoretically speaking, we can divide the feature selection procedure into two main parts: i) the ranking of variables and ii) the selection of the optimal number of variables [9,12,15,21]. This paper focuses on the first component of this process, where we evaluate various ranking methods (RMs) to establish a quantifiable scale that determines the relative performance of these methods. Specifically, we mainly compare RMs based on wrapper methods, originally introduced in [7]. In these methods, features are selected by directly optimizing a given cost function to enhance model performance, typically through a *sequential* evaluation of feature subsets. We also study a non-sequential wrapper method, which in the literature is called *leave-one-covariate-out (LOCO)*. Since it is non-sequential, LOCO could also be parallelized. It is worth noting that LOCO is closely related to the *Shapley importance values*, which are widely studied in the literature nowadays [6,11,17,22]. Finally, we also include a comparison with an embedded method in our study, which particularly fits with our experiments.

For a fair comparison, all the rankings have been tested in a controlled scenario. We generate a synthetic dataset of normalized variables or features. The dataset is created in such a way as to ensure the presence of correlated variables, dependencies, and different distributions. The observations are generated following a linear model so that we can easily obtain the ground-truth with the correct ranking of the variables. Then, we test each RM, obtaining a ranking for the variables. Finally, we give a score to the rankings obtained for each RM to also provide a ranking of the RMs. Our results indicate that the RM that achieves the best performance is the backward elimination method, which minimizes the external cost by sequentially removing variables. However, the embedded method achieved the highest overall score as expected, since it is the best possible embedded method, as it utilizes the same linear (true) model. All evaluated RMs consistently rank the top six variables correctly; however, their behavior diverges when considering the full feature set. A key contribution of our work is the identification of two wrapper methods, a) backward elimination, which removes variables to minimize external cost, and b) forward selection, which adds the least relevant variables to maximize external cost, as the most effective and broadly applicable strategies. Unlike the embedded method, which is tied to specific model assumptions, these wrapper methods can be employed across a wide range of regression and classification problems.

The rest of this paper is organized as follows. Section 2 describes the elements that form a generic wrapper method. Section 3 is devoted to describing different specific ranking methods for feature selection. The experiments are provided in Sect. 4. Finally, Sect. 5 presents concluding remarks.

2 Elements in Wrapper Methods

2.1 Parametric Model

Let us consider a set of R variables $\mathbf{x} = [x_1, ..., x_R]^\top$ (input vector) and a related variable y (output). In many real-world applications, we observe a dataset of N pairs $\{\mathbf{x}_n, y_n\}_{n=1}^N$ where $\mathbf{x}_n = [x_{n,1}, ..., x_{n,R}]^\top$. In this work, we consider the case $R \leq N$. The relationship between inputs and output is given by an unknown transformation/model. Let us define the $N \times R$ matrix of inputs

$$\mathbf{X} = [\mathbf{x}_1, \ldots, \mathbf{x}_N]^\top = [\mathbf{z}_1, \ldots, \mathbf{z}_R],$$

where

$$\mathbf{z}_r = [x_{1,r}, ..., x_{N,r}]^\top, \quad r = 1, ..., R,$$

represents the $N \times 1$ column vector of r-th variable/feature. Given also the vector of outputs $\mathbf{y} = [y_1, ..., y_N]^\top$, we can try to approximate the true model by a parametric model $\mathbf{g}$ with parameters, $\boldsymbol{\beta} = [\beta_1, ..., \beta_R]^\top$, i.e.,

$$\mathbf{y} = \mathbf{g}_\beta(\mathbf{X}) = \mathbf{g}(\mathbf{X}; \boldsymbol{\beta}), \tag{1}$$

where $\mathbf{g}(\mathbf{X}; \boldsymbol{\beta}) = [g(\mathbf{x}_1; \boldsymbol{\beta}), ..., g(\mathbf{x}_N; \boldsymbol{\beta})]^\top$ is a vectorial function that maps $\mathbf{X}$ to the output vector $\mathbf{y}$. Some common examples of scalar g are:

- classical linear regression with $g(\mathbf{x}_n, \boldsymbol{\beta}) = \beta_1 x_1 + \beta_2 x_2 + \cdots + \beta_R x_R = \boldsymbol{\beta}\mathbf{x}_n$;
- logistic regression $g(\mathbf{x}_n, \boldsymbol{\beta}) = \frac{1}{1+\exp(-\boldsymbol{\beta}\mathbf{x}_n)}$.

The parametric models above can be easily extended by adding an additional constant parameter β_0.

2.2 Internal Procedure for Training

The optimal vector of parameters $\widehat{\boldsymbol{\beta}}$ can be obtained by minimizing an (internal) cost function,

$$\widehat{\boldsymbol{\beta}} = \arg \min_{\boldsymbol{\beta}} C_{\texttt{int}}(\mathbf{y}, \mathbf{g}(\mathbf{X}; \boldsymbol{\beta})) \tag{2}$$

where $C_{\texttt{int}}$ can be any L_p distance, $C_{\texttt{int}} = ||\mathbf{y} - \mathbf{g}(\mathbf{X}; \widehat{\boldsymbol{\beta}})||_p$, or any alternatives for regression and classification. We denote the predictive output vector as

$$\widehat{\mathbf{y}} = \mathbf{g}(\mathbf{X}; \widehat{\boldsymbol{\beta}}).$$

2.3   External Evaluation of the Model Performance

For our purpose of building a feature ranking, we need an external evaluation of the model performance independent of the internal procedure used to learn the optimal parameter vector $\widehat{\boldsymbol{\beta}}$. Considering the vector of observed outputs $\mathbf{y}$ and the predicted outputs $\widehat{\mathbf{y}}$, we define a general external cost function $C_{\text{ext}}(\mathbf{y},\widehat{\mathbf{y}})$: $\mathbb{R}^N \times \mathbb{R}^N \to \mathbb{R}$. For instance, in regression, we can consider

$$C_{\text{ext}}(\mathbf{y},\widehat{\mathbf{y}}) = \|\mathbf{y} - \widehat{\mathbf{y}}\|_\alpha = \left(\sum_{n=1}^{N} |y_n - \widehat{y}_n|^\alpha \right)^{\frac{1}{\alpha}}, \tag{3}$$

or in classification, we can consider the cross-entropy, i.e.,

$$C_{\text{ext}}(\mathbf{y},\widehat{\mathbf{y}}) = - \sum_{n=1}^{N} [y_n \log \widehat{y}_n + (1 - y_n) \log(1 - \widehat{y}_n)]. \tag{4}$$

A relevant point is that, in the wrapper methods, we have that C_{ext} can differ from C_{int}. We have this flexibility that in certain circumstances can be useful and can increase the robustness of the obtained results.

3   Ranking Methods for Feature Selection

In this work, we describe six different RMs. The first five ranking procedures are based on wrapper methods and the last one is based on embedded methods. To describe briefly the RMs, we need to define models considering only a subset of the R possible variables, as inputs.

3.1   Model with a Subset of Variables

For the analysis of these methods, we define sub-models based on a set of m selected indices of features, $\mathcal{V}_m = \{k_1,\ldots,k_m\}$, where $m \leq R$ and $k_i \in \{1,\ldots,R\}$, with $k_i \neq k_j$ for $i \neq j$. Using this vector of indices, we construct the $N \times m$ matrix,

$$\mathbf{V}_m = [\mathbf{z}_{k_1}, \mathbf{z}_{k_2}, \ldots \mathbf{z}_{k_m}], \quad \text{where} \quad k_j \in \mathcal{V}_m,$$

which includes the selected features in $\mathcal{V}_m$, with $\mathbf{z}_{k_j} = [x_{1,k_j}, \ldots, x_{N,k_j}]^\top$. The corresponding coefficient vector is

$$\widehat{\boldsymbol{\beta}}_m = [\widehat{\beta}_{k_1}, \ldots, \widehat{\beta}_{k_m}]^\top \tag{5}$$

This vector is obtained as

$$\widehat{\boldsymbol{\beta}}_m = \arg \min_{\beta} C_{\text{int}}(\mathbf{y}, \mathbf{g}(\mathbf{V}_m; \boldsymbol{\beta}_m)). \tag{6}$$

This leads to the following sub-model and predictions,

$$\widehat{\mathbf{y}}_m = \mathbf{g}(\mathbf{V}_m; \widehat{\boldsymbol{\beta}}_m) = \mathbf{g}([\mathbf{z}_{k_1}, \mathbf{z}_{k_2}, \ldots \mathbf{z}_{k_m}]; \widehat{\boldsymbol{\beta}}_m), \tag{7}$$

where $\widehat{\mathbf{y}}_m$ is the $N \times 1$ vector of predictions.

3.2   Examples of Wrapper Ranking Methods

RM0 *Leave-One-Covariate-Out (LOCO).* This method introduced in [8] evaluates the importance of individual features by assessing their impact on the cost function C_{ext}, in a parallelized manner. Specifically, for each feature $\mathbf{z}_r$, $r \in \{1, \dots, R\}$, LOCO defines a sub-model that excludes $\mathbf{z}_r$, being the corresponding model and prediction as follows

$$\widehat{\mathbf{y}}_{-r} = \mathbf{g}(\mathbf{X}_{-r}; \widehat{\boldsymbol{\beta}}_{-r}),$$

where $\mathbf{V}_{R-1} = \mathbf{X}_{-r}$ corresponds to the feature matrix that includes all features except $\mathbf{z}_r$, and $\widehat{\boldsymbol{\beta}}_{-r}$ denotes the coefficient vector estimated for the sub-model excluding $\mathbf{z}_r$. Namely, the matrix $\mathbf{V}_{R-1}$ is build as

$$\mathbf{V}_{R-1} = \mathbf{X}_{-r} = [\mathbf{z}_1, \mathbf{z}_2, \dots \mathbf{z}_{r-1}, \mathbf{z}_{r+1}, \dots \mathbf{z}_R].$$

LOCO evaluates C_{ext} for each sub-model, measuring the performance impact of excluding $\mathbf{z}_r$ on the maximization of $C_r = C_{\text{ext}}(\mathbf{y}, \widehat{\mathbf{y}}_{-r})$. The obtained cost value C_r for the exclusion of each feature is then ordered to determine the feature's importance. Specifically, the feature corresponding to $\max_r C_r$ is considered the most significant (as its exclusion leads to the worst performance, hence we have lost the greatest amount of information) and is denoted by $\mathbf{z}_{k_1}$. Conversely, the feature corresponding to $\min_r C_r$ is considered the least significant (since removing it produces just a small increase in the error) and is denoted by $\mathbf{z}_{k_R}$. The final ranking, in decreasing order of importance, is $\mathbf{z}_{k_1}, \mathbf{z}_{k_2}, \dots, \mathbf{z}_{k_R}$. It is important to remark that LOCO is quite related to the famous Shapley values that determine the variable importance, as both aim to assess the contribution of each feature to a model's prediction, with Shapley values providing a fair, game-theoretic approach to distribute the prediction among features based on their individual contributions [22].

RM1 *Forward Selection Adding Variables "Forward" Minimizing The External Cost.* This method iteratively adds one feature at a time. In each iteration, the feature that achieves the best performance in terms of minimizing the cost function C_{ext} is selected. Thus, RM1 adds the most significant feature at each step. At the initial step, we denote the most significant selected feature as $\mathbf{z}_{k_1}$. Fixing $\mathbf{z}_{k_1}$ and considering a model with 2 features, at the second step, RM1 selects another feature denoted as $\mathbf{z}_{k_2}$. At the m-th iteration, fixing the previous $m - 1$ selected features, RM1 selects the most significant feature as $\mathbf{z}_{k_m}$, and so on. The sequence of models and predictions would be

$$\widehat{\mathbf{y}}_1 = \mathbf{g}(\mathbf{z}_{k_1}; \widehat{\boldsymbol{\beta}}_1), \quad \widehat{\mathbf{y}}_2 = \mathbf{g}([\mathbf{z}_{k_1}, \mathbf{z}_{k_2}]; \widehat{\boldsymbol{\beta}}_2), \dots,$$
$$\widehat{\mathbf{y}}_m = \mathbf{g}([\mathbf{z}_{k_1}, \mathbf{z}_{k_2}, \dots, \mathbf{z}_{k_m}]; \widehat{\boldsymbol{\beta}}_m),$$

etc. The final ranking (in a decreasing order of importance) would be $\mathbf{z}_{k_1}, \mathbf{z}_{k_2}, \dots, \mathbf{z}_{k_R}$ (where $\mathbf{z}_{k_1}$ is the best and $\mathbf{z}_{k_R}$ is the worst).

RM2 *Backward Elimination Removing Variables "Backward" Minimizing the External Cost.* This method iteratively removes one feature at a time. In each iteration, the feature that contributes the least to minimizing the cost function C_{ext} is removed. Thus, RM2 removes the least significant feature at each step. At the initial step, RM2 includes all R features in the model and removes the least significant feature, denoted as $\mathbf{z}_{k_1}$ (that is the least important variable). Fixing the remaining $R - 1$ features, RM2 removes the next feature denoted as $\mathbf{z}_{k_2}$, reducing the model to $R - 2$ features. At the m-th iteration, RM2 removes the feature that contributes the least among the $R - m$ remaining features, denoted as $\mathbf{z}_{k_m}$, and so on. The final ranking (in decreasing order of importance) would be $\mathbf{z}_{k_R}, \mathbf{z}_{k_{R-1}}, \ldots, \mathbf{z}_{k_1}$ (where $\mathbf{z}_{k_R}$ is the best and $\mathbf{z}_{k_1}$ is the worst).

RM3 *Backward Elimination Removing the Best Variable, "Backward" Maximizing the External Cost.* As RM2, this method also removes one feature at a time, but now the cost function C_{ext} is maximized. The method starts again considering the complete model. Then, RM3 removes the variable that produces the higher increase of the error in prediction (this variable would be the most important) and continues iteratively. Denote the first variable removed as $\mathbf{z}_{k_1}$ (the most relevant variable), the second removed variable $\mathbf{z}_{k_2}$, and so on. The final ranking (in decreasing order of importance) is $\mathbf{z}_{k_1}, \mathbf{z}_{k_2}, \ldots, \mathbf{z}_{k_R}$.

RM4 *Forward Selection Adding the Worst Variable, Maximizing the External Cost.* This method is similar to RM1, but instead of adding the most significant feature at each iteration, RM4 adds the feature that achieves the worst performance, i.e., the feature that maximizes the external cost function C_{ext}. In each iteration, RM4 selects the feature that causes the largest increase in the cost function (that is the worst variable). At the initial step, RM4 selects the feature, denoted as $\mathbf{z}_{k_1}$, that maximizes C_{ext} (then $\mathbf{z}_{k_1}$ is the worst variable). Fixing $\mathbf{z}_{k_1}$, RM4 then selects the next feature with a model of dimension 2, denoted as $\mathbf{z}_{k_2}$, which again maximizes C_{ext}. Note that, RM4 creates a sequence of variables from the worst to the best variable (i.e., increasing their relevance). The final ranking (in decreasing order of importance) would be $\mathbf{z}_{k_R}, \mathbf{z}_{k_{R-1}}, \ldots, \mathbf{z}_{k_1}$.

3.3 Ranking by an Embedded Method (RM-E)

Embedded techniques rank features directly inside their model learning process, i.e., the ranking is an *internal* procedure that is performed during the learning process of the parameters. Embedded techniques are clearly an alternative to the wrapper method. As an example of an embedded method, we consider one that perfectly matches the model employed in our experiment. In generalized linear models, such as multiple linear regression and logistic regression, we have that $E(\mathbf{Y} \mid \mathbf{X}) = \boldsymbol{\mu} = f(\mathbf{X}\boldsymbol{\beta})$, and we obtain the estimation $\widehat{\boldsymbol{\beta}} = [\widehat{\beta}_1, \ldots, \widehat{\beta}_R]^{\top}$, then the ranking is built according to the decreasing order of the absolute values of $\widehat{\beta}_j$, i.e., $|\widehat{\beta}_j|$ (i.e., the most important variable has the greatest $|\widehat{\beta}_j|$).

4   Experiments

4.1   Synthetic Dataset Generation

We generate a synthetic dataset to evaluate the performance, under controlled conditions, of the different RMs described in Sect. 3.2, i.e., knowing the ground-truth. The data structure follows a linear model with specific variables included and excluded based on predefined criteria. Specifically, the dataset consists of $N = 5000$ observations and $R = 20$ variables/features, $\mathbf{x} = [x_1, \ldots, x_{20}]$, and the variables are defined as shown in Table 1.

Table 1. Feature generation: sampling from a distribution

Variables	Generation/Distribution
$x_1,\ x_2,\ x_5,\ x_7,$ $x_{15},\ x_{16},\ x_{18},\ x_{19}$	$\mathcal{N}(0,1)$
$x_3,\ x_4,\ x_8,\ x_9,$ $x_{10},\ x_{13},\ x_{20}$	$\mathcal{U}\left(\left[-\frac{\sqrt{12}}{2}, \frac{\sqrt{12}}{2}\right]\right)$
x_6	x_2^2
x_{11}	$z = 0.5x_8 + \epsilon, \quad \epsilon \sim \mathcal{N}(0,1),$ $\frac{z - \mathrm{mean}(z)}{\mathrm{std}(z)}$
x_{12}	$z = 0.5x_{10} + \epsilon, \quad \epsilon \sim \mathcal{N}(0,1),$ $\frac{z - \mathrm{mean}(z)}{\mathrm{std}(z)}$
x_{14}	$z = x_5 + \epsilon, \quad \epsilon \sim \mathcal{N}(0,1),$ $\frac{z - \mathrm{mean}(z)}{\mathrm{std}(z)}$
x_{17}	$z = 0.2x_2 + u, \quad u \sim \mathcal{U}([0,1]),$ $\frac{z - \mathrm{mean}(z)}{\mathrm{std}(z)}$

Note that for all the input variables we have:

$$\mathbb{E}[x_i] = 0 \quad \mathrm{var}[x_i] = 1, \tag{8}$$

so that they are all normalized in terms of power of the signal.

True Model: The corresponding observations were generated as follows

$$\begin{aligned}
y_n =\ & 0.6x_2 + 0.6x_3 - 0.2x_4 + 0.1x_5 - 0.3x_7 + 0.1x_8 \\
& + 0.8x_9 - 0.3x_{11} + 0.3x_{12} + 0.3x_{14} + 0.5x_{15} + 0.9x_{16} \\
& + 0.2x_{17} - 0.3x_{18} - 0.5x_{19} + 0.6x_{20}.
\end{aligned} \tag{9}$$

Note that in this experiment we have not added noise in the generation of y. It is important to remark that the model in Eq. (9) excludes explicitly the following features:

$$x_1,\ x_6,\ x_{10},\ \text{and}\ x_{13}.$$

However, x_6 is included as a transformation of x_2, i.e., $x_6 = x_2^2$. Moreover, some variables present linear correlation: x_8 and x_{11}, x_{10} and x_{12}, x_5 and x_{14}, x_2 and x_{17}. Indeed, x_{11}, x_{12}, x_{14}, and x_{17} are obtained with a linear transformation of another variable plus noise as shown in Table 1. Furthermore, some variables have identical coefficients in the true model but have different distributions: x_2 and x_3 share the same regression coefficient but follow different distributions. Similarly, x_7 and x_9 have identical coefficients but have been drawn from different distributions. This design ensures a mix of collinearity, and redundant information, making the dataset a robust frame for assessing model performance. Note that the realizations of the variables are drawn from normalized distributions, ensuring zero mean and unit variance.

4.2   Parametric Model Used by the RMs

Next, we define the parametric model that is used to assess the different RMs for the dataset described in Sect. 4.1. The relationship between inputs and outputs is studied using the linear parametric model,

$$\mathbf{y} = \mathbf{X}\boldsymbol{\beta}, \tag{10}$$

The regularized least squares (LS) estimator is

$$\widehat{\boldsymbol{\beta}} = \left(\mathbf{X}^\top \mathbf{X} + \lambda \mathbf{I}\right)^{-1} \mathbf{X}^\top \mathbf{y}, \tag{11}$$

where $\lambda = 0.5$ and $\mathbf{I}$ is a diagonal unit matrix. Hence, the predicted output according to the model is

$$\widehat{\mathbf{y}} = \mathbf{X}\widehat{\boldsymbol{\beta}} = \mathbf{X}\left(\mathbf{X}^\top \mathbf{X} + \lambda \mathbf{I}\right)^{-1} \mathbf{X}^\top \mathbf{y}. \tag{12}$$

To evaluate the model performance, we use (3) with $\alpha = 2$, i.e., the Euclidean-norm. Note that the parametric model is linear as the true model. Then, we remove the issue of model mismatch and we can focus on the comparison of the RMs.

4.3   A Score for Each Ranking Method

In this section, we describe the method used to compare the different RMs. With this aim, we first define the ground-truth of our true model in (9).

Ground-Truth: The variables, considering only their subindices, are ranked in descending order of importance (obtained sorting in decreasing order the absolute values of the coefficients in the true model) as detailed below

$$16 \rightarrow 9 \rightarrow (2, 3, 20) \rightarrow (15, 19) \rightarrow (7, 11, 12, 14, 18)$$
$$\rightarrow (4, 17) \rightarrow (5, 8) \rightarrow (1, 6, 10, 13),$$

where indices within the parentheses $(\cdot, \ldots, \cdot)$ indicate ties in the ranking, meaning the variables inside the parentheses have the same importance in the model.

Any permutation of these variables will be considered a correct ranking. To evaluate the RMs, we use the following algorithm that provides a score associated with each RM according to the "distance" with respect to the ground-truth. Let us define S as the score for a given ranking method. The idea is the following:

- Each time that a variable is placed in the ranking in a correct position, the score of the ranking is increased by 1, i.e., $S \leftarrow S + 1$. Otherwise, the score remains invariant. Note that as a correct position we consider also the draw positions: for instance, variable 2 is considered well placed in the third, fourth, or fifth position.

Clearly, this scoring algorithm could be improved in several ways. For instance, perhaps we should assign more weight to errors in the first positions of the ranking than to errors in the last positions of the ranking. Moreover, in each error, we should also consider the distance from the correct position. Here, in Table 2, we consider the *total score* over all the 20 variables and a *partial score* over the first 7 most relevant variables.

4.4 Results

This section presents the ranked features obtained from the different RMs for $\alpha = 2$ in the external cost. Table 2 summarizes the rankings across RMs, allowing for a comparative analysis of feature importance. By examining these rankings, we can observe how similar or different the ranking methods are in prioritizing features. The table shows only the indices of the ranked variables. The total and partial scores are numerical values derived from the procedure described above, with a maximum value of 20 (for the total score) or 7 (for the partial score), when all features are ranked correctly, and the minimum score is 0.

Table 2. Comparison of the different RMs: final results.

RMs	Ranking	Scores
RM0	16 9 2 20 3 15 7 17 12 11 14 4 8 5 10 1 6 13 18 19	11 - 6
RM1	16 9 2 3 20 15 18 14 12 7 17 11 4 8 5 10 1 6 13 19	15 - 6
RM2	16 9 2 3 20 15 19 14 12 7 17 11 4 8 5 10 1 6 13 18	16 - 7
RM3	16 9 2 20 3 15 17 12 7 14 5 11 4 10 8 6 13 1 18 19	14 - 6
RM4	16 9 2 20 3 15 19 18 14 5 17 12 7 11 4 10 8 6 13 1	13 - 7
RM-E	16 9 2 20 3 15 12 14 7 17 11 18 19 4 8 5 10 1 6 13	17 - 6
	Maximum possible score values:	20 - 7

In Table 2, we observe that all RMs correctly rank the first six variables. There is a clear agreement among all the RMs on at least the first three most important variables, 16, 9, and 2. For the fourth and fifth positions, they agree to choose 3 and 20, and in the sixth position the variable 15. RM2 ranks the top 10 variables correctly, and the variable in position 15. RM-E performs better, ranking the first and the last six variables correctly. However, RM-E is an embedded method designed exactly for the multiple linear model (true model). Moreover, the score difference between RM-E and RM2 is minimal, with scores of 17 and 16, respectively, which shows the robustness of RM2. Additionally, RM-E is not the best in the ranking when considering the partial score. RM2 and RM4 are the highest-ranked according to the partial score. RM4 provides the correct variables in the first seven positions, similar to RM2, but accumulates more errors in the central part of the ranking. In fact, RM4 was able to correctly detect 3 of the 4 worst variables as well. Perhaps another scoring system could easily detect the benefits of RM4. Surprisingly, Table 2 shows that the worst performance is achieved by LOCO (RM0), which is quite related to Shapley importance values [22]. As a final consideration, note that both RM2 and RM4 start the sequential construction of the ranking by identifying the worst variable and end by selecting the best variable.

5   Conclusions

In this work, we test different ranking methods by designing a systematic and fair comparison. Specifically, we have compared six different wrapper methods and one embedded method that can be employed for generalized linear models. All the tested ranking methods correctly rank the first six variables, but various behaviors emerge when considering the full set of features. The two sequential procedures, which start deciding the worst feature at each iteration, denoted as RM2 and RM4, provide the best results (jointly with the specific embedded method RM-E, which can be applied only for generalized linear models). Unlike RM-E, RM2 and RM4 can be employed in any regression and classification scenario. Another interesting result is that the famous method called LOCO (RM0) provides the worst performance. This is quite relevant since LOCO is computationally and theoretically related to the so-called Shapley values [22].

Acknowledgments. This work has been supported in part by Comunidad de Madrid within the 2023–2026 agreement with Universidad Rey Juan Carlos for the granting of direct subsidies for the promotion and encouragement of research and technology transfer, Line of Action A Emerging Doctors, under Project OrdeNGN (Ref. F1177), by the "PIAno di inCEntivi per la RIcerca di Ateneo 2024/2026" (UPB 28722052159) and PIACERI Starting Grant BA-GRAPH (UPB 28722052144) of the University of Catania, and by project POLI-GRAPH - Grant PID2022-136887NB-I00 funded by MCIN/AEI/10.13039/501100011033.

References

1. Blesch, K., Watson, D.S., Wright, M.N.: Conditional feature importance for mixed data. AStA Adv. Stat. Anal. **108**(2), 259–278 (2024)

2. Chandrashekar, G., Sahin, F.: A survey on feature selection methods. Comput. Electr. Eng. **40**(1), 16–28 (2014). https://doi.org/10.1016/j.compeleceng.2013.11.024. 40th-year commemorative issue

3. Guyon, I.M., Elisseeff, A.: An introduction to variable and feature selection. J. Mach. Learn. Res. **3**, 1157–1182 (2003)

4. Heinze, G., Wallisch, C., Dunkler, D.: Variable selection - a review and recommendations for the practicing statistician. Biom. J. **60**(3), 431–449 (2018)

5. Kim, B., Shah, J.A., Doshi-Velez, F.: Mind the gap: a generative approach to interpretable feature selection and extraction. In: Advances in Neural Information Processing Systems, vol. 28 (2015)

6. Kim, S.H., Park, S.Y., Seo, H., Woo, J.: Feature selection integrating shapley values and mutual information in reinforcement learning: an application in the prediction of post-operative outcomes in patients with end-stage renal disease. Comput. Methods Programs Biomed. **257**, 108416 (2024). https://doi.org/10.1016/j.cmpb.2024.108416

7. Kohavi, R., John, G.H.: Wrappers for feature subset selection. Artif. Intell. **97**(1–2), 273–324 (1997)

8. Lei, J., G'Sell, M., Rinaldo, A., Tibshirani, R.J., Wasserman, L.: Distribution-free predictive inference for regression. J. Am. Stat. Assoc. **113**(523), 1094–1111 (2018). https://doi.org/10.1080/01621459.2017.1307116

9. Martino, L., Millán-Castillo, R.S., Morgado, E.: Spectral information criterion for automatic elbow detection. Expert Syst. Appl. **231**, 120705 (2023)

10. Martino, L., Morgado, E., Millán-Castillo, R.S.: An index of effective number of variables for uncertainty and reliability analysis in model selection problems. Signal Process. **227**, 109735 (2025). https://doi.org/10.1016/j.sigpro.2024.109735

11. Meepaganithage, A., Rath, S., Nicolescu, M., Nicolescu, M., Sengupta, S.: Feature selection using the advanced Shapley value. In: 2024 IEEE 14th Annual Computing and Communication Workshop and Conference (CCWC), pp. 0207–0213 (2024). 10.1109/CCWC60891.2024.10427665

12. Morgado, E., Martino, L., Millán-Castillo, R.S.: Universal and automatic elbow detection for learning the effective number of components in model selection problems. Digital Signal Processing **140**, 104103 (2023)

13. Pries, J., Berkelmans, G., Bhulai, S., van der Mei, R.: The Berkelmans-pries feature importance method: a generic measure of informativeness of features. arXiv preprint: arXiv:2301.04740 (2023)

14. Pudjihartono, N., Fadason, T., Kempa-Liehr, A.W., O'Sullivan, J.M.: A review of feature selection methods for machine learning-based disease risk prediction. Front. Bioinform. **2**, 927312 (2022)

15. San Millán-Castillo, R., Martino, L., Morgado, E.: A variable selection analysis for soundscape emotion modelling using decision tree regression and modern information criteria. IEEE Access (2024)

16. San Millán-Castillo, R., Martino, L., Morgado, E., Llorente, F.: An exhaustive variable selection study for linear models of soundscape emotions: rankings and Gibbs analysis. IEEE/ACM Trans. Audio, Speech, Lang. Process. **30**, 2460–2474 (2022)

17. Sebastián, C., González-Guillén, C.E.: A feature selection method based on Shapley values robust for concept shift in regression. Neural Comput. Appl. **36**(23), 14575–14597 (2024)

18. Servera, J.V., Martino, L., Verrelst, J., Rivera-Caicedo, J.P., Camps-Valls, G.: Multioutput feature selection for emulation and sensitivity analysis. IEEE

Trans. Geosci. Remote Sens. **62**, 1–11 (2024). https://doi.org/10.1109/TGRS.2024. 3358231

19. Tang, Z., Westling, T.: Nonparametric assessment of variable selection and ranking algorithms. arXiv preprint: arXiv:2308.11593 (2023)
20. Theng, D., Bhoyar, K.K.: Feature selection techniques for machine learning: a survey of more than two decades of research. Knowl. Inf. Syst. **66**, 1575–1637 (2023)
21. Thorndike, R.L.: Who belongs in the family? Psychometrika **3**, 267–276 (1953)
22. Verdinelli, I., Wasserman, L.: Feature importance: a closer look at Shapley values and LOCO. Stat. Sci. **39**(4), 623–636 (2024). https://doi.org/10.1214/24-STS937
23. Wood, D., Papamarkou, T., Benatan, M., Allmendinger, R.: Model-agnostic variable importance for predictive uncertainty: an entropy-based approach. Data Min. Knowl. Disc. **38**(6), 4184–4216 (2024)

UI Component Detection in Graphical Interfaces: A YOLO-Based Comparative Analysis

Mădălina Dicu[(✉)]

Faculty of Mathematics and Computer Science, Babeş-Bolyai University, Str. Mihail Kogălniceanu nr. 1, 400084 Cluj-Napoca, Romania
`madalina.dicu@ubbcluj.ro`

Abstract. In recent years, detecting visual components in graphical user interfaces (GUIs) has become increasingly relevant for modern applications that rely on automatic UI understanding. Unlike natural images, UI layouts present unique challenges, such as small, densely packed elements and high visual variability, which can affect standard object detectors. To explore this problem, we conduct a comparative analysis of six recent YOLO (You Only Look Once) versions, ranging from YOLOv5 to YOLOv12, for the task of detecting graphical components in user interfaces. We target three major platforms (mobile, web, desktop) and use representative datasets: VINS (mobile), GEN-GUI (web), and UICVD (desktop). All annotations were standardized to include the same three core UI classes: `Text`, `Button`, and `Icon`. Each YOLO model was trained and evaluated separately on each dataset, using the lightweight `nano (n)`, `tiny (t)`, and `small (s)` variants to analyze the trade-off between accuracy and efficiency. The results show noticeable performance differences across versions and interface types. YOLOv9s and YOLO11s performed best in complex scenarios, while YOLOv5s excelled on mobile data. `Button` was the easiest class to detect, `Text` remained stable, and `Icon` proved the most challenging. Several models, including YOLOv9s and YOLO11s, achieved mean Average Precisions (mAP) above 93% across desktop, web, and mobile interfaces, confirming that detection quality depends strongly on both the architecture and the interface type.

Keywords: Object detection · UI detection · YOLO · Mobile · Web · Desktop

1   Introduction

Modern graphical user interfaces (GUIs), found in mobile, web, and desktop applications, have become increasingly complex, reflecting the growing diversity of use cases and user expectations. In this context, the automatic extraction of a UI's visual structure, by identifying and locating graphical components, has

C. Chira et al. (Eds.): InnoComp 2025, CCIS 2793, pp. 123–139, 2026.
https://doi.org/10.1007/978-3-032-12478-4_9

become essential for tasks such as automated testing [18] or code generation [1]. However, object detection models designed initially for natural images face significant limitations in this domain: graphical interfaces often feature dense layouts, small component sizes, low contrast, and frequent overlapping between elements [4].

YOLO (You Only Look Once) models have become widely used in recent years as reliable solutions for real-time object detection, valued for their favorable balance between accuracy and computational efficiency [17]. Starting with YOLOv5 [12] and continuing through successive versions such as YOLOv8 [13], YOLOv9 [24], YOLOv10 [22], YOLOv11 [14], and the recent YOLOv12 [21], numerous architectural improvements have been introduced. Versions YOLOv6 [15] and YOLOv7 [23] are omitted from this study due to limited adoption or architectural overlap with other implementations not maintained by the main Ultralytics development line. These improvements include more efficient backbones [13], advanced spatial aggregation mechanisms [24], removal of post-processing steps such as Non-Maximum Suppression [22], and fast spatial attention modules [21]. While promising in theory and validated on general-purpose datasets (e.g., COCO, PASCAL VOC), these advancements do not necessarily guarantee strong performance in UI component detection.

Existing studies often treat YOLO versions as part of a linear progression, where each newer model is implicitly assumed to outperform its predecessors [10]. However, comprehensive comparative studies in the specific context of graphical interfaces are still lacking. Moreover, findings from [9] suggest that YOLO performance can vary significantly depending on visual properties, such as object density, size, and background, raising valid questions about the model's adaptability to non-standard scenarios, like UI component detection.

To address these issues, we propose an extensive experimental analysis of several modern YOLO versions, selected from the most relevant recent releases: v5, v8, v9, v10, v11, and v12. These are applied to the task of detecting graphical components in user interfaces. Each version is evaluated in two or three standard configurations, namely `nano (n)`, `tiny (t)`, and `small (s)`, to highlight the trade-off between model size and prediction accuracy.

The evaluation is conducted on three complementary datasets chosen to represent the most frequent visual interaction platforms: VINS for mobile interfaces [3], GENGUI for synthetic web applications [7], and UICVD for realistic desktop interfaces [6]. All datasets include annotated images with the most common UI component classes: `Text`, `Button`, and `Icon`, enabling a detailed performance analysis across classes and platforms.

These research questions were motivated by gaps observed in the literature. Most existing studies compare only a limited number of YOLO versions in specific settings, without examining cross-version behavior or class-level differences. Addressing these aspects is important for guiding model selection in UI component detection tasks.

Specifically, this study aims to answer the following research questions:

- **RQ1:** How does the performance of different YOLO versions vary in detecting UI components across mobile, web, and desktop platforms?
- **RQ2:** Are there consistent performance differences between the n/t and s variants of each YOLO version?
- **RQ3:** Which types of UI components (Text, Button, Icon) are harder to detect, and which models perform best for each?

The paper is structured as follows: Sect. 2 reviews related work, Sect. 3 describes the experimental methodology, and Sects. 4 and 5 present and analyze the results. Finally, Sect. 6 summarizes the conclusions and outlines directions for future research.

2   Related Work

The detection of visual components in graphical user interfaces (GUIs) has attracted increasing attention in recent years as mobile, web, and desktop applications have become increasingly complex. Early work in this area proposed rule-based and hierarchical approaches. Later, deep learning methods were successfully applied to extract visual structure automatically. A notable example is Beltramelli [1], who introduced the idea of converting UI images into source code using neural networks. Follow-up studies, such as those by Chen et al. [4] and Sun et al. [20], applied CNNs to detect UI components in mobile applications.

YOLO (You Only Look Once) models have become a popular choice for object detection tasks due to their balance between accuracy and computational efficiency. From YOLOv1 [17] to more recent versions like YOLOv12 [21], the series has undergone significant evolution in terms of architecture, introducing improvements related to gradient flow, multi-scale capabilities, and attention mechanisms. However, these models are generally evaluated on standard datasets such as COCO and Pascal VOC, which do not include UI components and fail to reflect the specific challenges of graphical interfaces, such as high element density and stylistic variation.

When it comes to UI detection, the literature remains limited and fragmented. Selçuk and Serif [19] compare the performance of YOLOv5 and YOLOv8 in a mobile scenario using a subset of the VINS dataset. They observe a modest performance improvement for YOLOv8s (3.3% mAP), along with a decrease in the detection of small elements such as icons and text. However, the study is limited to a single context (mobile) and compares only two YOLO versions.

Other studies [5,25,26] have explored UI component detection using hybrid architectures (YOLOv3, Faster R-CNN, Mask R-CNN) or modified YOLOv5 variants. Still, these rely on custom datasets with no standardized class definitions and no comparison across multiple YOLO versions. For instance, YOLOv5-MGC [5] introduces changes aimed at detecting small UI elements; however, it is tested only on mobile data without being extended to other types of interfaces.

Jiang and Zhong [10] present a broad analysis of YOLOv5 to YOLOv11 models in the ODVerse33 benchmark, which includes 33 diverse datasets. Their findings show that newer YOLO versions do not consistently outperform older ones, depending on the task. However, their study focuses exclusively on natural objects and does not include user interface (UI) components.

Despite recent progress in object detection architectures, comparative evaluations that explicitly target graphical user interfaces (GUIs) remain extremely limited. In many applied studies, YOLO models are treated as part of a linear evolution in which newer versions are assumed to perform better by default, without contextual validation. This assumption can be problematic for UI-related tasks, where specific challenges such as small component sizes, frequent overlaps, and cross-platform variation can significantly impact detection performance. The lack of dedicated analyses under controlled conditions leaves open the question of how robust these models are in GUI-oriented detection scenarios.

3   Methodology

This section outlines the methodology used in our study, including dataset selection, annotation standardization, YOLO model configurations, and the training and evaluation pipeline. Our goal is to evaluate the performance of various YOLO architectures in detecting standard UI components across mobile, web, and desktop platforms under a consistent experimental setup.

3.1   Dataset Description

This study uses three datasets that cover the major categories of graphical user interfaces: mobile, desktop, and web. The selection was guided by practical relevance, visual diversity, and the availability of accurate annotations. The essential characteristics of each dataset are described below.

VINS (Mobile UI). [3] is a dataset focused on mobile user interfaces, from which we used the Android subset, consisting of 2000 images in `.jpg` format. These are screenshots from real applications, spanning categories such as shopping, communication, and system settings. Annotations are provided in Pascal VOC format (`.xml`), with each file containing the bounding box coordinates and class labels for all visual components in the image. The dataset defines 11 UI-related classes and reflects the compact design typical of mobile apps, with a relatively low number of components per image. A representative example is shown in Fig. 1a.

GENGUI (Web UI). [7] consists of 250 `.png` images automatically generated using a large language model capable of producing realistic web interface layouts. Annotations are stored in `.csv` files, where each row corresponds to an individual component. The dataset contains 13 classes and includes interfaces in multiple languages (English, French, and German). Due to its synthetic nature, GENGUI provides controlled diversity and wide coverage of visual scenarios. A sample image is shown in Fig. 1b.

UICVD (Desktop UI). [6] contains 121 `.png` images collected from real desktop applications, including domains such as productivity tools and development environments. Annotations are provided in `.csv` format, with each row corresponding to one UI element, specifying its class and position. The dataset comprises 16 classes and is characterized by dense layouts and rigid component distribution, which is representative of traditional desktop software. An example from this dataset is shown in Fig. 1.

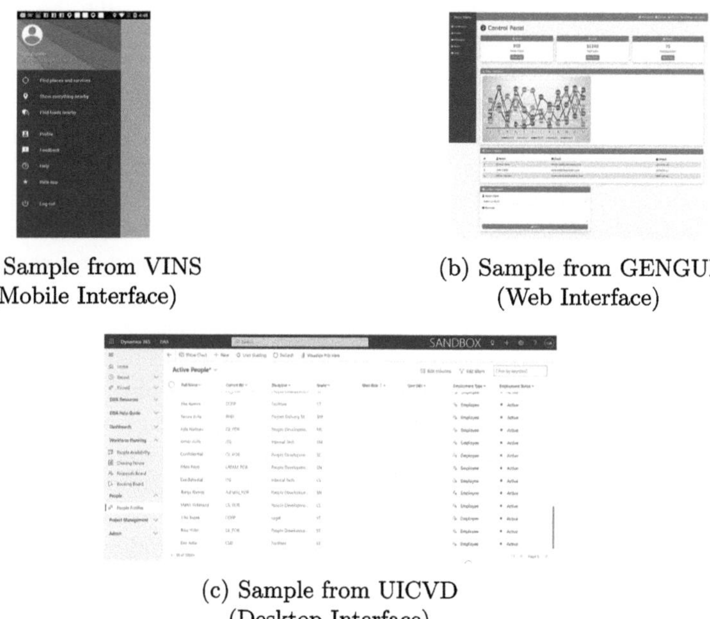

(a) Sample from VINS
(Mobile Interface)

(b) Sample from GENGUI
(Web Interface)

(c) Sample from UICVD
(Desktop Interface)

Fig. 1. Example screenshots from the three datasets: mobile and web interfaces (top), and a desktop interface (bottom).

3.2   Datasets and Preprocessing

To ensure a meaningful comparison between object detection models across different visual domains, we established a standard set of target classes. Although each dataset was initially created with its own set of labels, they all include a shared group of user interface components, namely: text, buttons, and icons. These were chosen because they are commonly used, essential for user interaction, and appear in all three datasets.

A manual inspection of class labels was carried out for each dataset, and all semantically equivalent labels were mapped to the same standard category: `Text`, `Button`, or `Icon`. For instance, labels such as `TextLabel`, `SectionTitle`, `textForNavigationBar`, and `statusLabel` (found in GENGUI and UICVD) were all mapped to `Text`. Similarly, labels like `TextButton` were grouped under `Button`. Icon labels showed consistent naming across datasets and were retained as-is. This normalization step helped resolve ambiguities and ensured semantic consistency during training.

After aligning the class names, we converted all annotations into a unified format suitable for YOLO-based training pipelines. Each image was associated with a .txt annotation file, where each row represents one object, formatted as: class index (0 for Text, 1 for Button, 2 for Icon), followed by the normalized coordinates of the bounding box (center x, center y, width, height, all relative to image dimensions). In datasets originally annotated using Pascal VOC (e.g., VINS), absolute coordinates were converted to the normalized YOLO format, and class names were translated into the standard indices. The conversion was automated using custom scripts with validation steps to ensure the integrity of the format across datasets.

To manage both annotation density and computational constraints, we selected a fixed-size subset from each dataset: 200 images from GENGUI, 100 from UICVD, and 310 from VINS (Android subset). These values were chosen to strike a balance between data diversity and feasible training times. The selection was made randomly without enforcing class distribution constraints; however, we later verified the balance to avoid significant biases.

Each subset was then divided into three parts: training (Train), validation (Val), and testing (Test). Instead of proportional splits (e.g., 80/10/10), we opted for fixed image counts per partition, which enables more controlled comparisons, especially in cases where datasets differ in size and annotation density. For instance, UICVD contains densely annotated screenshots, while VINS images typically contain fewer elements. The detailed image and annotation distributions are summarized in Tables 1 and 2.

Table 1. Image and annotation counts for each dataset across the training, validation, and test splits.

Dataset	Number of Images			Number of Annotations			
	Total	Train	Val	Test	Total	Train	Val	Test
VINS	310	250	30	30	2438	1967	232	239
GENGUI	200	160	20	20	11585	9188	1209	1188
UICVD	100	80	10	10	8921	7071	939	911

While the class distribution is generally acceptable, we note a significant imbalance in the VINS dataset, where the Button class is underrepresented compared to Text and Icon. This may influence per-class performance and should be considered when interpreting results.

3.3 YOLO Model Selection and Overview

This study examines a set of advanced YOLO (You Only Look Once) models for the task of detecting user interface components. YOLO is a widely used

Table 2. Annotation distribution per class (Text, Button, Icon) for each dataset and split.

Dataset	Text				Button				Icon			
	Total	Train	Val	Test	Total	Train	Val	Test	Total	Train	Val	Test
VINS	1665	1348	154	163	257	202	31	24	516	417	47	52
GENGUI	7133	5684	730	719	826	634	113	79	3626	2870	366	390
UICVD	3624	2915	367	342	629	494	69	66	4668	3662	503	503

object detection architecture known for its balance between speed and accuracy, making it well-suited for real-time applications [17]. Its lightweight nature, competitive precision, and ability to operate on resource-constrained systems [22] make YOLO particularly attractive for UI-related use cases, where fast response time and the ability to detect visually similar elements of varying scale and position are essential.

We selected six YOLO versions for evaluation, covering a progression of architectural improvements from YOLOv5 and YOLOv8 up to the most recent YOLOv12. The selected YOLO versions were chosen to represent the most actively maintained and widely used releases from the Ultralytics line, each introducing key architectural innovations that are relevant for UI detection tasks. These versions represent a coherent sequence of innovations in model design, efficiency, and performance. A brief description of each version follows:

YOLOv5. [12], developed by Ultralytics as a PyTorch-native implementation, introduced several practical enhancements, including scalable model variants (n, s, m, l, x), automatic learning rate tuning, and the Mosaic data augmentation technique. Although not officially part of the original YOLO series, its accessibility, modular design, and consistently strong performance have made it one of the most widely adopted YOLO versions in the computer vision research and practitioner communities.

YOLOv8. [13] represents a complete architectural redesign featuring a modular structure with a C2f-based backbone and a flexible detection head. It introduced a unified interface for training, validation, and deployment and demonstrated strong performance across heterogeneous datasets, particularly in real-world object detection tasks.

YOLOv9. [24] introduced a new backbone architecture called GELAN (Gradient Efficient Layer Aggregation Network), designed to enhance gradient flow and multi-scale feature representation. In addition, it integrated a PGI (Proto-Guided Interaction) module that leverages prototype-based distillation to improve semantic consistency across detection layers. These changes significantly improve detection performance on visually complex inputs such as user interface layouts and cluttered scenes.

YOLOv10. [22] introduced a fully end-to-end object detection framework by eliminating Non-Maximum Suppression (NMS) and employing a dual label assignment strategy. Its redesigned detection head adopts a one-to-one matching paradigm inspired by transformer-based models, allowing for direct output of final predictions. These innovations significantly reduce inference latency while maintaining high accuracy, making YOLOv10 particularly effective for real-time applications.

YOLOv11. [14] introduced the C3k2 block to replace C2f in the neck and head, improving gradient flow and inference speed. It also integrated modules like SPPF (Spatial Pyramid Pooling Fast) and C2PSA (Cross-Stage Partial Spatial Attention) for improved multi-scale representation.

YOLOv12. [21] is the first YOLO architecture to center attention mechanisms, integrating Area Attention and FlashAttention for improved spatial understanding with minimal overhead. This makes it especially effective in tasks with subtle visual features or limited data, such as detecting UI components.

For each YOLO version, we tested lightweight configurations optimized for small-scale datasets, such as ours (200310 images per subset). Specifically, we used the `nano` (n) and `small` (s) variants for YOLOv5, v8, v10v12, and the `tiny` (t) and `small` (s) variants for YOLOv9, which follows a different naming convention.

To ensure consistency in comparison and reporting, we treat all first-level compact models uniformly as **n** (nano), even if the official YOLOv9 release uses the name `tiny`. This normalization simplifies cross-version analysis while preserving the distinction between smaller (**n**) and larger (**s**) variants throughout the paper.

These versions are widely recommended for low-resource environments, offering a favorable trade-off between training speed, memory usage, and generalization [8]. All configurations preserve the full architectural design of their respec-

Table 3. Summary of YOLO model configurations and architectural characteristics

Model	Submodel	Params (M)	Key Features
YOLOv5	yolov5n	2.6	Mosaic, CSPNet, auto-lr
	yolov5s	9.1	
YOLOv8	yolov8n	3.2	C2f backbone, modular head
	yolov8s	11.2	
YOLOv9	yolov9t	2.0	GELAN, PGI
	yolov9s	7.2	
YOLOv10	yolov10n	2.3	End-to-end, NMS-free, dual-label head
	yolov10s	7.2	
YOLOv11	yolo11n	2.6	C3k2, SPPF, C2PSA modules
	yolo11s	9.4	
YOLOv12	yolo12n	2.6	Area Attention, FlashAttention
	yolo12s	9.3	

tive YOLO versions, differing only in network depth and width, thus making them structurally comparable while varying in learning capacity.

Table 3 summarizes the YOLO models evaluated in this study, listing the tested submodels, parameter counts, and a brief description of the key architectural features for each version.

3.4 Experimental Setup and Evaluation Metrics

All YOLO models evaluated in this study were trained using the Google Colaboratory platform [2], with access to a `T4 GPU` accelerator. We employed the official Ultralytics implementations [11] for each version, using the `model.train()` API command to conduct training. To ensure consistent and fair comparisons, all models were trained for 100 epochs with a batch size of 2 and an input resolution of 640 pixels. Early stopping was enabled with the patience of 10 epochs, and the optimizer used was `AdamW`, with an initial learning rate of approximately `0.0014`, which corresponds to the default configuration in the Ultralytics framework. These hyperparameter values were selected following the official Ultralytics defaults, as they are generally well-suited for small-scale datasets and offer a good trade-off between training stability and performance without requiring manual tuning.

The training process included automatic checkpointing to retain the best-performing model based on the highest `mAP@0.5`, minimizing the risk of overfitting. Additionally, the default data augmentation strategies provided by the Ultralytics pipeline were preserved in all experiments, including light transformations such as blur, grayscale conversion, and contrast enhancement.

Model performance was evaluated using a standardized set of metrics applied uniformly across all architectures. Specifically, scores were computed with a fixed Intersection over Union (`IoU`) threshold of `0.5` and a confidence threshold of `0.25`. The primary evaluation metric was mean Average Precision (`mAP`), computed following the Pascal VOC protocol using the public implementation proposed by Padilla et al. [16]. For each class (`Text`, `Button`, `Icon`), the Average Precision (`AP`) was computed individually. The final `mAP` score represents the average of these values.

3.5 Training Strategy

To ensure a consistent and comprehensive evaluation, each YOLO configuration (a total of 12 models) was independently trained on each of the three datasets: GENGUI, VINS, and UICVD. This resulted in 36 distinct training runs. Each training was conducted from scratch, without cross-dataset transfer, and followed the same experimental setup described previously.

This one-to-one modeldataset setup enables us to evaluate how effectively each architecture accommodates the visual and structural differences among mobile, web, and desktop user interfaces. By isolating the behavior of each model on each dataset, we aim to highlight architectural robustness and dataset-specific performance, rather than domain adaptation effects. The results of these 36 experiments form the basis of the comparative analysis discussed in the next section.

4   Results

This section presents the performance of YOLO models evaluated on the three datasets using the unified training and evaluation setup described earlier. The analysis is structured by visual domain (mobile, web, and desktop) in order to highlight the particular challenges of each interface type.

4.1   Mobile Interfaces

For the VINS dataset, which includes screenshots of real mobile applications characterized by compact layouts and relatively few components, the results are summarized in Table 4.

Table 4. YOLO models performance on the VINS dataset, using an IoU threshold of 0.5 and a confidence threshold of 0.25. The table reports overall mAP and per-class AP for Button, Icon, and Text. Best values are highlighted in bold.

Model	Submodel	mAP	Button	Icon	Text
YOLOv5	yolov5n	93.31	95.14	89.18	95.60
	yolov5s	**93.63**	95.32	90.37	95.21
YOLOv8	yolov8n	89.88	**95.83**	80.83	92.98
	yolov8s	91.22	95.66	81.50	**96.50**
YOLOv9	yolov9t	91.30	95.66	83.46	94.79
	yolov9s	93.56	**95.83**	**90.93**	93.91
YOLOv10	yolov10n	77.11	87.12	62.48	81.73
	yolov10s	78.11	70.37	81.52	82.43
YOLOv11	yolo11n	90.52	90.46	88.08	93.01
	yolo11s	91.98	94.78	85.62	95.54
YOLOv12	yolo12n	91.07	91.06	90.59	91.55
	yolo12s	90.65	**95.83**	81.29	94.83

The YOLOv5s model achieves the highest mAP, suggesting strong adaptability to the dataset characteristics. Detection of Button components is consistently good across all models, with YOLOv8n showing a slight edge in accuracy. For Text, the best results are obtained by YOLOv8s, likely due to its modular architecture, which improves textual feature representation. Icon components, being smaller and more visually diverse, are generally harder to detect. However, YOLOv5s maintains balanced performance across all three classes, suggesting overall robustness in this context.

4.2   Web Interfaces

The GENGUI dataset, composed of automatically generated web interfaces with dense layouts and rigid structure, yields the results shown in Table 5.

Table 5. YOLO models performance on the GENGUI dataset, using an IoU threshold of 0.5 and a confidence threshold of 0.25. The table reports overall mAP and per-class AP for `Button`, `Icon`, and `Text`. Best values are highlighted in bold.

Model	Submodel	mAP	Button	Icon	Text
YOLOv5	yolov5n	84.94	98.79	67.32	88.71
	yolov5s	85.99	99.58	66.39	**92.00**
YOLOv8	yolov8n	84.47	99.59	63.96	89.86
	yolov8s	**86.27**	**99.76**	66.26	88.78
YOLOv9	yolov9t	84.38	99.35	**70.41**	83.38
	yolov9s	85.20	99.70	66.34	89.55
YOLOv10	yolov10n	85.36	99.64	66.20	90.25
	yolov10s	84.88	99.73	65.48	89.43
YOLOv11	yolo11n	84.77	99.53	65.45	89.34
	yolo11s	85.10	99.71	64.78	90.80
YOLOv12	yolo12n	85.46	99.56	64.26	92.57
	yolo12s	85.92	99.65	67.66	90.45

`YOLOv8s` delivers the best overall mAP, highlighting its balanced architecture and generalization capacity in structured, synthetic environments. The `Button` class is consistently detected with high accuracy across all models, likely due to its standardized shape and high contrast in synthetic layouts.

For `Icon`, the best performance is achieved by `YOLOv9t`, suggesting the GELAN backbone is effective for small and visually weak components. In the case of `Text`, the top results are obtained from `YOLOv5s`, followed closely by `YOLOv12n` and `YOLO11s`, indicating that both mature and newer models can effectively handle textual detection in structured environments.

4.3   Desktop Interfaces

The **UICVD** dataset contains real desktop application screenshots characterized by dense layouts and a wide variety of components. The results are shown in Table 6.

The highest mAP on this dataset is obtained by `YOLOv9s`, confirming its strong performance in visually complex environments. `Button` components are accurately detected by nearly all models, with `YOLOv9s` again standing out, suggesting its architecture effectively captures standardized control elements.

Table 6. YOLO models performance on the UICVD dataset, using an IoU threshold of 0.5 and a confidence threshold of 0.25. The table reports overall mAP and per-class AP for `Button`, `Icon`, and `Text`. Best values are highlighted in bold.

Model	Submodel	mAP	Button	Icon	**Text**
YOLOv5	yolov5n	87.07	92.87	85.40	82.94
	yolov5s	92.56	97.50	90.41	89.78
YOLOv8	yolov8n	90.15	95.65	86.89	87.90
	yolov8s	92.60	98.35	90.59	88.86
YOLOv9	yolov9t	91.76	97.04	90.45	87.79
	yolov9s	**93.00**	**98.63**	**91.55**	88.81
YOLOv10	yolov10n	91.11	97.20	88.36	87.78
	yolov10s	92.77	98.24	90.98	89.09
YOLOv11	yolo11n	91.10	97.10	88.14	88.06
	yolo11s	92.88	98.27	91.00	89.36
YOLOv12	yolo12n	91.17	97.01	88.59	**89.91**
	yolo12s	92.81	98.41	90.68	89.35

`Icon` detection also yields strong results across models, with `YOLOv9s` leading, reinforcing its robustness for small and frequent elements. For `Text`, `YOLOv12n` achieves the best score, likely due to the attention mechanisms introduced in this version that help isolate overlapping or dense textual content.

Overall, the results indicate that newer YOLO versions, especially v9 and v12, offer notable advantages in handling complex interfaces, ensuring reliable detection across all three component types.

5   Discussion

The results reveal notable differences across the YOLO versions evaluated, which are shaped by both the architecture of each model and the size of its subvariants, as well as the visual characteristics of the tested interfaces. A comparison of the three data domains (mobile, web, and desktop) shows that performance varies significantly with context, underscoring the importance of selecting the appropriate model based on the application and the visual nature of the data.

Figure 2 summarizes the variation in detection accuracy across different YOLO submodel sizes and dataset types, providing a consolidated view of model behavior in diverse interface settings.

Regarding **RQ1**, the performance of YOLO versions differs substantially across domains. No single model consistently dominates all three, but recent versions, particularly `YOLOv9s` and `YOLO11s`, demonstrate stronger adaptability to diverse visual complexity. These models benefit from architectural improvements that support more effective multi-scale feature representation and enhanced

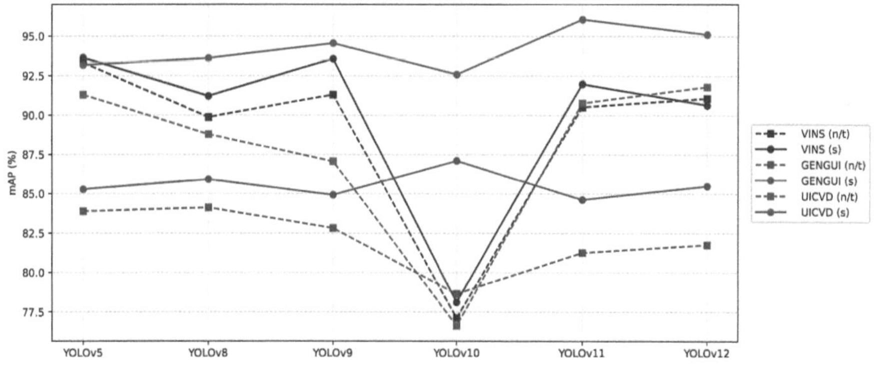

Fig. 2. Comparison of YOLO `nano`/`tiny` and `small` submodels across the VINS (mobile), GENGUI (web), and UICVD (desktop) test sets. Solid lines and circle markers represent `small` variants, while dashed lines and square markers indicate `nano` or `tiny` variants. The chart highlights accuracy trends across model capacities and interface domains.

information propagation, which are especially valuable in cluttered or low-contrast interface environments.

Mobile interfaces (VINS) are handled effectively even by smaller models, likely due to the predictable structure and clear visual separation of elements. On the UICVD dataset, where interfaces are denser and more varied, high-performing models capitalize on recurring visual patterns to achieve strong results. In contrast, although GENGUI interfaces are regular, their artificial uniformity appears to limit the model's ability to leverage distinctive cues, resulting in smaller performance gaps across versions.

Thus, no model clearly outperforms all others in every scenario. Choosing the optimal model depends on the visual characteristics of the interface. For instance, `YOLO11s` shows solid versatility across domains, while models like `YOLOv10` appear more sensitive to interface structure. This underscores that architectural compatibility with the data distribution is just as important as global mAP scores.

In response to **RQ2**, there is a clear trend in favor of the **s** variants, which consistently outperform their **n** counterparts within the same version. The visualization in Fig. 2 supports this observation: the curves for the **s** submodels consistently lie above those of the compact variants across all datasets. This advantage is even more pronounced in visually complex scenarios, such as GENGUI and UICVD, where richer feature representation translates into higher performance.

On the VINS dataset, however, the differences are less significant, suggesting that in mobile contexts, which typically feature clean layouts and clearly defined components, smaller models can be competitive while offering reduced computational cost. Local exceptions, such as `YOLOv12n` on VINS, also show that the base architecture can compensate for reduced size under favorable visual conditions.

In summary, s variants generally provide a more reliable choice for robust performance, particularly in complex user interfaces. However, in resource-constrained environments or more uniform datasets (such as mobile layouts), the n variants can serve as efficient alternatives.

Regarding **RQ3**, which examines the relative difficulty of detecting different UI components, a clear hierarchy emerges among the three classes: `Button`, `Text`, and `Icon`. When deliberately excluding the YOLOv10 models, which show the most significant deviations from the others, with lower and more inconsistent scores across classes, a more coherent pattern is observed. `Button` components are consistently the easiest to detect, achieving the highest AP scores across all datasets. This is plausible given their repetitive appearance, standardized shapes, and high contrast across most interfaces and platforms.

In parallel, `Text` proves to be the most stable class, with minimal variation between models and datasets, suggesting strong generalization capabilities, even though it does not always yield the top scores. This balance makes it a reliable category, especially for recent models such as `YOLOv8s`, `YOLO11s`, and `YOLO12s`.

Conversely, `Icon` consistently emerges as the most challenging class to detect. Its lower performance can be attributed to small size, limited contrast, and high visual diversity, all of which complicate localization and classification. While some models, such as `YOLOv9s` and `YOLO11s`, show promising results for this class, scores generally remain below those achieved for `Button` and `Text`.

When reincluding YOLOv10 in the analysis, the overall picture becomes more heterogeneous. This version introduces the highest class-level variation, with inferior performance on `Button` and `Text` across all datasets. Interestingly, certain submodels, such as `YOLOv10s` on GENGUI and UICVD, achieve relatively strong scores for `Icon`, indicating a potential architectural sensitivity to this class. However, this does not compensate for the underperformance observed in other areas.

Therefore, we conclude that `Button` is, on average, the most accessible class for YOLO models, `Text` the most consistent, and `Icon` the most challenging. These insights emphasize that global mAP alone may not sufficiently reflect model suitability and that detection performance on specific classes should also guide model selection based on the application's priorities.

To complement the quantitative results presented above, we include visual examples that illustrate typical detection difficulties encountered across interface types. Figure 3 shows samples from the test sets of VINS (mobile), GENGUI (web), and UICVD (desktop), with predictions and ground-truth annotations overlaid. The `YOLOv10s` model was selected for this illustration due to its architectural distinctiveness and mixed quantitative performance, which makes it well-suited for highlighting both successful and failed cases. Correct detections are marked in blue, missed components (i.e., present in the ground truth but not detected) are shown in red, and incorrect predictions (i.e., model detections with no matching ground-truth component) appear in orange.

In the VINS image (Fig. 3a), several predicted elements lack corresponding annotations in the ground truth, which suggests that the original dataset may include incomplete labeling. This highlights an important consideration when interpreting evaluation scores, particularly in mobile scenarios. In the GEN-GUI and UICVD samples, cropped to focus on visually dense regions, icons and overlapping text often result in missed detections or confusion between classes. These examples qualitatively support the observations made earlier regarding the higher difficulty of detecting small or ambiguous elements, especially in synthetic or desktop environments.

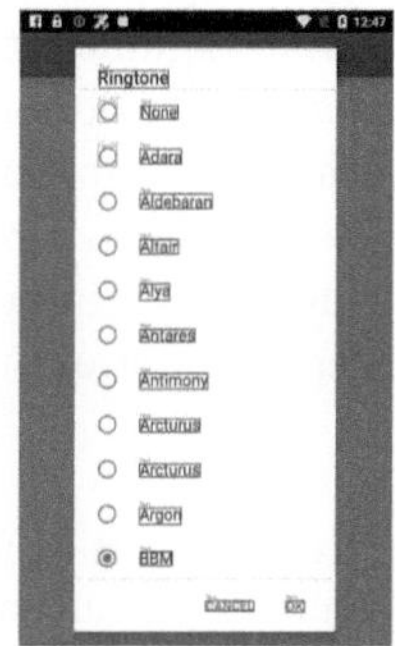
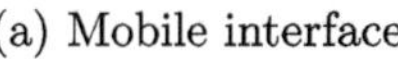

(a) Mobile interface

(b) Web interface

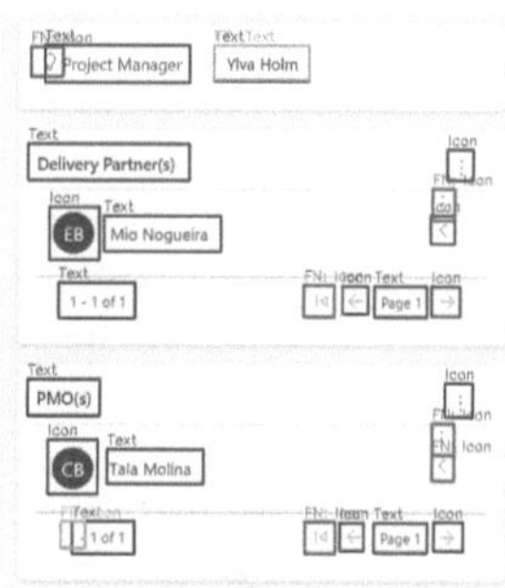

(c) Desktop interface

Fig. 3. Qualitative detection examples using the YOLOv10s model. Blue boxes indicate correct predictions, red boxes represent missed components (present in the ground truth but not detected), and orange boxes correspond to incorrect predictions (detections with no matching ground-truth element).

6   Conclusions and Future Work

In this paper, we conducted a comparative analysis of six modern versions of YOLO models, ranging from YOLOv5 to YOLOv12, for the task of detecting graphical components in user interfaces across three platforms: mobile, web, and desktop. We selected a standard set of UI classes (Button, Text, and Icon) to ensure consistent comparison between models. For each YOLO version, we evaluated the lightest available configurations (nano/tiny and small), which are often used in resource-constrained environments. The results show that performance varies significantly depending on the YOLO architecture, model size, and the visual characteristics of the target interface.

Newer models, such as YOLOv9s and YOLO11s, demonstrated better adaptability in visually complex scenarios, while earlier versions, like YOLOv5s, remained competitive in simpler cases. Regarding the UI classes, Button was generally the easiest to detect, thanks to its standardized shapes and high contrast with the background. The Text class performed consistently well across experiments, indicating that YOLO models can generalize effectively for this type of element, even in varied visual conditions. In contrast, `Icon` was the most challenging class to detect due to its small size, diverse appearance, and frequent occurrence in cluttered layouts.

Although the `small (s)` variants generally delivered the best results, compact models like `YOLOv5n` and `YOLOv12n` achieved competitive performance on mobile interfaces. This suggests that model selection should consider both the visual complexity of the application and the available computational resources.

Given the promising results of compact models, a future research direction could be to explore intermediate variants (such as `medium` or `large`) to see whether a higher parameter count leads to better accuracy without significantly affecting efficiency. Another interesting path would be to explore multi-domain training strategies, such as pretraining on synthetic layouts followed by fine-tuning on real-world interfaces. This approach may improve generalization, especially for small datasets. Lastly, combining visual data with structural metadata (e.g., UI component hierarchies) could support a more advanced understanding of user interfaces in future applications.

Acknowledgment. This research is supported by the project "Romanian Hub for Artificial Intelligence - HRIA", Smart Growth, Digitization and Financial Instruments Program, 2021–2027, MySMIS no. 334906.

References

1. Beltramelli, T.: pix2code: generating code from a graphical user interface screenshot. In: Proceedings of the ACM SIGCHI Symposium on Engineering Interactive Computing Systems, pp. 1–6 (2018)
2. Bisong, E.: Google Colaboratory. Apress (2019)
3. Bunian, S., Li, K., Jemmali, C., Harteveld, C., Fu, Y., El-Nasr, M.S.: VINS: visual search for mobile user interface design (2021)
4. Chen, J., et al.: Object detection for graphical user interface: old fashioned or deep learning or a combination? In: Proceedings of the 28th ACM Joint Meeting on European Software Engineering Conference and Symposium on the Foundations of Software Engineering, pp. 1202–1214 (2020)
5. Cheng, J., Tan, D., Zhang, T., Wei, A., Chen, J.: YOLOv5-MGC: GUI element identification for mobile applications based on improved YOLOv5. Mob. Inf. Syst. **2022**(1), 8900734 (2022)
6. Dicu, M., Sterca, A., Chira, C., Orghidan, R.: UICVD: a computer vision UI dataset for training RPA agents. In: ENASE, pp. 414–421 (2024)
7. Dicu, M., González, E.G., Chira, C., Villar, J.R.: GenGUI: a dataset for automatic generation of web user interfaces using ChatGPT. In: Proceedings of the 17th International Conference on Agents and Artificial Intelligence - Volume 3: ICAART, pp. 707–714. INSTICC, SciTePress (2025)
8. Hussain, M.: YOLOv1 to v8: unveiling each variant-a comprehensive review of YOLO. IEEE Access **12**, 42816–42833 (2024)
9. Jegham, N., Koh, C.Y., Abdelatti, M., Hendawi, A.: YOLO evolution: a comprehensive benchmark and architectural review of YOLOv12, YOLO11, and their previous versions. YOLO11, and Their Previous Versions (2024)

10. Jiang, T., Zhong, Y.: ODverse33: is the new YOLO version always better? A multi domain benchmark from yolo v5 to v11. arXiv preprint: arXiv:2502.14314 (2025)
11. Jocher, G., Chaurasia, A., Qiu, J.: YOLO by Ultralytics (2023). https://github.com/ultralytics/ultralytics
12. Jocher, G., et al.: YOLOv5 - Ultralytics (2020). https://github.com/ultralytics/yolov5
13. Jocher, G., et al.: YOLOv8 - Ultralytics (2023). https://github.com/ultralytics/ultralytics
14. Jocher, G., et al.: YOLOv11 - Ultralytics (2024). https://github.com/ultralytics
15. Li, C., et al.: YOLOv6: a single-stage object detection framework for industrial applications. arXiv preprint: arXiv:2209.02976 (2022)
16. Padilla, R., Passos, W., Dias, F., Netto, S., da Silva, E.: A survey on performance metrics for object-detection algorithms. arXiv preprint: arXiv:2006.02607 (2020)
17. Redmon, J., Divvala, S., Girshick, R., Farhadi, A.: You only look once: unified, real-time object detection. In: Proceedings of the IEEE Conference on Computer Vision and Pattern Recognition, pp. 779–788 (2016)
18. Said, K.S., Nie, L., Ajibode, A.A., Zhou, X.: GUI testing for mobile applications: objectives, approaches and challenges. In: Proceedings of the 12th Asia-Pacific Symposium on InternetWare, pp. 51–60 (2020)
19. Selcuk, B., Serif, T.: A comparison of YOLOv5 and YOLOv8 in the context of mobile UI detection. In: International Conference on Mobile Web and Intelligent Information Systems, pp. 161–174. Springer (2023)
20. Sun, X., Li, T., Xu, J.: UI components recognition system based on image understanding. In: 2020 IEEE 20th International Conference on Software Quality, Reliability and Security Companion (QRS-C), pp. 65–71. IEEE (2020)
21. Tian, Y., Ye, Q., Doermann, D.: YOLOv12: attention-centric real-time object detectors. arXiv preprint: arXiv:2502.12524 (2025)
22. Wang, A., et al.: YOLOv10: real-time end-to-end object detection. In: Advances in Neural Information Processing Systems, vol. 37, pp. 107984–108011 (2024)
23. Wang, C.Y., Bochkovskiy, A., Liao, H.Y.M.: YOLOv7: trainable bag-of-freebies sets new state-of-the-art for real-time object detectors. In: Proceedings of the IEEE/CVF Conference on Computer Vision and Pattern Recognition, pp. 7464–7475 (2023)
24. Wang, C.Y., Yeh, I.H., Mark Liao, H.Y.: YOLOv9: learning what you want to learn using programmable gradient information. In: European Conference on Computer Vision, pp. 1–21. Springer (2024)
25. Xie, M., Feng, S., Xing, Z., Chen, J., Chen, C.: UIED: a hybrid tool for GUI element detection. In: Proceedings of the 28th ACM Joint Meeting on European Software Engineering Conference and Symposium on the Foundations of Software Engineering, pp. 1655–1659 (2020)
26. Zhang, T., Liu, Y., Gao, J., Gao, L.P., Cheng, J.: Deep learning-based mobile application isomorphic GUI identification for automated robotic testing. IEEE Softw. **37**(4), 67–74 (2020)

Embedding-Efficient Brain-to-Image Reconstruction Using Diffusion Models

Ioana Gabor[(⊠)]

Department of Computer Science, Faculty of Mathematics and Computer Science,
Babes-Bolyai University, Cluj-Napoca, Romania
`ioana.gabor.cj@gmail.com`

Abstract. Visual brain decoding is the process of reconstructing the images a person sees by analyzing their brain activity, as measured using functional MRI, electroencephalography, or magnetoencephalography. Recent advances in generative AI, particularly diffusion models, have significantly improved the semantic quality of the reconstructed images. In this work, we investigate the potential for improving the computational efficiency of brain-to-image models by focusing on a reduced number of informative embeddings. Specifically, we explore whether high-quality image reconstructions can be achieved by predicting a small, selected subset of CLIP-derived embeddings. We perform evaluations using the Natural Scenes Dataset, applying both a simple single-subject ridge regression model and a multi-subject federated learning framework adapted from a recent approach. The results demonstrate that embedding-efficient decoding can achieve competitive performance while substantially reducing model size, although with a decrease in quality on certain metrics. Code available at: https://github.com/IoanaGabor/efficient-vbd.

Keywords: Braincomputer interfaces · functional magnetic resonance imaging (fMRI) · image reconstruction · diffusion models · neural decoding · CLIP embeddings · federated learning · ridge regression

1 Introduction

Reconstructing visual stimuli from human brain activity is a core challenge at the intersection of cognitive neuroscience and artificial intelligence. Visual brain decoding refers to the task of reconstructing images that a person has perceived, using recorded neural activity, such as functional Magnetic Resonance Imaging (fMRI), electroencephalography (EEG), or magnetoencephalography (MEG). Among these, fMRI is particularly suited for high-resolution visual decoding due to its superior spatial accuracy [1].

Recent advancements in generative modeling, in particular latent diffusion models, have revolutionized the fidelity of image generation from neural signals. These models iteratively refine noisy inputs into realistic images, having outperformed earlier approaches based on Variational Autoencoders (VAEs) and Generative Adversarial Networks (GANs) [5]. This is partially due to the fact that

C. Chira et al. (Eds.): InnoComp 2025, CCIS 2793, pp. 140–156, 2026.
https://doi.org/10.1007/978-3-032-12478-4_10

OpenAI's Contrastive LanguageImage Pretraining (CLIP) embeddings [21] have proven to be an effective intermediate representation, as they correlate well with higher-level semantic patterns encoded in the visual cortex. This alignment has enabled a new class of brain decoding pipelines that map neural activity directly to the CLIP space, subsequently generating images via latent diffusion models [19].

While these models have demonstrated state-of-the-art performance, their embedding strategies often rely on predicting a few hundreds of high-dimensional tokens, 257 visual and 77 textual embeddings in most cases [19,24]. This might lead to extended training times, and potential overfitting, as claimed by Han et al. [11]. Additionally, they rely on Versatile Diffusion [35], which might add unwanted artifacts, as opposed to Stable Diffusion [11].

The main contributions of this paper lie in extending the efficient embedding approach introduced by Han et al. [11] through the integration of novel, lightweight architectures aimed at reducing model complexity. We explore the potential of using a smaller, processed subset of CLIP-derived embeddings to enable high-quality brain-to-image reconstruction, with the goal of enhancing computational efficiency while maintaining reconstruction accuracy. Specifically, we extract just 16 informative visual embeddings using a pretrained IPAdapterPlus model [36] and evaluate the performance within two reconstruction pipelines: (i) a simple ridge regression single-model-per-subject approach, and (ii) an adapted federated learning model based on BrainGuard [31]. Our experiments are conducted using the Natural Scenes Dataset (NSD) [1], which has become a well-established benchmark for fMRI-based decoding.

The results suggest that competitive reconstruction performance can be obtained while significantly reducing model complexity. Additionally, the good performance obtained under federated learning conditions indicates potential for future practical applications in privacy-sensitive and clinical environments, serving as a starting point for future research in this area.

2   Related Work

The field of visual brain decoding has seen substantial evolution, moving from early traditional methods to deep learning and generative models [22]. This section reviews key milestones, categorizing them by their methodology and underlining their contributions to the current state-of-the-art.

2.1   Non-generative Methods

Early efforts in visual brain decoding primarily focused on classification, identification, or direct reconstruction using non-generative approaches [22]. Incipient techniques involved learning linear mappings between fMRI voxel activity and hand-engineered image features, such as Gabor filters [7] or specific image structures. These methods, while foundational, were limited in their ability to capture non-linear relationships between brain signals and visual stimuli. The advent of deep learning, particularly Convolutional Neural Networks (CNNs), marked a significant leap forward. CNNs, with their multi-layered architectures, proved more effective in learning intricate mappings from fMRI data to visual features. Encoder-decoder models based on CNNs, like those proposed by Beliy et al. [2] and Gaziv et al. [8], enabled more abstract feature learning. These models typically involved a two-stage training process, where an encoder mapped stimulus images to brain activity, and a decoder performed the inverse, reconstructing images from neural signals.

2.2   Generative Models

The release of large-scale fMRI datasets like the Natural Scenes Dataset (NSD) [1] led to increased interest in generative brain decoding. Mind Reader, presented at NeurIPS 2022, was among the first to utilize the NSD, adopting a multimodal approach [17]. They mapped fMRI signals to CLIP embeddings (both visual and textual, utilizing Common Objects in Context (COCO) [18] captions) and then used these embeddings to condition a Generative Adversarial Network (GAN), specifically StyleGAN2 [32], for image generation. While Mind Reader laid the groundwork for multimodal brain-to-image generation, it was constrained to single-subject analysis and relied on GANs, which can suffer from training instability and mode collapse [9].

2.3   Diffusion Models

More recently, diffusion models have emerged as the leading generative framework for high-quality image synthesis, demonstrating improved performance in reproducing fine details and overall image quality compared to GANs [5, 12, 23].

One Model Per Subject Approaches. Brain Diffuser, introduced by Ozcelik et al. in 2023, published by Nature, proposed a two-stage pipeline for brain decoding [19]. The first stage generated low-level image reconstructions using a Very Deep Variational Autoencoder (VDVAE) [4] by mapping fMRI signals to its latent space. The second stage trained additional regression models to map fMRI data to both CLIP-Vision and CLIP-Text features, which, combined with the low-level reconstruction, guided a Versatile Diffusion [35] model for final image generation. Brain Diffuser achieved state-of-the-art results across various quantitative metrics, largely due to its multi-stage reconstruction and multimodal guidance. Similarly, MindEye [24] utilized two parallel pipelines: a high-level one mapping fMRI voxels to CLIP ViT-L/14 image space and a low-level one mapping voxels to a Variational Autoencoder's image embedding space [24]. Both pipelines shared a residual Multilayer Perceptron (MLP) backbone and task-specific submodules, with optimal performance achieved through contrastive and mean squared error losses.

Multi-subject Approaches. A significant challenge in brain decoding has been the inter-subject variability, often necessitating individual models for each subject, which limits scalability. Recent advancements have focused on multi-subject models to overcome this limitation. MindBridge (2024) [33] introduced a cross-subject framework featuring an aggregation function, a "brain embedder", a "brain builder", and a "brain translator". It incorporated a cyclic fMRI reconstruction mechanism and a "reset-tuning" strategy for adapting to new subjects. Though it did not consistently outperform existing methods across all benchmarks, MindBridge demonstrated comparable performance with a single model, highlighting its potential for scalable cross-subject brain decoding. Psychometry (2024) [20] proposed an "omnifit" model designed to reconstruct images from fMRI data across different subjects by integrating an "Omni Mixture-of-Experts (Omni MoE)" module and a retrieval-enhanced inference strategy called "Ecphory". Its architecture featured a Vision Transformer with an "Omni MoE" layer within the transformer block, trained with a bidirectional contrastive learning loss [20]. Brainguard (2025) [31] introduced a global-local architecture that enhances privacy by training individual models on local subject data while a shared global model captures cross-subject patterns, eliminating the need for fMRI data aggregation. The global model integrates parameters from independently trained individual models, which then dynamically synchronize with the merged global parameters.

3   NSD Dataset

The Natural Scenes Dataset comprises high-resolution (1.8-mm isotropic, 1.6-s sampling rate) whole-brain fMRI measurements obtained from 8 healthy adult subjects [1]. These measurements were collected while the subjects viewed thousands of diverse color natural scenes, taken from the Microsoft Common Objects in Context dataset [18], across 3040 individual scan sessions. Data acquisition was performed at ultra-high-field (7T) strength at the Center of Magnetic Resonance Research at the University of Minnesota. During the scanning sessions, subjects engaged in a continuous recognition task, where they reported whether they had previously seen each presented image. Out of those 8 subjects, only 4 completed all the scanning sessions, therefore most research in visual reconstruction focuses on them [19,24,31]. Using the same split as in related research, for each subject, the training set consists of 8,859 unique image stimuli and 24,980 corresponding fMRI trials, while the test set includes 982 image stimuli and 2,770 fMRI trials, containing only images that were shared across subjects.

Data Preprocessing. As in other studies [19,24,31,33], we based our analysis on time-collapsed fMRI data represented by beta values. The preprocessing included fitted hemodynamic response functions, along with denoising and regularization using GLMDenoise [15] and ridge regression [13,14]. To restrict the analysis to relevant brain areas, the NSDGeneral Region-of-Interest (ROI) mask was applied [1]. This ROI includes a wide range of visual regions, from early visual cortex to higher-level visual areas, covering [15724, 14278, 13039, 12682] voxels for the four subjects, respectively [1]. Please refer to the original Natural Scenes Dataset publication [1] for additional information.

4   Methodology

This section presents the methodology upon which the performed experiments are based. First, we formalize the problem and present the main stages of reconstruction, then proceed with describing them individually.

4.1   Problem Formalization

The objective is to reconstruct the original image viewed by a subject based on recorded fMRI signals. After preprocessing (detailed in the *Data Preprocessing* section), the fMRI data is represented as a one-dimensional vector of voxel intensities. The length of this vector, denoted d_s, varies with each subject $s \in \{1, 2, \ldots, S\}$, where S is the number of subjects.

Formally, let $X_{s,n} \in \mathbb{R}^{d_s}$ represent the preprocessed fMRI signal recorded from subject s when viewing image I_n. The goal is to map $X_{s,n}$ to a corresponding feature vector $F_n \in \mathbb{R}^{c \times t}$, where F_n represents features extracted directly from the image I_n, c denotes the number of features, and t denotes the dimensionality of each feature. These features serve as intermediate representations that are subsequently used as input for an image reconstruction model, which generates an approximation $\hat{I}_n$ of the original image I_n.

This process can be described as a composition of functions:

$$X_{s,n} \xrightarrow{f_s} \hat{F}_n \xrightarrow{g} \hat{I}_n,$$

where $f_s : \mathbb{R}^{d_s} \to \mathbb{R}^{c \times t}$ is a subject-specific feature prediction model, and $g : \mathbb{R}^{c \times t} \to \mathcal{I}$ is a fixed image reconstruction model, with $\mathcal{I}$ denoting the space of possible images.

Each component of this pipeline is addressed in the following sections. The three main components are: feature extraction from the original images, the brain mapping module that predicts features from fMRI data, and the image reconstruction model that generates images from predicted features.

4.2   Feature Extraction

This stage aims to extract informative visual embeddings from each image, which will later be used as prediction targets for the brain-to-feature mapping module. These embeddings encode both semantic content and spatial structure, serving as an intermediate representation between fMRI signals and final image reconstruction. In particular, they are used to condition diffusion models, which iteratively generate high-fidelity images that align with the predicted visual features.

We utilize the IP-AdapterPlus [36] architecture, which was first used in this context by Han et al. [11]. Unlike earlier approaches that relied on high-dimensional CLIP embeddings (257×768 visual, 77×768 textual) [19,24,31,33], their approach proved that the IP-AdapterPlus can provide a more compact and effective alternative, functioning as an output to a ViT-inspired architecture. Specifically, brain signals are mapped to a 16×768 feature space, reducing size and the risk of overfitting [11]. The fixed number of 16 embeddings is determined by the pretrained model provided by [36]. We adopt this model following the setup in [11], as it offers a favorable balance between compactness and representational capacity.

The IP-Adapter [36] architecture is originally designed for conditioning pre-trained text-to-image diffusion models with image prompts. Its key innovation lies in a decoupled cross-attention mechanism, which separates the processing of image and text features [36]. IP-AdapterPlus, a better version of the original adapter, improves detail preservation by using fine-grained features. It extracts grid-level activations from the penultimate layer of the CLIP image encoder [36]. These grid features are then processed by a lightweight transformer-based query network with 16 learnable tokens, which converts them into 16 embeddings of size 768. These embeddings capture both semantic and spatial characteristics of the image more effectively than global representations used by IP-Adapter [36].

In our setup, these fine-grained embeddings serve as the target outputs for the brain-to-feature mapping model.

4.3   Brain Mapping Module

As outlined before, we performed experiments on two visual reconstruction frameworks. We will describe them individually in this section.

Single-Subject Framework. This framework is constrained to a single-subject paradigm, as training is performed on each subject individually, and the commonalities are not taken into account. A source of inspiration for this framework was BrainDiffuser [19], in terms of using ridge regression; however their approach is more complex, involving two stages, the first one involving a Very Deep Variational Autoencoder [4], the second one using Versatile Diffusion [35] with both textual and visual embeddings [19], whereas ours has only one stage. For each subject, 16 ridge regressors, one per predicted embedding, were trained. After generating predictions, the outputs were aligned to the training distribution, similar to the Brain Diffuser approach. The model architecture is shown in Fig. 1. Ridge regression was chosen for these initial experiments to model simple, linear mappings between the embedding space and brain representations, providing a baseline for future comparisons. This approach becomes impractical in broader scenarios, as it effectively amounts to overfitting a model to the data of a single subject.

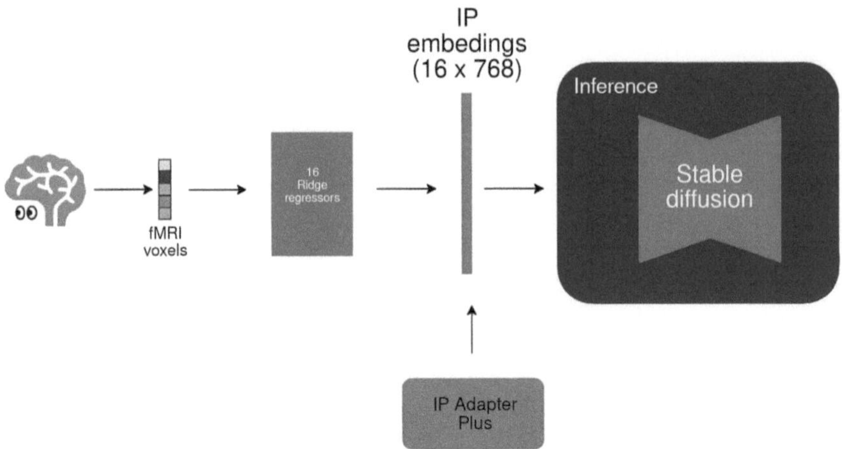

Fig. 1. Simple single-subject Ridge Regression diagram.

Multi-subject Framework. The second approach builds on the federated learning method introduced in the BrainGuard paper [31]. There are three major differences: (i) we used IP-Adapter embeddings, as opposed to their visual and textual CLIP embeddings, reducing parameter count by 7 times, as outlined in Table 9, (ii) we replaced the Versatile Diffusion model with Stable Diffusion (iii) we used a prediction strategy that is independent of the training data, whereas their method involves averaging predicted embeddings with similar ones found during training. In terms of similarities, both approaches use the same training algorithm, synchronization strategy and network architecture. The training process consists of repeated cycles of local iterations, performed by each client, followed by an aggregation and synchronization of the model weights, on the central server [31]. The advanced layers are updated using a "Dynamic Fusion Learner" module [37], the intermediate layers are synchronized via parameter copying, while those foundational remain local [31]. The network architecture consists of MLPs with residual blocks, followed by a projector head, using GELU for activation. The loss is calculated by considering two components: MSE loss and SoftCLIP loss [31]. For more technical details regarding the BrainGuard framework, please consult the original paper, or the adapted codebase. The architecture can be viewed in Fig. 2, with the differences between the frameworks highlighted in orange.

4.4   Image Generation

The final stage of the pipeline reconstructs images from the predicted visual features using a text-to-image diffusion model. Specifically, we use a Stable Diffusion pipeline [27] configured with a Denoising Diffusion Implicit Models (DDIM) scheduler [26] and a separately loaded VAE decoder [28]. This design mirrors the approach proposed by MindFormer [11].

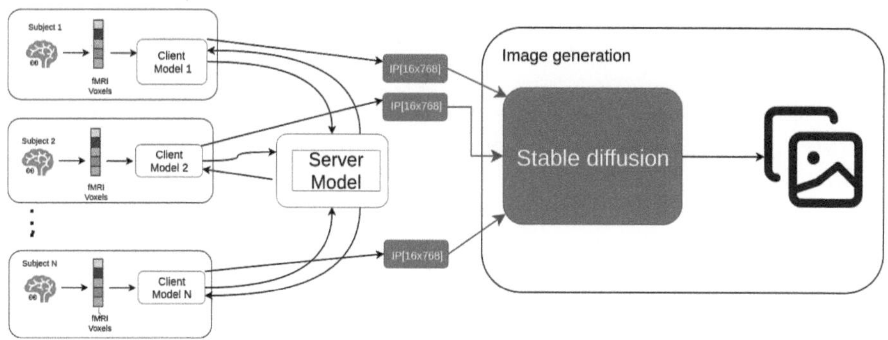

Fig. 2. Multi-subject Federated Learning diagram, adapted from BrainGuard. (Color figure online)

Our specific hyperparameters were based on open-source IP-Adapter-Plus [36] examples. These parameter values are listed in Table 1.

Table 1. Stable Diffusion Reconstruction Configuration

Component	Value
DDIM Timesteps	1000
Beta Start	0.00085
Beta End	0.012
Beta Schedule	`scaled_linear`
Clip Sample	False
Alpha to One	False
Steps Offset	1
Torch Dtype	`float16`
Inference Steps	100

4.5 Evaluation Metrics

We evaluate reconstruction quality using eight image similarity metrics, chosen to capture both low-level and high-level visual characteristics, following evaluation protocols from related work [24].

Low-level metrics focus on fundamental visual attributes such as structure and pixel alignment:

- *PixCorr*: Measures pixel-wise correlation between reconstructed and ground-truth images.
- *SSIM* [34]: Evaluates structural similarity, emphasizing luminance, contrast, and structure.
- AlexNet features [16]: Features were extracted from the second and fifth convolutional layers, denoted *Alex(2)* and *Alex(5)* respectively.

High-level metrics assess semantic similarity using feature representations from deeper networks:

- *Inception-v3* [29]: Features were extracted from the final pooling layer.
- *CLIP-Vision* [21]: Uses output embeddings from the CLIP vision encoder.
- *EfficientNet-B1* [30] and *SwAV-ResNet50* [3]: Provide correlation-based distances computed from their respective feature spaces.

For *Alex(2)*, *Alex(5)*, *Inception-v3*, and *CLIP*, we compute the Pearson Correlation Coefficient and apply two-way identification [24] over the test dataset to quantify recognition accuracy. For all metrics except *EffNet-B1* and *SwAV*, higher values indicate better reconstruction quality.

5   Experiments and Results

5.1   Single Subject Framework

We performed a quantitative analysis of the α hyperparameter used in ridge regression. In ridge regression, α controls the strength of the L_2 penalty applied to the model weights. A larger α imposes a stronger constraint on the magnitude of the weights, helping prevent overfitting. Conversely, smaller α values allow the model to fit the training data more closely, but at the risk of reduced generalization.

The results obtained by averaging across subjects, for one run each, can be visualized in the Table 2. As it can be observed, α values of 20000 and 30000 achieve the best performance. For lower values of α (e.g., 1000), decreased performance across all metrics is observed. The first column contains the α value. A per-subject breakdown for a sample run, with $\alpha = 20000$, is provided in Table 3.

Table 2. Quantitative results for various α values using the 8 evaluation metrics

Alpha	PixCorr↑	SSIM↑	Alex (2) ↑	Alex (5) ↑	Incept. ↑	CLIP ↑	EffNet-B ↓	SwAV ↓
1000	0.185	0.301	88.5%	94.6%	87.5%	84.1%	0.774	0.449
5000	0.202	0.312	91.7%	96.9%	92.2%	89.1%	0.717	0.408
10000	0.214	0.318	92.7%	97.3%	92.9%	90.2%	0.707	0.400
20000	0.220	0.322	93.1%	97.5%	93.3%	90.5%	0.703	0.399
30000	0.221	0.324	93.0%	97.4%	93.1%	90.5%	0.709	0.401
60000	0.219	0.326	92.6%	97.2%	92.5%	89.7%	0.723	0.411

5.2   Multiple Subject Framework

Following the approach presented by BrainGuard, we performed experiments to better understand how the frequency of synchronization across the client architectures influences the performance. After these experiments, we found that the best results, for all metrics, are obtained when global iterations happen after exactly one local iteration for all the client models. Relevant values, averaged for all subjects, for a single run, are presented in Table 4.

Table 3. Quantitative results for $\alpha = 20000$ for each subject

Subject	PixCorr ↑	SSIM ↑	Alex (2) ↑	Alex (5) ↑	Incept. ↑	CLIP ↑	EffNet-B ↓	SwAV ↓
1	0.248	0.328	94.45%	98.13%	93.51%	90.98%	0.698	0.394
2	0.225	0.324	93.35%	97.91%	93.26%	89.48%	0.706	0.399
5	0.209	0.323	93.35%	97.81%	94.52%	92.18%	0.685	0.387
7	0.200	0.315	91.23%	96.22%	91.88%	89.43%	0.723	0.415

Table 4. Effect of number of local iterations on reconstruction quality

Iter.	PixCorr ↑	SSIM ↑	Alex(2) ↑	Alex(5) ↑	Inc. ↑	CLIP ↑	Eff ↓	SwAV ↓
1	0.263	0.335	0.951	0.985	0.959	0.939	0.611	0.334
5	0.256	0.330	0.949	0.983	0.957	0.937	0.614	0.336
10	0.254	0.335	0.948	0.982	0.955	0.939	0.618	0.335
20	0.252	0.331	0.946	0.983	0.956	0.937	0.618	0.339

Additional experiments were performed by variating the hidden layer size. The results, averaged for all subjects, for a single run, can be viewed in Table 5. These results not only offer insight into the performance impact that the hidden layer has, but also showcase the relationship between the metrics in evaluating the same images. EffNet-B has the largest difference in score, followed by Alex(2) and PixCorr. On the opposite side of the spectrum, SSIM has the smallest difference in score.

Table 5. Effect of hidden layer size on reconstruction quality

Hidden	Pix ↑	SSIM ↑	A(2) ↑	A(5) ↑	Inc. ↑	CLIP ↑	Eff ↓	SwAV ↓
64	0.091	0.248	0.759	0.854	0.798	0.802	0.837	0.489
128	0.134	0.293	0.839	0.927	0.883	0.884	0.741	0.426
256	0.188	0.320	0.891	0.956	0.918	0.918	0.683	0.389
2048	0.263	0.334	0.951	0.985	0.959	0.937	0.611	0.334

One component of the BrainGuard framework [31] is the "Dynamic Fusion Learner (DFL) module" [37], which fine-tunes the deeper layers of the network responsible for capturing high-level features essential to interpreting individual brain activity. Rather than fully replacing model parameters, DFL adaptively fuses global and subject-specific parameters on a layer-by-layer basis to balance shared patterns with personalized representations. The performance of this module is sensitive to the number of synchronized layers, with the original Brain-Guard study identifying 8 as the optimal setting. We replicated this analysis in our modified framework, with the results presented in Table 6, averaged across subjects, for a single run. In our case the best results seem to be obtained by synchronizing only 4 layers.

Table 6. Effect of number of synchronized layers on reconstruction quality

Layers	Pix ↑	SSIM ↑	A(2) ↑	A(5) ↑	Inc. ↑	CLIP ↑	Eff ↓	SwAV ↓
1	0.251	0.330	0.947	0.982	0.954	0.942	0.616	0.339
2	0.262	0.332	0.949	0.983	0.952	0.939	0.622	0.341
3	0.247	0.328	0.943	0.982	0.954	0.942	0.612	0.337
4	0.248	0.330	0.945	0.983	0.958	0.951	0.608	0.335
5	0.249	0.331	0.948	0.982	0.957	0.942	0.613	0.338
6	0.255	0.333	0.947	0.982	0.957	0.943	0.610	0.338
7	0.250	0.329	0.945	0.982	0.955	0.943	0.613	0.338
8	0.252	0.331	0.945	0.982	0.955	0.942	0.611	0.338
10	0.247	0.331	0.944	0.981	0.953	0.943	0.614	0.339

In Table 7, we provide a per-subject breakdown for one run of an architecture having a hidden layer size of 2048, 4 synchronized layers, with synchronization happening after each local iteration.

Table 7. Quantitative results for each subject, federated approach

Subject	PixCorr ↑	SSIM ↑	Alex (2) ↑	Alex (5) ↑	Incept. ↑	CLIP ↑	EffNet-B ↓	SwAV ↓
1	0.286	0.340	95.84%	98.66%	96.00%	93.96%	0.608	0.331
2	0.265	0.342	95.38%	98.42%	95.10%	92.45%	0.620	0.340
5	0.243	0.330	94.63%	98.49%	96.68%	94.70%	0.590	0.325
7	0.228	0.326	92.54%	96.87%	94.13%	91.96%	0.637	0.355

Figure 3 presents some examples of good and of bad reconstructions for the federated approach. The original image and reconstructions for the first three subjects are displayed. One noteworthy observation, regarding the "Low quality" reconstructions, is that generally, the framework struggles with encoding information related to the multiplicity of the objects. This is a known shortcoming of CLIP [21]. Additionally, due to the limited number of embeddings used, it may happen that details are not fully reconstructed, as it can be observed for the image featuring a cake.

5.3   Comparison with Other Works

Quantitative Comparison. Table 8 provides a detailed comparison of our best performing configuration for both methods, single-subject (*ours (RR)*) and multi-subject (*ours (BG)*), with several recent state-of-the-art approaches, on the same test split, using the average of the scores obtained for all 4 subjects, across 7 repetitions. The external metrics and images are provided by [11,25,31]. Across all evaluation metrics, ours (BG) demonstrates strong and consistent performance, outperforming earlier works such as Mind-Vis [17], Gu et al. [10], and MindBridge [33]. In terms of pixel-level similarity, as measured by PixCorr

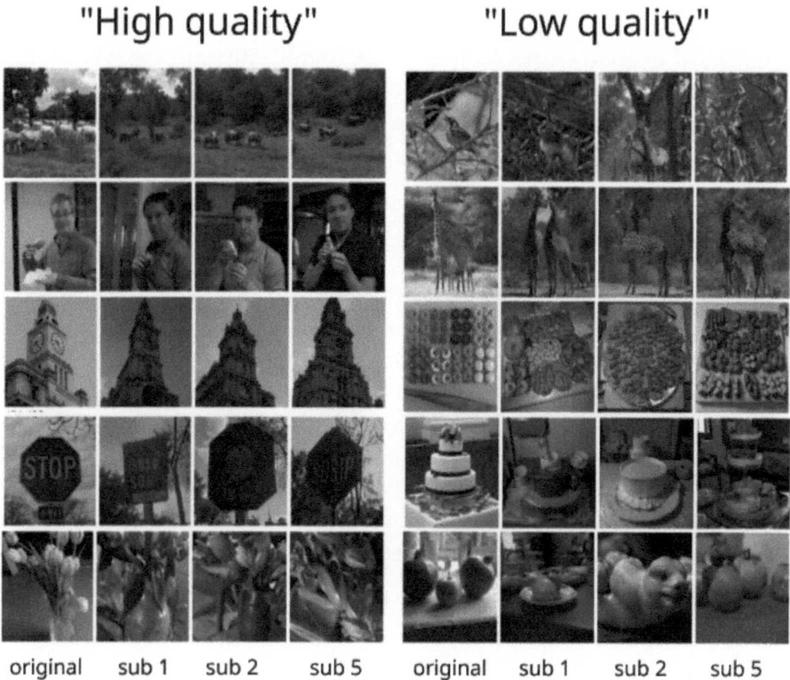

Fig. 3. Qualitative results for the federated approach; "sub" labels indicate reconstructions per subject.

(0.253) and SSIM (0.332), *ours (BG)* achieves results that exceed those of more complex frameworks, such as MindBridge.

Regarding model-based similarity metrics, including AlexNet layers (2 and 5), Inception, and CLIP, our method attains scores of 94.38% and 98.06% on AlexNet (2) and (5), and 95.37% on Inception. These figures place *ours (BG)* among the leading methods, closely following top performers such as MindEye2 [25] and BrainGuard, which show higher values. Importantly, on the EffNet-B metric, where lower values indicate better performance, *ours (BG)* surpasses, with an average of 0.615, the best score of MindEye2, reflecting good feature representation quality. Additionally, our SwAV score of 0.338 is outperforming many prior models and approaching the best results reported. When comparing to MindFormer [11], the only other approach that uses IP-Adapter [36] embeddings, we can observe slight improvements across five metrics. This could indicate that the underlying federated learning framework might be more performant in sharing cross-subject commonalities than their ViT method [6]. Taken together, these results indicate that *ours (BG)* achieves a good balance between low-level pixel accuracy and high-level semantic fidelity. The method's performance across diverse metrics highlights its effectiveness, despite the limited number of tokens used.

The single-subject framework, *ours (RR)* performs competitively across most metrics, achieving good scores such as 92.9% and 97.39% on AlexNet (2) and (5), respectively. While *ours (RR)* generally falls behind *ours (BG)* on all metrics, it still outperforms several earlier works like Mind-Vis [17] and Gu et al. [10].

Qualitative Comparison. As illustrated in Fig. 4, our reconstructed images closely match the semantic content of the originals across diverse categories, using subject 1 data. The leftmost image represents the image viewed by the subject and it is followed by reconstruction sourced from comparisons presented by the referred papers. Most prior approaches struggle to reconstruct the flower vase, often failing to capture both its identity and spatial placement. In contrast, our method successfully recovers not only the correct object category, but also its approximate position within the frame, including the occlusion at the top where the flowers are cut off. Spatial placement is reconstructed successfully for the picture of the man sitting on the bench, although the man does not look similar. In the case of the pictures featuring the motorcycle and the birds, our method performs worse than Brainguard or Mindbridge, because of missing or added objects.

Table 8. Comparison of prior work and our variants, including standard deviations

Methods	PixCorr ↑	SSIM ↑	Alex (2) ↑	Alex (5) ↑	Incept. ↑	CLIP ↑	EffNet-B ↓	SwAV ↓
Mind-Reader, NeurIPS 2022	–	–	–	–	78.2%	–	–	–
Mind-Vis, CVPR 2023	0.080	0.220	72.1%	83.2%	78.8%	76.2%	0.854	0.491
Takagi et al. CVPR 2023	–	–	83.0%	83.0%	76.0%	77.0%	–	–
Gu et al., MIDL 2023	0.150	0.325	–	–	–	–	0.862	0.465
Brain-Diffuser, Nature 2023	0.254	0.356	94.2%	96.2%	87.2%	91.5%	0.775	0.423
MindEye, NeurIPS 2023	0.309	0.323	94.7%	97.8%	93.8%	94.1%	0.645	0.367
MindBridge, CVPR 2024	0.148	0.259	86.9%	95.3%	92.2%	94.3%	0.713	0.413
MindFormer, arxiv 2024	0.243	0.345	93.5%	97.6%	94.4%	94.4%	0.648	0.350
MindEye2, ICML 2024	**0.322**	**0.431**	**96.1%**	**98.6%**	95.4%	93.0%	0.619	**0.333**
Psychometry, CVPR 2024	0.295	0.328	94.5%	96.8%	94.9%	95.3%	0.632	0.361
BrainGuard, AAAI 2025	0.313	0.330	94.7%	97.8%	**96.1%**	**96.4%**	0.624	0.353
ours (RR), 2025	0.219 ± 0.002	0.324 ± 0.001	$92.90\% \pm 0.20$	$97.39\% \pm 0.09$	$93.08\% \pm 0.20$	$90.49\% \pm 0.22$	0.710 ± 0.002	0.414 ± 0.001
ours (BG), 2025	0.253 ± 0.003	0.332 ± 0.002	$94.38\% \pm 0.27$	$98.06\% \pm 0.13$	$95.37\% \pm 0.21$	$93.51\% \pm 0.24$	$\mathbf{0.615} \pm 0.002$	0.338 ± 0.002

Efficiency Comparison. From an efficiency and scalability perspective, our frameworks are significantly lighter than the other models listed, both at the single-subject and multi-subject levels. This reduction in parameter count allows for faster training and inference, making it more practical for large-scale or resource-constrained applications without sacrificing performance. The data presented in Table 9 is sourced from [11] and [24]. We only performed computations for our models and for the BrainDiffuser model. For single subject approaches, we estimated the number of parameters from the third column by multiplying with 4. Our model achieves competitive performance while using over 15× fewer parameters than BrainDiffuser and over 25× fewer than MindEye.

Fig. 4. Qualitative comparison between our (BG) and SOTA methods.

Table 9. Parameter count of different models

Model	One Subject	All Subjects
Ours (BG, 2048 hidden)	81,539,328	326,157,312
Ours (BG, 1024 hidden)	35,534,080	142,136,320
Takagi et al.	~487,000,000	~1,948,000,000
BrainGuard	589,308,928	~2,357,235,712
BrainDiffuser	1,445,693,600	5,782,774,400
MindEye	2,099,141,060	8,396,564,344
MindBridge	–	693,579,264
MindFormer	304,782,336	765,607,680

6 Conclusion

In this study, we examined the effectiveness of an embedding-efficient brain-to-image reconstruction approach that uses only a small set of visual features extracted via the IP-AdapterPlus model. Our experiments, conducted across both single-subject and federated multi-subject frameworks using the Natural Scenes Dataset, demonstrate that comparable reconstruction performance can be achieved using significantly fewer embeddings than in traditional setups.

These results suggest that the widely adopted practice of predicting hundreds of CLIP-derived embeddings may not be strictly necessary for high-quality reconstructions on this specific dataset. Instead, they highlight the potential of more targeted representations, motivating further exploration into alternative feature selection strategies for neural decoding. At the same time, our findings point to a broader limitation: the dataset used, while well-established, may not fully capture the variability of real-world visual experiences. This points to the need for more representative datasets, as well as evaluation methodologies that better align with real-life scenarios.

Taken together, our findings support the promise of embedding-efficient decoding as a step toward more scalable and efficient brain-to-image reconstruction systems. Future work could further investigate embedding selection, alternative brain imaging modalities, real-time approaches and reconstruction evaluation under more realistic visual conditions.

Acknowledgment. This work relies on the Natural Scenes Dataset. Collection of the NSD dataset was supported by NSF IIS-1822683 and NSF IIS-1822929. Further information is available at https://naturalscenesdataset.org/. We also thank the authors of BrainGuard, MindEye, BrainDiffuser, and IPAdapter for generously sharing their codebases, which served as foundations for our experiments.

References

1. Allen, E.J., et al.: A massive 7T fMRI dataset to bridge cognitive neuroscience and artificial intelligence. Nat. Neurosci. **25**(1), 116–126 (2022)
2. Beliy, R., Gaziv, G., Hoogi, A., Strappini, F., Golan, T., Irani, M.: From voxels to pixels and back: self-supervision in natural-image reconstruction from fMRI. In: Advances in Neural Information Processing Systems, vol. 32 (2019)
3. Caron, M., Misra, I., Mairal, J., Goyal, P., Bojanowski, P., Joulin, A.: Unsupervised learning of visual features by contrasting cluster assignments. Adv. Neural. Inf. Process. Syst. **33**, 9912–9924 (2020)
4. Child, R.: Very deep VAEs generalize autoregressive models and can outperform them on images. arXiv preprint arXiv:2011.10650 (2020)
5. Dhariwal, P., Nichol, A.: Diffusion models beat GANs on image synthesis. Adv. Neural. Inf. Process. Syst. **34**, 8780–8794 (2021)
6. Dosovitskiy, A., et al.: An image is worth 16×16 words: transformers for image recognition at scale. arXiv preprint arXiv:2010.11929 (2020)
7. Gabor, D.: Theory of communication. J. Inst. Electr. Eng.-Part III: Radio Commun. Eng. **93**(26), 429–441 (1946)
8. Gaziv, G., et al.: Self-supervised natural image reconstruction and large-scale semantic classification from brain activity. Neuroimage **254**, 119121 (2022)
9. Goodfellow, I.J., et al.: Generative adversarial nets. In: Advances in Neural Information Processing Systems, vol. 27 (2014)
10. Gu, Z., Jamison, K., Kuceyeski, A., Sabuncu, M.R.: Decoding natural image stimuli from fMRI data with a surface-based convolutional network. In: Medical Imaging with Deep Learning (MIDL) (2023)
11. Han, I., Lee, J., Ye, J.C.: Mindformer: a transformer architecture for multi-subject brain decoding via fMRI. arXiv preprint arXiv:2405.17720 (2024)
12. Ho, J., Jain, A., Abbeel, P.: Denoising diffusion probabilistic models. Adv. Neural. Inf. Process. Syst. **33**, 6840–6851 (2020)

13. Hoerl, A.E., Kennard, R.W.: Ridge regression: applications to nonorthogonal problems. Technometrics **12**(1), 69–82 (1970)
14. Hoerl, A.E., Kennard, R.W.: Ridge regression: biased estimation for nonorthogonal problems. Technometrics **12**(1), 55–67 (1970)
15. Kay, K.N., Rokem, A., Winawer, J., Dougherty, R.F., Wandell, B.A.: Glmdenoise: a fast, automated technique for denoising task-based fMRI data. Front. Neurosci. **7**, 247 (2013)
16. Krizhevsky, A., Sutskever, I., Hinton, G.E.: Imagenet classification with deep convolutional neural networks (2012)
17. Lin, S., Sprague, T., Singh, A.K.: Mind reader: reconstructing complex images from brain activities. Adv. Neural. Inf. Process. Syst. **35**, 29624–29636 (2022)
18. Lin, T.-Y., et al.: Microsoft COCO: common objects in context. In: Fleet, D., Pajdla, T., Schiele, B., Tuytelaars, T. (eds.) ECCV 2014. LNCS, vol. 8693, pp. 740–755. Springer, Cham (2014). https://doi.org/10.1007/978-3-319-10602-1_48
19. Ozcelik, F., VanRullen, R.: Natural scene reconstruction from fMRI signals using generative latent diffusion. Sci. Rep. **13**(1), 15666 (2023)
20. Quan, R., Wang, W., Tian, Z., Ma, F., Yang, Y.: Psychometry: an omnifit model for image reconstruction from human brain activity. In: Proceedings of the IEEE/CVF Conference on Computer Vision and Pattern Recognition, pp. 233–243 (2024)
21. Radford, A., et al.: Learning transferable visual models from natural language supervision. In: International Conference on Machine Learning, pp. 8748–8763. PmLR (2021)
22. Rakhimberdina, Z., Jodelet, Q., Liu, X., Murata, T.: Natural image reconstruction from fMRI using deep learning: a survey. Front. Neurosci. **15**, 795488 (2021)
23. Rombach, R., Blattmann, A., Lorenz, D., Esser, P., Ommer, B.: High-resolution image synthesis with latent diffusion models. In: Proceedings of the IEEE/CVF Conference on Computer Vision and Pattern Recognition, pp. 10684–10695 (2022)
24. Scotti, P., et al.: Reconstructing the mind's eye: fMRI-to-image with contrastive learning and diffusion priors. Adv. Neural. Inf. Process. Syst. **36**, 24705–24728 (2023)
25. Scotti, P.S., et al.: Mindeye2: shared-subject models enable fMRI-to-image with 1 hour of data. arXiv preprint arXiv:2403.11207 (2024)
26. Song, J., Meng, C., Ermon, S.: Denoising diffusion implicit models. arXiv preprint arXiv:2010.02502 (2020)
27. StabilityAI: Realistic vision v4.0 novae stable diffusion model (2020). https://huggingface.co/SG161222/Realistic_Vision_V4.0_noVAE. Accessed 29 May 2025
28. StabilityAI: Sd-vae-ft-mse vae model (2025). https://huggingface.co/stabilityai/sd-vae-ft-mse. Accessed 29 May 2025
29. Szegedy, C., et al.: Going deeper with convolutions. In: Proceedings of the IEEE Conference on Computer Vision and Pattern Recognition, pp. 1–9 (2015)
30. Tan, M., Le, Q.: Efficientnet: rethinking model scaling for convolutional neural networks. In: International Conference on Machine Learning, pp. 6105–6114. PMLR (2019)
31. Tian, Z., Quan, R., Ma, F., Zhan, K., Yang, Y.: Brainguard: privacy-preserving multisubject image reconstructions from brain activities. In: Proceedings of the AAAI Conference on Artificial Intelligence, vol. 39, pp. 14414–14422 (2025)
32. Viazovetskyi, Y., Ivashkin, V., Kashin, E.: Stylegan2 distillation for feed-forward image manipulation. In: European Conference on Computer Vision. pp. 170–186. Springer, Cham (2020)

33. Wang, S., Liu, S., Tan, Z., Wang, X.: Mindbridge: a cross-subject brain decoding framework. In: Proceedings of the IEEE/CVF Conference on Computer Vision and Pattern Recognition, pp. 11333–11342 (2024)
34. Wang, Z., Bovik, A.C., Sheikh, H.R., Simoncelli, E.P.: Image quality assessment: from error visibility to structural similarity. IEEE Trans. Image Process. **13**(4), 600–612 (2004)
35. Xu, X., Wang, Z., Zhang, G., Wang, K., Shi, H.: Versatile diffusion: text, images and variations all in one diffusion model. In: Proceedings of the IEEE/CVF International Conference on Computer Vision, pp. 7754–7765 (2023)
36. Ye, H., Zhang, J., Liu, S., Han, X., Yang, W.: IP-adapter: text compatible image prompt adapter for text-to-image diffusion models. arXiv preprint arXiv:2308.06721 (2023)
37. Zhang, J., et al.: Fedala: adaptive local aggregation for personalized federated learning. In: Proceedings of the AAAI Conference on Artificial Intelligence, vol. 37, pp. 11237–11244 (2023)

Music Emotion Recognition with Deep Learning Techniques

Denis-Angel Moldovan[ID] and Claudia-Ioana Coste[(✉)][ID]

Babeş-Bolyai University, 400084 Cluj-Napoca, Romania
`denis.angel.moldovan@stud.ubbcluj.ro`,
`claudia.coste@ubbcluj.ro`
`https://www.cs.ubbcluj.ro/`

Abstract. Music is considered an *universal language*, a way to reduce language and cultural barriers between people from around the globe. The present paper proposes to dive into the subjective domain of emotion recognition, aiming to decipher the relationship between musical elements and human emotions. In the present paper, deep learning methods are leveraged to classify the mood induced by a piece of music. For feature modelling, multiple attributes are used. Spotify features are employed, such as rhythm, tone, danceability, energy, loudness, valence, speechiness, and others. Other features, such as Mel-Frequency Cepstral Coefficients, Spectral Centroid, Chroma Energy Normalized Statistics, and the Mel Spectrogram, are concatenated into an image. The conducted experiments employed deep learning methods, including a ResNet18 model, which had not been used in this context before. The best-performing model, with 86% accuracy, proves to be a hybrid one, based on ResNet18 architecture and enriched with two Bidirectional Long-Short Term Memory layers, outperforming by 6%-11% other deep learning architectures that the present paper experimented with. The results are rather similar to other studies from the literature, even though the approaches and the datasets are different.

Keywords: music emotion recognition · deep learning · information retrieval

1   Introduction

Music can induce emotions and even behaviors through its tune, tempo, pitch, and lyrics. Music emotion recognition (MER) is a topical domain, complex with applications in different fields. How the brain reacts to music can advance the research into neuroscience and psychology fields. In computer science, music can contribute to improving user-friendliness in applications, leading to a pleasant user experience. In meditation applications, for example, music can soothe, relax, and add a calming effect. Moreover, in marketing, sound can help build a brand identity.

C. Chira et al. (Eds.): InnoComp 2025, CCIS 2793, pp. 157–174, 2026.
https://doi.org/10.1007/978-3-032-12478-4_11

Studies presented in [10] consider that moods are influenced by music in two dimensions, arousal and valence. This is called the circumplex model proposed by Russell [28]. Valence indicates how much pleasure is released, and arousal refers to the activation level, the energy transfer. Music can induce certain moods based on brain reflexes, different associations between music genres and emotions, contagion, visual effects generated from listening to a piece of music, memory associations, and anticipation [10]. Music can evoke certain memories in humans, which then induce emotions. Additionally, it was proved that music can help individuals boost their cognitive abilities through Mozart's effect documented in [26]. Furthermore, Lang et al. [12] experimented with multiple people where each of them had to listen to three different types of music, represented by rhythmic, non-rhythmic, and white noise. After listening to the audio pieces, they were required to complete a cognitive task. It was proved that the people who have been listening to rhythmic music reached the highest productivity level compared to the other groups. Therefore, music can motivate, immerse, and engage people in different activities.

However, extracting moods from songs can be especially subjective, depending on the mixed feelings induced, cultural differences, social, economic, and educational status, and even whether the human subject is female or male. In [35] are proposed for future work fine-tuned models for mood prediction in music according to each demographic group, adjusted to cultural variations.

Spotify is the best application providing music streaming services [30] to over 675 million consumers. Spotify offers 100 million songs, 6.5 million podcasts, and 350K audiobooks [32]. The Spotify app provides an API for developers [31] where a large variety of statistics about music can be retrieved, such as listening patterns, features computed for each song relevant in performing MER.

The present paper proposes to advance research in the MER field by applying deep learning techniques to predict the mood of a song. The experiments are run on a 4-label annotated dataset with 1,200 music samples. The main contributions can be summarized to the following:

- Developing neural network (NN) models for both Spotify and other types of features (e.g., Mel-Frequency Cepstral Coefficients, Spectral Centroid, Chroma Energy Normalized Statistics, and Mel spectrogram);
- Features are combined in an image, which will represent the input data;
- Developing four NN models: artificial neural network (ANN), Convolutional Neural Network (CNN) + a Long-Short Term Memory (LSTM) layer, ResNet18, and ResNet18 + Bidirectional LSTM (biLSTM);
- Applying ResNet18 model in a novel MER context;
- Obtained a 86% accuracy with ResNet18 + biLSTM model.

The present study is structured in five sections. The first one is the *Introduction* 1 where we explain the domain and the motivation behind current research. The following segment, 2, reports the most relevant systems developed in MER previously published in the literature. The third section, entitled *Empirical Methodology* 3 formally describes the proposed approach, with the dataset selected, the feature extracted, and the classification algorithms used. The fourth section, named *Results and Discussion* 4 presents the computations, comparisons with the depicted models and other solutions from literature. Finally, the last part lays out the conclusions and explains new ideas of research for the future 5.

2 Previous Work

MER solutions address the problem as a classification task or as a regression one. Moreover, the features depicted consist of the input for the intelligent methods, which can be manually engineered or automatically produced by deep learning techniques [5]. Some deep learning strategies used for MER are, for example, CNN, VGGNet, Deep Boltzmann Machine, and others. Regarding machine learning systems, one could recall: K-Nearest Neighbor (KNN), Support Vector Machine (SVM), Decision Tree (DT), Logistic Regression (LR), Naive Bayes (NB), etc. The following paragraphs introduce relevant solutions proposed in the MER context.

Yang et al. [35] consider regression and predict values for arousal and valence. Their regression approach is based on Thayer's Cartesian system as presented in Fig. 1, very similar to Russell's circumplex one. The dataset used contains 195 songs in Chinese or Japanese. Features are extracted by using third-party services (i.e., PsySound [3], Marsyas [34], Spectral Contrast [17], DWCH [14]). From the collected characteristics, one could enumerate: spectral centroid, loudness, volume, multiplicity (count of pitches), chord (simultaneous musical pitches), tonality (whether it is major or minor), etc. Additionally, some features are extracted by using two independent algorithms. In total, there are 114 attributes used, and then a feature selection step is performed. The most relevant information considering the prediction for arousal values is the tonality, multiplicity, etc. The most important features for valence are Spectral dissonance, roughness of all spectrum components, tonality, and others. For classification, Yang et al. [35] are using Machine-Learning Ranking (MLR) and Support Vector Regressors (SVRs).

Similarly, Han et al. [4] are using an SVR that predicts values for arousal and valence to be represented in Thayer's Cartesian system and the polar system. Authors develop suitable formulas to map coordinates from one representation system to another. The dataset in this case has 165 songs from the All Music dataset [1]. For features, authors choose tonality, tempo, intensity, rhythm, and harmonies. The best classification is done by the SVR, achieving an accuracy of 94.55% in the polar system and 92.73% using the Gaussian Mixture Model (GMM) with the polar system. The training is done using cross-validation, and it was observed that peace or boring classes were often confused with calm.

To perform classification for four emotions: happiness, fear, sadness, and neutrality, Rajesh et al. [25] developed a Recurrent Neural Network (RNN). It achieved 85.6% accuracy for happy, 91.0% for sad, 87.3% for neutral, and 93.4% for fear. The input features were Mel Frequency Cepstral Coefficients (MFCCs). For comparisons, another set of music characteristics was used, such as Chroma Energy Normalized Statistics (CENS), Chroma Short Time Fourier Transform (STFT), Spectral Centroid, Spectral Bandwidth, Spectral Rolloff, and Zero Crossing Rate (ZCR). In total, Rajesh et al. [25] used 39 input values. The dataset included 800 15-second audio files, balanced against all four

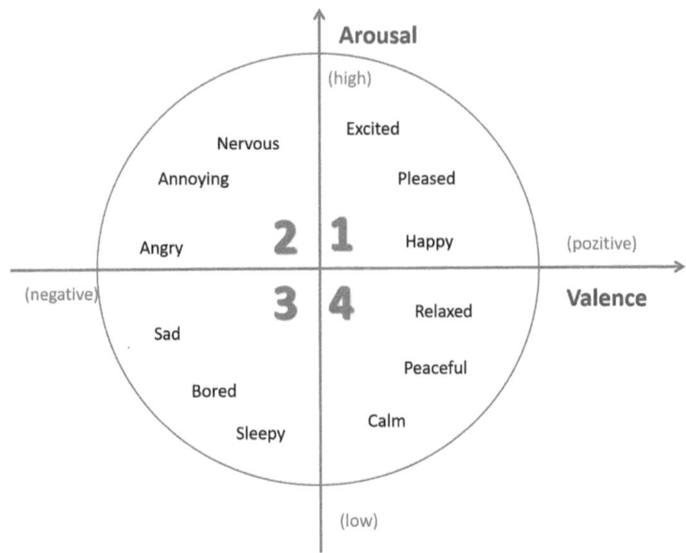

Fig. 1. The arousal-valence cartesian system proposed by Thayer from [35].

classes. Each music piece was instrumental music (i.e., piano, violin, flute, or trumpet) in the *wav* format. The final RNN model was represented by a neural network model with two LSTM layers added. Additionally, authors observed a correlation between instruments and emotions, such as the violin would induce sadness, the piano, happiness, the flute, neutrality, and the trumpet, fear.

Likewise, using deep learning ideas, Hizlisoy et al. [7] propose a Convolutional Recurrent NN to classify 124 Turkish songs of 30-second length. All songs were annotated by music specialists with arousal and valence scores. The final labels are computed based on the average and a categorization into one of the following three classes: happy, sad, and angry. For features, Hizlisoy et al. [7] use Mel Coefficients (MFCCs) and Log-mel Filterbank Energies. Each song was processed in three seconds, each section containing 10 s. Additionally, relevant data was extracted by using a CNN model from the MFCCs and Log-mel Filterbank Energies features. Other standard attributes were gathered by using OpenS-MILE, MIRtoolbox, or jAudio [18]. For feature selection, the authors used a Correlation-based method. The classification model is a Deep NN enriched with LSTM layers. The best accuracy of 91.93% was obtained on all input features, and by adding the selection methods, accuracy increased to 99.19%. Relevant comparisons are done with RF, SVM, or KNN models. What is unique about this approach is that it investigates a novel cultural context in the MER field.

Advancing the deep learning research, Huang et al. [9] propose an Attention-based Deep Feature Fusion model for predicting the mood from music clips. The samples are collected from the PMEmo database [36], where each music fragment from a total of 794 was annotated by 10 international volunteers. The annotation process should blur the cultural differences and output an accurate label. The proposed deep model makes use of different intelligent methods to form the final system. First, the input data consists of the Mel Spectrogram, then it uses VGGNet to extract features, and in the following step, they are spatially represented. Next, the temporal attention model named squeeze-and-excitation is run. Finally, the proposed idea achieves a performance increase up to 10.43% for valence and 4.82% for arousal compared to State-of-the-Art solutions. Since the authors are solving a regression problem, the metrics are R^2 scores and Root Mean Squared Error (RMSE). For generating the Mel diagram, the librosa [15] library is employed, and comparisons are made with other deep feature fusion variants, as well.

Delivering a novel approach by making use of Spotify features, Panda et al. [22] proposed to investigate how relevant these characteristics are in MER. There are 12 features provided by Spotify, and with an SVM classification model, they reach just 58.5%, which is far below other State-of-the-Art MER systems. However, by enriching previous literature models with Spotify information, an increase in performance was observed by 2.3% for the TOP 5 features and by 1.1% for the TOP 10. Experiments included a 704-record dataset depicted from AllMusic [1] spread across all four quadrants of Russell's model. The most relevant Spotify features proved to be: energy, valence, and acousticness.

All in all, there are not many ideas in the MER field considering deep learning models having as input multiple types of audio features. This is why, present approach focuses on this research niche, by applying novel deep learning architectures to MER.

3   Methodology

The present section details the methodology followed during experiments. The dataset is presented, together with the features used and the deep learning models.

3.1   Dataset

The present study is based on the Moodify dataset provided in [21]. It contains 1200 radio songs annotated with four classes: sad, happy, energetic, and calm. Moodify database has already extracted the Spotify features (i.e., acousticness, danceability, energy, tempo, liveness, loudness, speechiness, valence, and instrumentalness), and a description of them is presented in Table 1. Unfortunately, Spotify API [31] does not explain how these attributes are computed. Based on the samples in the dataset, we propose to retrieve the associated audio files by using the YouTube API.

Table 1. Features extracted using the Spotify API [21].

Feature	Description	Values
Acousticness	Measures how acoustic the song is	[0.0–1.0]
Danceability	States how suitable is the song for dancing	[0.0–1.0]
Energy	The intensity of the song is measured, e.g., songs with high energy are usually loud and upbeat	[0.0–1.0]
Instrumentalness	The value should denote if the music contains vocals or not, e.g., if the value is close to 1.0, then there is a low probability for that piece of music to contain vocals	[0.0–1.0]
Liveness	High values indicate that the music was recorded in a live setting	[0.0–1.0]
Loudness	Measures the power of the music in decibels (dB)	[-60, 0]
Speechiness	Predicts the probability for a song to contain lyrics	[0.0–1.0]
Valence	Describes the feelings given by the melody. For example, while a high value may denote feelings of happiness, euphoria, a low value can indicate feelings of sadness, anxiety, anguish.	[0.0–1.0]
Tempo	Beats per minute (BPM)	[0, 300]

3.2   Characteristics Extracted from Audio Files

The following step of the proposed methodology involves extracting features from the audio file. In the MER literature, there is an emphasis on extracting the most relevant information from the musical file. These characteristics should summarize musical attributes that will be important as input for an intelligent algorithm. The features we focus on have managed to achieve a good importance score in the literature. Such audio features are: Mel-frequency cepstral coefficients (MFCC), Spectral Centroid, Chroma Energy Normalized Statistics (CENS), and the Mel Spectrogram. All these characteristics are gathered from the 30-second audio file, which was then split into three segments of ten seconds. Then, for each segment, feature extraction is run and the results are concatenated and normalized to fall in the [0,1] interval with Standard Scaler [23].

In the following paragraphs, there are details about these features. During experiments, all features were retrieved by using the Python library named Librosa [15].

Mel-Frequency Cepstral Coefficients (MFCC). Mel-frequency cepstral coefficients are a set of attributes proposed in [16]. These coefficients are extracted in a multi-step process as tackled in [11]. The main steps are:

1. Sound conversion to 8–16 kHz;
2. Emphasising the high frequency audio segments (i.e., audio segments having voice) to compute a higher energy score;
3. Applying the Fourier Discrete Transformation;
4. Run Mel-Filter Bank formula to process the sound and modify it to have a suitable frequency for humans;

5. Applying the logarithm;
6. The inverse transformation from item 3;
7. Extracting features (a total of 39 coefficients).

Spectral Centroid. Statistical measurements on the center of mass of the signal spectrum represent spectral centroids. These should encapsulate information on the location where the center of energy regarding frequency is situated. There are two concepts here worth mentioning: the frequency of the spectral centroid and its amplitude. This characteristic is obtained according to the methodology presented in [13]. Firstly, the square of the Fourier transformation magnitude is computed. Then, the frequency of the spectral centroid is calculated based on the weighted average of all frequencies for each audio band. Then, the amplitude of the spectral centroid results from the weighted mean of the spectral magnitude from each audio band.

In Librosa library [15] version 0.11.0, the spectral centroid is computed taking into consideration the mathematical expression detailed in Eq. 1, where t is the temporal index of the current frame, S represents the magnitude of the spectrogram, and $freq$ is the frequency array, where values are measured in Hz and were obtained by applying the Fast Fourier Transform function.

$$centroid_t = \frac{\sum_k S[k,t] \cdot freq[k]}{\sum_j S[j,t]} \tag{1}$$

Chroma Energy Normalized Statistics (CENS). There are 12 CENS attributes [20] collected based on the 12 tones (i.e., Do, Do#, Re, ..., Si), and they should represent how the energy is distributed. Moreover, an Euclidean normalization can be applied to balance the differences in energy levels that might appear. After that, a quantization stage can be included, which should reduce the local fluctuations denoted by tempo, musical notes execution, and the joints in between. The quantization formula is presented in Eq. 2.

$$Q : [0,1] \rightarrow \{0,1,2,3,4\}; Q(a) = \begin{cases} 0 & 0 \le a < 0.05 \\ 1 & 0.05 \le a < 0.1 \\ 2 & 0.1 \le a < 0.2 \\ 3 & 0.2 \le a < 0.4 \\ 4 & 0.4 \le a < 1 \end{cases} \tag{2}$$

The Mel Spectrogram. The Mel Spectrogram is humanly readable, and it provides information on the frequency of the audio. The frequencies are represented with the Mel scale for human explainability. The creation of the Mel spectrogram is based on the stages described in [27], and it follows the following processes. First, the sound is represented as a waveform, then it is mapped employing the Fast Fourier Transform algorithm. Then, the logarithmic function is applied, and the amplitude is transformed into decibels. Finally, results are mapped to the Mel scale.

All the presented features are important, and they are extracted using the librosa library [15]. These features are merged by concatenation on the horizontal axis, resulting in a feature concatenation on each line of a matrix (i.e., image).

3.3   Classification Models

The experiments revolve around artificial intelligence algorithms, capable of supervised learning because they are trained on labelled data. We tried multiple variants of deep learning approaches, such as a simple ANN, then a CNN enriched with simple LSTM layers, and the most performant model is ResNet18 with Bidirectional LSTM layers. In the next subsections, all these deep learning models are presented together with their calibrated parameters. For implementation, multiple 3.9 Python libraries were used, such as: pytorch version 2.6.0 [33], matplotlib 3.10.1, numpy, pandas, scikit-learn 1.6.1 [23], etc.

ANN. The baseline ANN model takes as input the Spotify features already extracted in the used dataset [21]. The Spotify attributes were described in the previous section in Table 1. The data is first normalized with Standard Scaler and then, through successive experiments, the parameter calibration process is done. After the calibration, the neural network architecture is set to four hidden layers with 256 neurons, 128, 64, and 32 neurons. The input layer has 11 units corresponding to the Spotify features. The learning rate is 0.0002, then training is done over 100 epochs, and the batch size is set to 8. More details about the ANN structure are presented in Table 2, where the number of training parameters is displayed, as well. Additionally, it can be observed that the activation function is ReLU.

Table 2. The architecture for the ANN model.

Layer	Output size	No. parameters
Linear, Relu, Dropout	[256]	3,072
Linear, Relu, Dropout	[128]	32,896
Linear, Relu, Dropout	[64]	8,256
Linear, Relu, Dropout	[32]	2,080
Linear, Softmax	[4]	132
		Total: 46,436

CNN. Taking as input the combination of the audio features extracted by Librosa [15], we developed a more complex deep learning model by adding three Convolution Layers. The input data consists of three images, each of which represents features for ten seconds. After several runs, the simple CNN model was enriched with an LSTM layer, which should capture pertinent temporal patterns. The model's design is described in Table 3. It can be observed that the three Convolutional layers should extract relevant data, which is then linearized and fed to the LSTM layers. Finally, the classification is done through a Softmax layer. Additionally, the Conv2D layers use a kernel of 10×10 or 3×3, 3×3 depending on the order, and a 1×1 stride. Batch normalization is applied to reduce the algorithm's sensitivity to initial parameters. Moreover, Dropout layers with a 0.05 probability are used to prevent overfitting. ReLU is employed as the activation function, and MaxPool2D type of layers are added. There are two simple LSTM layers that use 256 neurons.

Table 3. The architecture for the CNN + LSTM model.

Layer	Output size	No. parameters
ConvBlock1	[64, 38, 105]	6,592
ConvBlock2	[128, 9, 25]	74,112
ConvBlock3	[128, 2, 7]	147,840
Linear	[256]	459,008
LSTM	[2, 128]	264,192
Linear, Softmax	[4]	516
		Total: 952,260

ResNet18. After multiple empirical studies and direct observations, the final developed model is represented by ResNet18 [6] combined with two Bidirectional LSTM layers. The original ResNet18 architecture can be observed in Fig. 2, and it was proposed in [6]. ResNet18 is characterized by 18 layers, from which the main ones are the four types of Convolutional layers, whose size increases from 64 to 128, 256, and 512 [6]. ResNet18 implements the skip connection mechanism to counteract the vanishing gradient problem. The main advantage of this architecture is that it is very compact and has fewer parameters for training compared with other models. In empirical studies, ResNet in general proved a higher performance compared to other approaches in different computer science fields.

Our proposed architecture consists of binding a ResNet18 with two Bidirectional LSTM layers. In our experiments, this configuration proved to be the most suitable for the mood detection task. The LSTM layers have 512 neurons. LSTM architecture is created based on Recurrent Neural Networks (RNNs), and they have a more complex unit with four components [8]. Additionally, the LSTM mechanism has three types of gates: forget gate, input gate, and output gate [8]. LSTM is a robust model mitigating problems such as vanishing gradient or exploding gradient [8]. A visual representation of the proposed architecture can be seen in Fig. 3. Moreover, in Table 4, there is a compilation of the number of trainable parameters and the output size of the components from our proposed method.

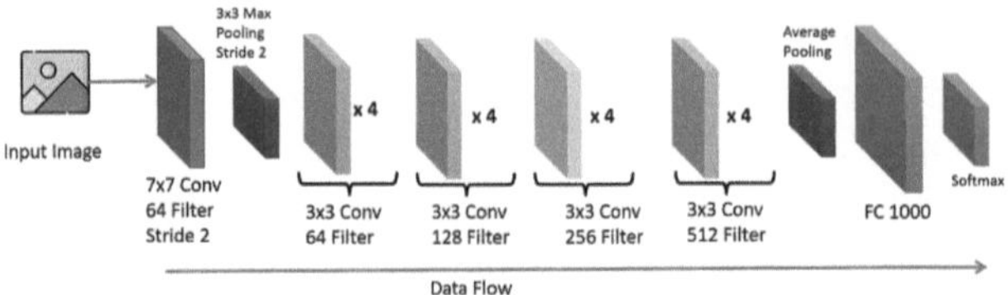

Fig. 2. The original architecture for ResNet18 [6]. Diagram inspired by [2] and [19].

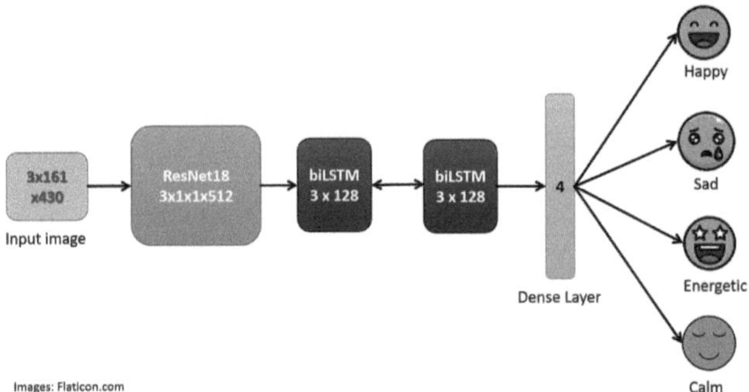

Fig. 3. The proposed architecture for the ResNet18 model combined with BiLSTM layers.

Table 4. The architecture for the ResNet18 + biLSTM model.

Layer	Output size	No. parameters
ResNet18	[512, 1, 1]	11,170,240
LSTM	[3, 128]	395,264
Linear, Softmax	[4]	516
		Total: 11,566,020

4   Results and Discussion

The current section will detail the results obtained for the experiments. These results are put into perspective by taking into account other relevant proposed work in the MER domain.

4.1   Dataset and Data Preprocessing

The audio file of all songs from the used dataset was retrieved using YouTube API (i.e., YouTube Search [29] version 2.1.2, PyTube [24] version 15.0.0, MoviePy [37] version 1.0.3). The audio file (i.e., MP3) was truncated to make use of just 30 s of the song starting from the first quarter of the tune. The time needed for this processing is approximately 6 s per sample. After the processing, the dataset distribution is not affected, and the final dataset of 1,162 songs is relatively

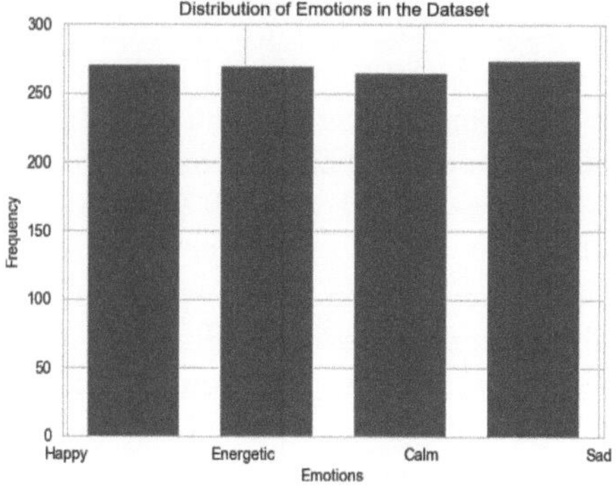

Fig. 4. Dataset distribution.

balanced across the four classes, as it can be observed from Fig. 4. The number of samples for each label is almost the same.

All in all, we use the Moodify dataset [21], where 1,100 songs are used for training and 100 of them for testing in the case of using Spotify features with the simple ANN. Each set is balanced. Then, for deep learning tasks, the training set contains 1,042 audio files and 120 samples for testing, all balanced.

The audio files are processed to obtain a spectrogram for each 10 s. The 30-second audio clip will output three spectrograms, which will be concatenated on the horizontal axis, normalized with Min-Max scaling, and fed into the deep learning models. Each spectrogram has a dimension of 161×430 pixels, and an example can be seen in Fig. 5. The input channel is set to 1 for all classification models.

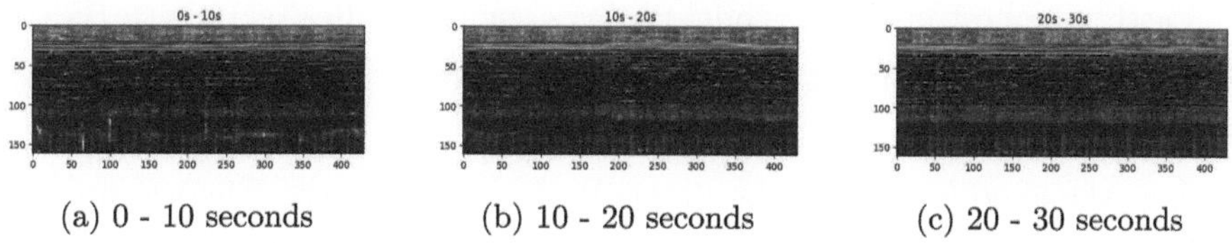

(a) 0 - 10 seconds   (b) 10 - 20 seconds   (c) 20 - 30 seconds

Fig. 5. Example of spectrograms for all three segments of a 30-second audio clip.

4.2   Experiments and Results

The performance metrics used in experiments are accuracy, precision, F1 score, confusion matrix from the scikit-learn library [23], and loss evolution from PyTorch [33]. All experiments were run on a laptop with Nvidia GeForce RTX 3050, Intel Core i7 11th generation, 16GB of RAM, and 1TB of SSD.

Table 5 presents the most relevant results achieved. Other results were obtained during the experimental phase when parameters were fine-tuned. The metrics are computed as an average of multiple runs. The ANN model uses as input features extracted using the Spotify API. ANN performed well with an accuracy of 75%. The ANN model was trained over 100 epochs, with Adamax optimizer and a learning rate of 0.0002. The hand-crafted convolutional model (CNN + LSTM) outperformed ANN with 80% accuracy. The CNN + LSTM model was fitted to 50 epochs, Adam optimizer and a 0.0001 learning rate. The training of a simple ResNet18 was conducted with cross-validation and over 25 epochs. The best results were obtained with the SGD optimizer, a learning rate of 0.0009, and the Cross-Entropy loss function. Finally, the model reaches an overall accuracy of 80%. By comparing ResNet18 and our designed model (CNN + LSTM), one could observe that both obtained the same performance. However, considering the computational resources used, ResNet18 consumes the most, followed by our convolutional model and the ANN model. Moreover, taking into account the number of parameters for ResNet18 and CNN+LSTM model, our crafted model (i.e., CNN + LSTM) has eleven times fewer as seen in Table 3.

Table 5. Models' performance. S - Sad, H- Happy, E - Energetic, C - Calm.

Model	Acc. (%)	Precision (%)				Recall (%)				F1 score (%)			
		S	H	E	C	S	H	E	C	S	H	E	C
ANN	75	66	70	72	93	92	56	52	100	77	62	60	96
CNN + LSTM	80	88	62	85	93	70	83	77	90	78	71	81	92
ResNet18	80	86	69	73	96	80	90	63	87	83	78	68	91
ResNet18 + biLSTM	**86**	**87**	**81**	**80**	**96**	**90**	**83**	**80**	**90**	**89**	**82**	**80**	**93**

Lastly, the best obtained model involves combining ResNet18 with Bidirectional LSTM layers. The final model was trained for 50 epochs, until the accuracy and the loss had stabilized, as seen in Fig. 6. The learning rate was set to 0.0001, the chosen optimizer was Adam, and the loss function was Cross-entropy, the same as the loss function for the simple ResNet18 algorithm. The learning rate is adjusted during training by the use of a scheduler that considers the loss function result. The obtained performance is 86% in terms of accuracy. There can be observed an increase of approximately 0.8% for all metrics in comparison with the other methods. Figure 6 represents how the loss value evolved during each epoch. One could observe that after the 20th epoch, the classification does not improve much over the next 30 epochs.

The confusion matrices from Figs. 7a and 7b are computed for the ANN model and the ResNet18 + biLSTM model. There is a predilection for the ANN model to misclassify the happy or energetic audios in approximately 30% of the cases. Moreover, as observed in Table 5, there are low precision scores for sad, happy, or energetic songs in comparison with the precision obtained for calm songs. While the ResNet18 + biLSTM system seems to solve the misclassification problem from the ANN, the ResNet-based model has difficulty in properly classifying

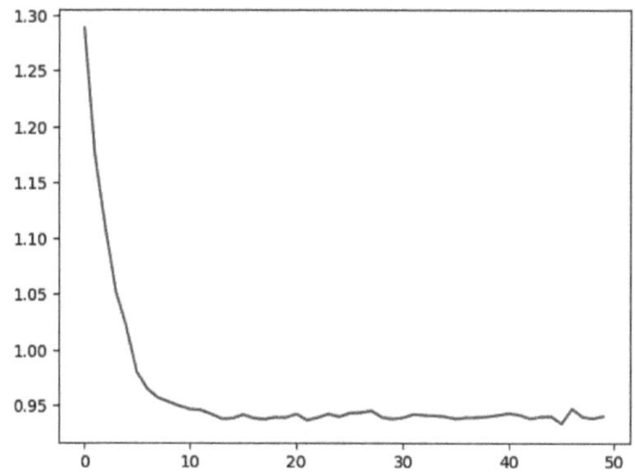

Fig. 6. Graphic representing the loss over 50 epochs for the ResNet18 + biLSTM model.

energetic songs, some of which are considered happy. This confusion can be explained because of the subjectivity of the domain; these two classes can be unclear even for human subjects.

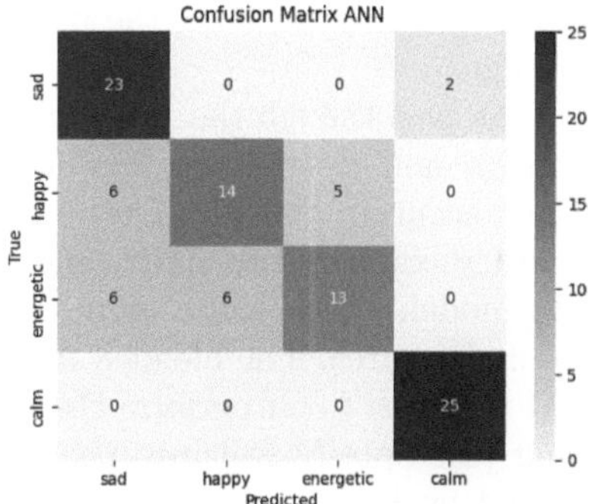

(a) Confusion matrix computed for the ANN model.

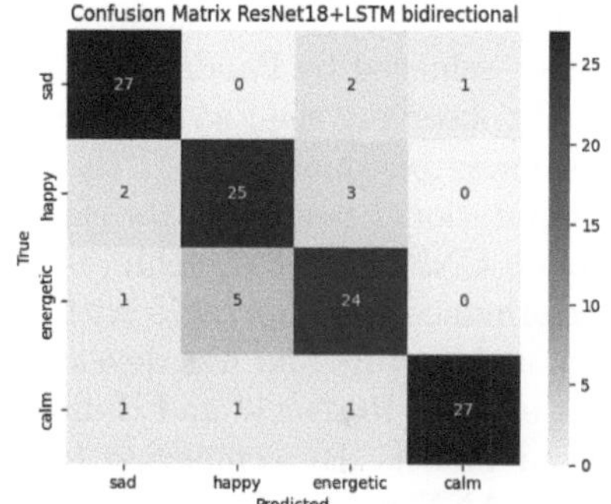

(b) Confusion matrix computed for the best model: ResNet18 + biLSTM.

Fig. 7. Confusion matrices for the ANN and the ResNet18 + biLSTM models.

4.3   Comparisons and Discussion

Most comparisons between different literature solutions are not very appropriate because the datasets are usually diverse. The majority of literature solutions have collected and annotated their internal and private dataset. Moreover, the annotation process is very subjective, and the research domain is still in its incipient stages.

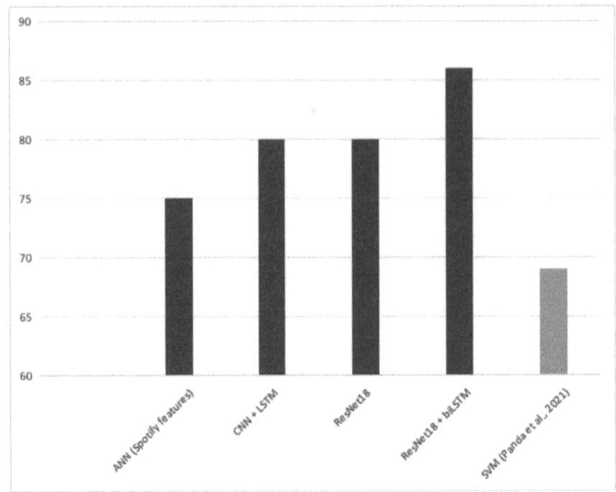

Fig. 8. Comparisons graphic with other literature approaches.

In Fig. 8 there are displayed our best models and the performance obtained by other relevant models from literature, such as [22]. The metric considered for performance is the accuracy obtained during testing, computed as an average over multiple runs. The accuracy obtained by us is colored in blue, the performance achieved by Panda et al. [22] is with orange.

By observing the Fig. 8, it can be observed that even though the ANN model has fewer parameters (i.e., the Spotify features), and its training has fewer computational resources, its classification is not accurate enough. The hybrid model, CNN + LSTM, achieves 5% more accuracy compared to ANN, and its performance is comparable with the ResNet18 model, which has a more complex architecture and was developed by Microsoft Research. The blended model crafted from ResNet18 and Bidirectional LSTM manages to outperform the rest of our models. By comparing the models using just Spotify features, the proposed ANN model outperforms the SVM created by Panda et al. [22] by 6%. However, the datasets are different; the one used in [22] has just 704 music pieces from AllMusic [1]. For the experiments, we trained multiple SVMs according to the configurations by Panda et al. [22] on our dataset. Then, the best performance of 69% accuracy on the testing set was chosen. It was obtained for RBF kernel with $C = 8$, and $gamma = 0.008$.

When comparing our accuracy of 86% attained by the ResNet18 + biLSTM model with other literature solutions such as the one proposed in [25], we must state that the datasets are different. Rajesh et al. [25] classifies with 89% accuracy, a 3% increase compared to our best model. This difference in accuracy may be because the dataset used in [25] is much more limited, having just 800 pieces of instrumental music. The authors manage to associate piano music with happiness, violin denotes sadness, flute signals neutral music, and trumpet implies fear. The accuracies obtained are 85.5% for piano, 91% for violin, 87.3% for flute, and 93.4% for trumpet. The violin and trumpet prove to have a very distinguishable sound, being labeled with a higher accuracy compared to the piano. The piano music is much more complex, and the algorithm has difficulties in properly annotating it. Similarly, the proposed model contains LSTM layers, and the input characteristics are equivalent to our proposed input features. They used MFCCs, and for comparisons, Rajesh et al. [25] tried to work with CENS, STFT, Spectral Centroid, Spectral Bandwidth, Spectral Rolloff, and Zero Crossing Rate (ZCR). By comparison, our model ResNet18 + biLSTM worked with MFCC, Spectral Centroid, CENS features, and the Mel Spectrogram.

Furthermore, by comparing our approach with the system proposed in [7], Hizlisoy et al. [7] obtained a considerably better performance represented by a 99% accuracy rate. The difference in their performance and ours is represented by the dataset used, which in [7]'s case contains just Turkish music, a rather narrow niche. In our case, the dataset is more culturally diverse. They also used the MFCCs features, and besides that, the authors added Log-mel energies and other characteristics extracted with the openSMILE software and other services. Moreover, another dissimilarity with our idea is the three labels used in the annotation process. On the other hand, a similarity comes from the structure of the CNN model, which integrates a bidirectional LSTM, but is much more complex than our model. A major improvement made by Hizlisoy et al. [7] was adding a feature selection algorithm. That led to an increase of 7% between their proposed models.

All in all, ResNet18 is a novel idea applied in the context of Music Emotion Recognition. Considering the above results, ResNet18 has good results, and we anticipate accurate results with other ResNet architectures as well. Additionally, the obtained accuracy for the ANN is outperforming the SVM proposed in Panda et al. [22]. Moreover, the ResNet18 + biLSTM method outperforms the ANN, positioning our approach among other relevant solutions from the literature, such as Hizlisoy et al. [7] and Rajesh et al. [25].

5   Conclusions and Future Work

To wind up, Music Emotion Recognition is a particularly subjective domain, taking into account cultural, social, and educational factors when labeling emotions suggested by music. The present approach proposed simple and hybrid deep learning techniques for song emotion detection. The experiments were run on a 1200-song dataset, which is diverse and contains radio-friendly tunes. The neural network models include a simple Artificial Neural Network using just Spotify features as a baseline model, and other more complex architectures, such as ResNet18 and other hybrid techniques: a Convolutional Neural Network or ResNet18 enriched with Long-Short Term Memory layers. The baseline model outperforms the State-of-the-Art approaches. Besides, the best performance was reached by the ResNet18 + biLSTM model with almost 86% accuracy. The obtained metrics are comparable with other literature models, even though the present approach does not outperform them. The input features are integrated in a novel way to form an image fed as input to the algorithm.

In light of future work, we propose experimenting with other, more complex ResNet architectures, such as ResNet-101, ResNet-50, or VGG architectures. These are expected to increase the performance. Moreover, the usage of Large-Language Models could be employed, either by using prompt engineering techniques or by fine-tuning and training specifically on the Music Emotion Recognition task. Additionally, incorporating features extracted from lyrics could lead to an improvement in performance, as sentiment analysis technology is relatively mature.

Acknowledgments. This research is supported by the project "Romanian Hub for Artificial Intelligence - HRIA", Smart Growth, Digitization and Financial Instruments Program, 2021–2027, MySMIS no. 334906.

Disclosure of Interests. The authors have no competing interests to declare that are relevant to the content of this article.

References

1. (2025). https://www.allmusic.com/. Link to AllMusic website. Accessed 25 Mar 2025
2. Atik, M.E., Duran, Z.: Deep learning-based 3D face recognition using derived features from point cloud. In: Innovations in Smart Cities Applications Volume 4: The Proceedings of the 5th International Conference on Smart City Applications, pp. 797–808. Springer, Cham (2021)
3. Cabrera, D., et al.: Psysound: a computer program for psychoacoustical analysis. In: Proceedings of the Australian Acoustical Society Conference, vol. 24, pp. 47–54. AASC Melbourne, Australia (1999)
4. Han, B., Rho, S., Dannenberg, R.B., Hwang, E.: Smers: music emotion recognition using support vector regression. In: ISMIR, pp. 651–656. Citeseer (2009)
5. Han, D., Kong, Y., Han, J., Wang, G.: A survey of music emotion recognition. Front. Comput. Sci. **16**(6), 1–11 (2022). https://doi.org/10.1007/s11704-021-0569-4
6. He, K., Zhang, X., Ren, S., Sun, J.: Deep residual learning for image recognition. In: Proceedings of the IEEE Conference on Computer Vision and Pattern Recognition, pp. 770–778 (2016)
7. Hizlisoy, S., Yildirim, S., Tufekci, Z.: Music emotion recognition using convolutional long short term memory deep neural networks. Eng. Sci. Technol. Int. J. **24**(3), 760–767 (2021)
8. Hochreiter, S., Schmidhuber, J.: Long short-term memory neural computation. Search In **8**(9) (1997)
9. Huang, Z., Ji, S., Hu, Z., Cai, C., Luo, J., Yang, X.: Adff: attention based deep feature fusion approach for music emotion recognition. arXiv preprint arXiv:2204.05649 (2022)
10. Hunter, P.G., Schellenberg, E.G.: Music and emotion. Music Perception, pp. 129–164 (2010)
11. Kiran, U.: Mfcc technique for speech recognition (2023). https://www.analyticsvidhya.com/blog/2021/06/mfcc-technique-for-speech-recognition/, Tutorial on MFCC. Accessed 15 Mar 2025

12. Lang, M., Shaw, D.J., Reddish, P., Wallot, S., Mitkidis, P., Xygalatas, D.: Lost in the rhythm: effects of rhythm on subsequent interpersonal coordination. Cogn. Sci. **40**(7), 1797–1815 (2016)
13. Le, P.N., Ambikairajah, E., Epps, J., Sethu, V., Choi, E.H.: Investigation of spectral centroid features for cognitive load classification. Speech Commun. **53**(4), 540–551 (2011)
14. Li, T., Ogihara, M.: Content-based music similarity search and emotion detection. In: 2004 IEEE International Conference on Acoustics, Speech, and Signal Processing, vol. 5, pp. V–705. IEEE (2004)
15. librosa development team: librosa / librosa: 0.11.0. https://doi.org/10.5281/zenodo.591533. https://librosa.org/doc/latest/index.html, Librosa API. Accessed 1 Mar 2025
16. Logan, B., et al.: Mel frequency cepstral coefficients for music modeling. In: Ismir, vol. 270, p. 11. Plymouth, MA (2000)
17. Lu, L., Liu, D., Zhang, H.J.: Automatic mood detection and tracking of music audio signals. IEEE Trans. Audio Speech Lang. Process. **14**(1), 5–18 (2005)
18. McKay, C.: jaudio: towards a standardized extensible audio music feature extraction system. Course Paper, McGill University, Canada (2009)
19. Mhasakar, P., Trivedi, P., Mandal, S., Mitra, S.K.: Handwritten digit recognition using Bayesian resnet. SN Comput. Sci. **2**(5), 399 (2021)
20. Müller, M., Kurth, F., Clausen, M.: Chroma-based statistical audio features for audio matching. In: Proceedings of the Workshop on Applications of Signal Processing (WASPAA), pp. 275–278. New Paltz, New York, USA (2005)
21. Orzan, A.: Orzanai/moodify: This application utilizes the lgbm model to accurately classify the emotions of songs and provide tailored song recommendations based on mood and cosine similarity (2023). https://github.com/orzanai/Moodify, Link to Moodify dataset on GitHub. Accessed 25 Mar 2025
22. Panda, R., Redinho, H., Gonçalves, C., Malheiro, R., Paiva, R.P.: How does the spotify API compare to the music emotion recognition state-of-the-art? In: Proceedings of the 18th Sound and Music Computing Conference (SMC 2021), pp. 238–245. Axea SAS/SMC Network (2021)
23. Pedregosa, F., et al.: Scikit-learn: machine learning in Python. J. Mach. Learn. Res. **12**, 2825–2830 (2011)
24. Pelykh, A.: Pytube/pytube: a lightweight, dependency-free Python Library (and command-line utility) for downloading YouTube videos (2023). https://github.com/pytube/pytube, Link to PyTube library on GitHub. Accessed 25 Mar 2025
25. Rajesh, S., Nalini, N.: Musical instrument emotion recognition using deep recurrent neural network. Procedia Comput. Sci. **167**, 16–25 (2020)
26. Rauscher, F.H., et al.: Music and spatial task performance: a causal relationship. ERIC (1994)
27. Roberts, L.: Understanding the mel spectrogram (2024). https://medium.com/analytics-vidhya/understanding-the-mel-spectrogram-fca2afa2ce53, Tutorial on Mel Spectrogram. Accessed 25 Mar 2025
28. Russell, J.A.: A circumplex model of affect. J. Pers. Soc. Psychol. **39**(6), 1161 (1980)
29. Saini, H.K.: Alexmercerind/YouTube-Search-Python: search for YouTube videos, Channels & playlists (2022). https://github.com/alexmercerind/youtube-search-python, Link to YouTube Search library on GitHub. Accessed 25 Mar 2025
30. Scarrott, B.: The best music streaming services 2025: spotify, apple music, tidal, qobuz, more (2023). https://www.techradar.com/audio/audio-streaming/

the-best-music-streaming-services#section-the-best-music-streaming-service-for-most-people, Link to article. Accessed 25 Mar 2025
31. Spotify: Web API - retrieve metadata from spotify content or control playback, Spotify Web API. Accessed 1 Mar 2025
32. Spotify: Spotify - about spotify (2025). https://newsroom.spotify.com/company-info/, Link to Spotify data. Accessed 25 Mar 2025
33. The Linux Foundation: Pytorch documentation (2025). https://pytorch.org/docs/stable/index.html, Link to PyTorch documentation. Accessed 25 Mar 2025
34. Tzanetakis, G., Cook, P.: Marsyas: a framework for audio analysis. Organised Sound **4**(3), 169–175 (2000)
35. Yang, Y.H., Lin, Y.C., Su, Y.F., Chen, H.H.: A regression approach to music emotion recognition. IEEE Trans. Audio Speech Lang. Process. **16**(2), 448–457 (2008)
36. Zhang, K., Zhang, H., Li, S., Yang, C., Sun, L.: The pmemo dataset for music emotion recognition. In: Proceedings of the 2018 ACM on International Conference on Multimedia Retrieval, pp. 135–142 (2018)
37. Zulko: Moviepy documentation (2024). https://zulko.github.io/moviepy/, Link to MoviePy documentation. Accessed 25 Mar 2025

Comparative Analysis of Activation Functions in Neural Networks for Computer Vision Applications

Iaz Andrei-Sebastian[(✉)] [ID], Bolojan Octavia-Maria[ID], and Neagu Robert Valentin[ID]

University of Oradea, Universității Street 1, 410087 Oradea, Romania
`{iaz.andreisebastian,neagu.robertvalentin}@student.uoradea.ro,`
`obolojan@uoradea.ro`

Abstract. Activation functions play a pivotal role in the performance and convergence of deep neural networks (DNNs) as well as convolutional neural networks (CNNs). In this study, a comprehensive comparative analysis is conducted on fourteen activation functions, including both classical functions, for example, ReLU, Sigmoid, Tanh and recent proposals such as SiLU, Mish, Serf, Nipuna. We apply these functions to a DNN architecture trained on the MNIST dataset, and then on a CNN trained on both MNIST and CIFAR-10. Results include a-ccuracy, training time, inference speed, and the cumulative loss area (computed via numerical integration). The results indicate that while some functions deliver high accuracy and an efficient convergence, others result in statistically significant decreases in accuracy and prolonged convergence times. This paper outlines the methodology, provides the mathematical framework, and presents and discusses the findings.

Keywords: Accuracy · Activation Function · Convergence · Deep Learning · Digit Recognition · MNIST · Neural Networks

1 Introduction

Activation functions are a critical component in the design of neural networks that provide the nonlinear transformations that make deep neural networks (DNNs) and convolutional neural networks (CNNs) capable of learning comprehensive, hierarchical features for various real-world applications, including but not limited to computer vision (CV) tasks. By introducing nonlinearity, they enable networks to learn complex representations of data, which is essential for addressing a wide range of challenges in computer vision.

Over the years, numerous activation functions have been proposed, from classic choices such as the ReLU or the Sigmoid, to more recent variants which have trainable para-meters. While advancements in the study area of NNs have led to explorations of activation functions having the potential to yield improvements in both inference accuracy and convergence, there still isn't a systematic and comprehensive comparison when

C. Chira et al. (Eds.): InnoComp 2025, CCIS 2793, pp. 175–188, 2026.
https://doi.org/10.1007/978-3-032-12478-4_12

considering relatively large architectures applied to CV. While some studies have high-lighted the benefits of adaptive or non-monotonic activation functions, their comparative impact hasn't been thoroughly quantified or organised.

Foundational work proved that feedforward networks with sigmoidal activations can universally approximate any continuous function [1, 2], but these saturating functions impede gradient flow and slow convergence as depth increases. The Rectified Linear Unit (ReLU) [3] overcame this bottleneck by remaining non-saturating for positive inputs, rapidly becoming the standard. Further refinements, such as sparse variants that enforce neuron-level separation, have improved stability in very deep models [4]. Despite this extensive list of proposals, no single study has benchmarked all leading activation functions side by side under identical conditions.

This paper addresses a notable gap in the literature by systematically evaluating four-teen activation functions across both deep neural networks (DNNs) and convolutional neural networks (CNNs), using the MNIST and CIFAR-10 datasets. The comparative analysis focuses on classification accuracy, convergence speed, and computational e-fficiency, all assessed under a consistent experimental framework. By employing stan-dardized training conditions and key performance indicators, such as accuracy, training duration, and loss trajectory, this research offers comprehensive and clear insights into the relative strengths and limitations of widely used and emerging activation functions, with a focus on which functions deliver superior performance, supporting informed decision-making for researchers in deep learning.

The remainder of this paper is structured as follows: Sect. 2 presents a brief historical and technical overview of activation functions in neural networks; Sect. 3 introduces the activation functions selected for comparison; Sect. 4 details the experimental method-ology, including model architecture and evaluation metrics; Sect. 5 presents the results on three network-dataset combinations: DNN on MNIST, CNN on MNIST, and CNN on CIFAR-10, respectively; Finally, Sect. 6 concludes the study and outlines directions for future work.

2   Current Research and Related Literature Case Studies

Since the beginning of humanity's exploration of the ability of neural networks to imitate the human mind, the evolution of activation functions has been pivotal in overcoming fundamental theoretical and practical challenges in the field of deep learning. Cybenko's early work [1] and Hornik et al. [2] established the fact that feedforward neural networks with sigmoidal activation functions are universal approximators, a fact now known in the community as the Universal Approximator Theorem. However, as network depths increased to accommodate ever-more difficult tasks, the limitations of the Sigmoid and Tanh functions became clear, especially since they have a tendency to saturate and produce vanishing gradients.

The introduction of the Rectified Linear Unit (ReLU) [3] was a significant break-through, as its simple formulation allowed for efficient gradient propagation. Glorot et al. [4] and subsequent studies [5, 6] demonstrated that networks with ReLU-based architectures achieved unprecedented performances, especially on large-scale tasks such as ImageNet classification tasks. It is widely acknowledged that image processing currently represents one of the most computationally intensive tasks for neural networks, however, real-time execution could provide a substantial advantage.

To further enhance learning dynamics, alternative functions such as ELU [7] and GELU [8] were developed. For one, ELU's incorporation of exponential components helped push mean activations closer to zero, which led to accelerated convergence, while GELU introduced stochastic regularization by leveraging the notion of Gaussian cumulative distribution. Ramachandran et al. [9] expanded the design space and possibilities for a neural network by looking for new activation functions. This led to the proposal of functions like Swish and Mish. Mish [10], in particular, has received attention for its self-regularizing properties and smooth non-monotonic behavior, both of which are much appreciated advantages.

Other studies have explored variations that address specific challenges. For instance, Elfwing et al. [11] proposed sigmoid-weighted linear units with the idea to improve function approximation in reinforcement learning (RL), while Barron's Squareplus [12] offered a differentiable alternative to ReLU. Klambauer et al. [13] introduced self-normalizing neural networks through SeLU, which demonstrated that proper scaling can preserve network stability across deep architectures, an increasingly important property as neural networks continue to grow in depth. Jin et al. [14], Basirat and Roth [15] as well as Godfrey and Gashler [16] examined the continuum of activation function behaviors, suggesting that hybrid forms might improve generalization even further. Researchers also focused on the impact of activation functions on universal approximation properties. More concretely, Kidger and Lyons [17] and Kratsios and Papon [18] provided theoretical guarantees for the approximation capabilities of deep networks under various activation schemes.

Recent survey work by Dubey et al. [19] provides a comprehensive overview and benchmark of many of these activation functions, which acted as a motivation for this very paper, while Kidger's latest work [20] underscores their ongoing significance in modern neural network design. Collectively, these studies illustrate that while ReLU, together with its variants, do indeed remain powerful baselines, there is a lot of room for innovation, particularly towards achieving a balance between computational efficiency, convergence speed, and overall performance.

In neural networks, two main challenges emerge as parameter counts grow. Firstly, training time can become a bottleneck, delaying development cycles in performance-critical applications. Secondly, networks demand vast computational resources. Although matrix multiplication is highly optimized in parallel computing, other operations still consume significant processing power and electricity.

To address the first issue, that of computational demand, Choi et al. [21] introduced an optical encoder during image capture and processing. Because color images are stored as three-dimensional tensors, holding both pixel values and their spatial relationships, any speedup at the acquisition stage can cascade through the entire pipeline, yielding substantial efficiency gains. For convergence speed, Wong et al. [22] modified weight-initialization strategies in ways that are directly related to the slected activation functions. Their results highlight how weights, biases, and activation functions together shape both the training process and final performance. By tweaking initial parameters, they achieved faster convergence without altering the overall architecture. Recently, deep ReLU networks have dominated tasks in CV and NLP due to their simplicity and empirical performance [23]. However, using the same activation function in every layer demands careful function choice, as its issues might get amplified. Selecting an activation function that maintains stable gradients across dozens, or even hundreds, of layers could reduce vanishing (or exploding) gradients and boost accuracy in dense architectures.

This overview sets the stage for our comparative analysis, where fourteen activation functions are quantitatively assessed on the MNIST dataset, measuring not only accuracy, but also training dynamics using metrics such as training time, inference time, and loss area.

3 Activation Functions Used

In this section, we present the activation functions analyzed in this study. Each function transforms an input scalar $x \in \mathbb{R}$ into an output value, denoted $f(x)$, which introduces nonlinearity into the neural network's processing layers. All activation functions were applied consistently in the hidden layers of each architecture network. The output layers remained linear, producing raw logits for classification. We categorize the functions into classical, modern, and experimental variants, and provide their mathematical definitions along with commentary on their properties.

Perceptron (Identity function)
This function represents the simplest activation function, defined as:

$$f(x) = x,$$

where $x \in \mathbb{R}$.

It passes input unchanged and serves as a neutral baseline. It lacks nonlinearity and it cannot model complex patterns on its own, but it helps illustrate the necessity of non-linear activation.

Sigmoid
This function maps the set of real numbers to the interval $(0, 1)$:

$$f(x) = \frac{1}{1 + e^{-x}},$$

where $x \in \mathbb{R}$.

It is being commonly used in binary classification problems to represent probabilities (since it's in the same numerical range as a probability). However, one problem it has is that it can suffer from vanishing gradients for large positive or negative inputs ($|x|$), which can hinder training in deep networks.

Hyperbolic Tangent (Tanh)
The Tanh function outputs values in the interval $(-1, 1)$, with zero-centered activation, being defined as:

$$f(x) = \frac{e^x - e^{-x}}{e^x + e^{-x}},$$

where $x \in \mathbb{R}$.

It generally performs better than Sigmoid in hidden layers but still suffers from vanishing gradient issues (unlike the Sigmoid function, Tanh is 0-centered, which contributes to faster convergence).

Rectified Linear Unit (ReLU)

This function is a widely used nonlinear function defined by:

$$f(x) = max(0, x),$$

where $x \in \mathbb{R}$.

It introduces sparsity and also mitigates the problem of the vanishing gradient, because its derivative is 1 for $x > 0$, which allows gradients to flow unchanged during the backpropagation step. While computationally efficient, it can cause neurons to "die" when they consistently receive negative inputs.

Leaky ReLU

The function is defined as follows:

$$f(x) = \begin{cases} x, & ifx \geq 0 \\ \alpha x, & ifx < 0 \end{cases},$$

where $x \in \mathbb{R}$ and $\alpha \in (0,1)$ is a given constant.

This variant addresses the dying ReLU problem by allowing a small, non-zero gradient for negative inputs (it solves the "dying ReLU" problem). The function is widely used in deep learning due to its computational efficiency and ability to mitigate the vanishing gradient problem, which allows for faster and more effective training in deep neural networks.

Exponential Linear Unit (ELU)

The function is defined as follows:

$$f(x) = \begin{cases} x, & ifx \geq 0 \\ \alpha(e^x - 1), & ifx < 0 \end{cases},$$

where $x \in \mathbb{R}$ and $\alpha \in (0,1)$.

ELU introduces an exponential component for negative inputs, which helps center activations toward a zero mean and accelerates network convergence by maintaining a more stable gradient flow.

Softplus

Softplus is a smooth approximation of ReLU and always differentiable, being defined as

$$f(x) = \ln(1 + e^x),$$

where $x \in \mathbb{R}$.

The Softplus activation function provides a smooth and continuous alternative to ReLU, avoiding abrupt transitions and ensuring differentiability, which can lead to more stable training in certain deep learning models.

Gaussian Error Linear Unit (GELU)

This function is defined as

$$f(x) = x \bullet \Phi(x),$$

where $x \in \mathbb{R}$ and $\Phi(x)$ is the cumulative distribution function (CDF) of the standard normal distribution. GELU introduces stochastic regularization effects and is widely used in transformer models.

Scaled Exponential Linear Unit (SELU)

SELU function is defined as

$$f(x) = \lambda \bullet \begin{cases} x, & if x \geq 0 \\ \alpha(e^x - 1), & if x < 0 \end{cases},$$

where $x \in \mathbb{R}$. With specific values for λ and α, SELU preserves mean and variance in self-normalizing networks, especially in deep architectures.

Sigmoid Linear Unit (SiLU or Swish)

SiLU is a smooth, non-monotonic, and differentiable activation function that combines the input with its sigmoid transformation, being defined as follows:

$$f(x) = x \bullet \frac{1}{1 + e^{-x}},$$

where $x \in \mathbb{R}$.

By allowing small negative values and supporting better gradient flow, it often improves convergence and leads to higher classification accuracy in computer vision tasks and deep vision models.

Mish

This function is defined as

$$f(x) = x \bullet \tanh(\ln(1 + e^x)),$$

where $x \in \mathbb{R}$.

Mish is a smooth, non-monotonic activation function known for its self-regularizing behavior and strong empirical performance across various deep learning tasks. By allowing small negative outputs and promoting smoother gradient flow, Mish facilitates better generalization and improved stability during training, particularly in deep neural networks.

Collapsing Linear Unit (CoLU)

A relatively new proposal, CoLU aims to reduce variance in deeper layers by "collapsing" neuron outputs toward fixed values during training. Its formulation is inspired by regularization techniques such as dropout and weight compression, aiming to improve stability and generalization in deep neural networks.

CoLU is defined as

$$f(x) = \frac{x}{1 - xe^{-(x+e^x)}},$$

where $x \in \mathbb{R}$. The function is smooth, differentiable, non-monotonic, and zero-centered, features that promote stable training in deep networks. The presence of nested exponentials results in higher computational complexity compared to classical alternatives such as ReLU.

Self-Regularizing Function (Serf)
Serf is a recent experimental activation function that introduces a modified sigmoid-like saturation combined with a self-regularizing mechanism. It is designed to reduce overfitting and gradient saturation, particularly in reinforcement learning and exploration-driven domains.

This function is defined as

$$f(x) = x \bullet \tanh\left(\frac{1}{1 + e^{-x}}\right),$$

where $x \in \mathbb{R}$. The result is a smooth, differentiable function that combines linear growth with controlled saturation.

Nipuna
Nipuna is a recent, experimental activation function that includes adaptive components designed to learn optimal scaling and translation per neuron. Its structure is typically of the form:

$$f(x) = a \bullet x + b \bullet tanh(c \bullet x),$$

where $x \in \mathbb{R}$ and a, b, c are trainable parameters.

The function combines a linear term with a saturating nonlinear component. It reduces overfitting by implicitly regularizing the activation distribution, supports stable training in deep architectures without explicit normalization layers, and dynamically adapts its shape based on task complexity and network depth.

The purpose of presenting the functions was to show that most of them rely either on properties of exponentials, logarithms, or linear functions of some type. To ensure a fair and consistent comparison of activation functions, in what follows we designed a rigorous experimental framework applied across multiple network architectures and datasets. Therefore, we outline the experimental protocol, including data handling, network architecture, training procedures, and evaluation metrics.

4 Methodology

In this section, we describe the experimental methodology employed, including the model architectures of the DNN and CNN, datasets, the training procedure and configurations, and finally the evaluation metrics used in comparing the activation functions, which are the results.

4.1 Common Experimental Setup

All three experiments were done in an identical training regime to ensure direct comparability. We split each dataset into five equally sized folds using a fixed random seed (42) and performed five-fold cross-validation, reporting the mean and standard deviation of all metrics. Each model was trained for 10 epochs with a batch size of 64 under the Adam optimizer with default parameters. To mirror modern production practices, we leveraged PyTorch's automatic mixed precision (AMP) by wrapping both forward and backward passes in autocast and scaling gradients with a GradScaler.

The code executed in Google Colab, on an NVIDIA A100 GPU. We measured training time by surrounding the loop of each fold with `time.perf_counter()`. For inference time, we performed one bulk forward pass over each validation fold, invoked `torch.cuda.synchronize()` in order to flush the GPU operations, and divided the elapsed seconds by the number of images, obtaining seconds-per-image values, which were then multiplied by 1000 to report milliseconds per image. We added the Perceptron not as a competitor, but as a "base measurement", the "neutral" activation function.

The cross-entropy loss function is used to compute the training loss. For a given input x with true label y and predicted probability distribution $\hat{y}$, the Cross-Entropy Loss is defined as:

$$\mathcal{L}CE = -\sum_{i=1}^{10} y_i log(\hat{y_i}) \tag{1}$$

where y_i is a one-hot encoded true label vector and $\hat{y_i}$ is the predicted softmax output.

4.2 Network Architectures

DNN on MNIST

Our first model is a DNN optimized for MNIST. Grayscale 28×28 images are flattened to a 784-dimensional vector and passed through 5 dense layers of sizes 2^8, 2^7, ..., 2^4. After each hidden layer, the activation function is applied. The output layer maps the 16-dimensional embedding to 10 logits, one per digit class. We bind activations dynamically at model construction, either from `torch.nn` or from our custom implementations.

CNN on MNIST

The second architecture is a "same-size" CNN. Inputs are $1 \times 28 \times 28$ images. Three convolutional blocks each consist of a 3×3 conv layer (channels $1 \rightarrow 16$, $16 \rightarrow 32$, $32 \rightarrow 64$), followed by the activation and a 2×2 max-pool. After the third block, feature maps of size $64 \times 7 \times 7$ are flattened and passed through a 72-unit dense layer with the same activation before the final logits.

CNN on CIFAR-10

Because MNIST is a relatively simple task to solve even for a DNN, our third model mirrors the MNIST CNN but takes as data $3 \times 32 \times 32$ color images from the CIFAR-10 dataset. The three conv blocks and pooling pattern remain identical, while the post-pool feature map is $64 \times 8 \times 8$ this time. We flatten to 4 096 dimensions, feed into a 55-unit dense layer with the test activation, ending with a 10-unit output (CIFAR-10 has the same number of classes as MNIST, just more complex data).

All 3 architectures were constructed to have very close to 250K trainable parameters, so model sizes are very similar.

4.3 Common Evaluation Metrics

To assess the performance of each activation function, we consider the following metrics:

- Accuracy (fraction): The ratio between correctly classified test samples to the sample count.
- Training Time (s): The average time taken to complete training across all 5 cross-validation folds.
- Inference Time (ms): The average time taken for a forward pass one image.
- Loss Area (unitless): A numerical approximation of the area under the loss curve over the training epochs, calculated using the trapezoidal rule `np.trapezoid()`), where $L(t)$ is the loss at epoch t:

$$\text{Loss Area} \approx \sum_{t=1}^{T-1} \frac{L(t) + L(t+1)}{2} \tag{2}$$

This reflects convergence behavior and optimization dynamics across epochs.

The next section presents our results, including metric comparisons for each activation function across the three network-dataset combinations.

5 Obtained Results and Comparative Analysis

This section presents the experimental results for all activation functions across the three evaluated architectures: a fully connected DNN on MNIST, a CNN on MNIST, and a CNN on CIFAR-10. Each function was tested under the same training configuration and dataset splits to ensure comparability. For each experiment, we report: Classification accuracy, Training time, Loss area, Inference time.

These metrics are reported as mean values over five-fold cross-validation. Outliers such as Sigmoid and Perceptron are discussed and omitted from some visualizations for improved interpretability.

5.1 DNN on MNIST

First, we look at Accuracy, arguably the most important metric. We use a scale of 0 to 1, and we can see that GELU obtains the best accuracy (≈ 0.9757). By training time, the fastest nonlinear function is ReLU (≈ 21.75 s). Intuitively, the computation time of the function itself matters as well. In terms of inference time/image, CoLU is the fastest (≈ 0.05014 ms). Finally, we look at loss area, where ELU converges the fastest (≈ 0.9936). Table 1 below illustrates the obtained results, with all numerical values rounded to four decimal places for consistency and clarity.

Table 1. Obtained results for DNN on MNIST

Activation function	Accuracy Mean ± STD	Training Time Mean ± STD	Loss Area Mean ± STD	Inference Time Mean ± STD
Perceptron	0.9134 ± 0.0017	20.9270 ± 0.3035	2.8339 ± 0.0788	0.0542 ± 0.0010
ReLU	0.9711 ± 0.0036	21.7539 ± 1.2067	1.1608 ± 0.1485	0.0509 ± 0.0005
Sigmoid	0.9618 ± 0.0025	21.8347 ± 1.2218	1.5511 ± 0.0904	0.0509 ± 0.0010
Tanh	0.9714 ± 0.0016	22.0179 ± 1.3763	1.0547 ± 0.0801	0.0543 ± 0.0056
LeakyReLU	0.9721 ± 0.0032	22.2965 ± 1.4427	1.1167 ± 0.1172	0.0548 ± 0.0071
ELU	0.9733 ± 0.0015	22.9611 ± 1.6536	0.9936 ± 0.0668	0.0522 ± 0.0019
Softplus	0.9715 ± 0.0021	22.9592 ± 1.6087	1.0443 ± 0.0818	0.0541 ± 0.0051
GELU	0.9757 ± 0.0016	23.0785 ± 1.5900	1.0268 ± 0.0561	0.0502 ± 0.0008
SELU	0.9714 ± 0.0025	22.9592 ± 1.5511	1.0645 ± 0.0610	0.0514 ± 0.0010
SiLU	0.9732 ± 0.0030	22.8255 ± 1.3991	1.1055 ± 0.1126	0.0505 ± 0.0007
Mish	0.9754 ± 0.0021	23.4559 ± 1.8940	1.0426 ± 0.1111	0.0561 ± 0.0057
CoLU	0.9744 ± 0.0013	23.3010 ± 2.1639	1.0664 ± 0.0675	0.0501 ± 0.0009
Serf	0.9751 ± 0.0016	22.9323 ± 1.6374	1.0230 ± 0.0490	0.0509 ± 0.0014
Nipuna	0.9734 ± 0.0020	23.4285 ± 2.4577	1.1014 ± 0.0781	0.0539 ± 0.0050

5.2 CNN on MNIST

In the following part, we present the results of the same activation functions applied to a CNN training on the MNIST dataset, consisting of 60000 training and 10000 test images using 5-fold cross-validation. We may even see an effect where the CNN can make better use of the activation function.

First, we look at accuracy. Here, Mish obtains the best score (≈ 0.989). By training time, the fastest nonlinear function is the Sigmoid (≈ 25.14 s). In terms of inference time / image, ReLU is the fastest (≈ 0.4152 ms). Finally, we look at loss area, where ReLU converges the fastest (≈ 0.4929). So as a conclusion, for these types of tasks, one should use Mish, having the best accuracy. The difference in training time is insignificant, and the difference in inference speeds is negligible. The loss area difference is to be taken into account, but slower convergence with higher accuracy is much more valuable than faster convergence with lower accuracy. Table 2 presents the results obtained, with all values rounded to four decimal places in order to facilitate accurate interpretation and standardized comparison across the evaluation, as in the previous case.

Table 2. Obtained results for CNN on MNIST

Activation function	Accuracy Mean ± STD	Training Time Mean ± STD	Loss Area Mean ± STD	Inference Time Mean ± STD
Perceptron	0.9785 ± 0.0021	26.7374 ± 5.9811	0.8133 ± 0.1100	0.4907 ± 0.0222
ReLU	0.9890 ± 0.0010	25.5734 ± 3.2972	0.4152 ± 0.0317	0.4929 ± 0.0245
Sigmoid	0.9806 ± 0.0025	25.1437 ± 2.2425	0.5692 ± 0.0542	0.4993 ± 0.0244
Tanh	0.9864 ± 0.0004	25.2439 ± 2.2254	0.4262 ± 0.0084	0.4993 ± 0.0249
LeakyReLU	0.9881 ± 0.0014	25.7611 ± 2.4433	0.4497 ± 0.0547	0.4996 ± 0.0244
ELU	0.9871 ± 0.0019	25.5203 ± 2.5953	0.5918 ± 0.0945	0.5011 ± 0.0257
Softplus	0.9842 ± 0.0007	25.6609 ± 2.6347	0.6218 ± 0.0235	0.5005 ± 0.0249
GELU	0.9887 ± 0.0007	25.5385 ± 2.5092	0.4652 ± 0.0234	0.4981 ± 0.0254
SELU	0.9862 ± 0.0006	25.4038 ± 2.3905	0.6240 ± 0.0841	0.5007 ± 0.0245
SiLU	0.9879 ± 0.0008	25.5610 ± 2.3314	0.5013 ± 0.0550	0.4948 ± 0.0236
Mish	0.9890 ± 0.0020	25.5862 ± 2.6852	0.4649 ± 0.0907	0.5012 ± 0.0244
CoLU	0.9885 ± 0.0003	25.8308 ± 3.1765	0.5121 ± 0.0770	0.4989 ± 0.0240
Serf	0.9882 ± 0.0024	25.4520 ± 2.5963	0.5152 ± 0.1177	0.4960 ± 0.0253
Nipuna	0.9887 ± 0.0005	25.7415 ± 3.1266	0.5123 ± 0.0237	0.4947 ± 0.0245

It is worth noting that the differences in inference time across activation functions are not significant. The fastest function would be ReLU ($\approx$ 0.4929 ms), while the slowest would be Mish ($\approx$ 0.5012 ms). This difference of 0.01 ms is insignificant for any practical application, as even software that has to work with subatomic, quasi-instant events records the images, cleans the data, and only afterwards performs the analysis. It can also be observed that Mish and ReLU achieve a similar accuracy mean ($\approx$ 0.9890), but the standard deviation associated with ReLU is less than half of that obtained for Mish, suggesting that ReLU provides enhanced stability. High-speed driving could also be a use case, but a hundredth-of-a-milisecond delay is insignifcant, so accuracy still remains the decisive factor for such functions.

5.3 CNN on CIFAR-10

As we kept the structure of the CNN as close to the original as possible, including the number of trainable parameters, but we increased the difficulty of the dataset (having to deal with coloured imaged of objects, not only digits), while not providing things like border detection, we expect the network to perform more poorly, further stressing the power of the activation function to determine the final output given by the network, and thus showing even more clearly which activation function is a better fit to be used in a CNN.

First, we look at accuracy. Here, Serf obtains the best score ($\approx$ 0.6817). Next, we look at Loss Area, where Tanh has the greatest value ($\approx$ 0.119). By training time, of course the Perceptron is the fastest, but of the nonlinear ones, ReLU is the fastest ($\approx$

9.0794). SeLU is the fastest (≈ 0.6418 s) in inference time/image. The results presented in Table 3 are also reported with numerical values rounded to four decimal places in order to ensure data uniformity and readability, as was done previously.

Table 3. Obtained results for CNN on CIFAR-10

Activation function	Accuracy Mean ± STD	Training Time Mean ± STD	Loss Area Mean ± STD	Inference Time Mean ± STD
Perceptron	0.6362 ± 0.0057	19.4784 ± 0.1119	9.7984 ± 0.1070	0.6597 ± 0.0146
ReLU	0.6692 ± 0.0105	19.8514 ± 0.1021	9.0794 ± 0.3139	0.6746 ± 0.0033
Sigmoid	0.4877 ± 0.0145	21.2565 ± 2.5109	12.8670 ± 0.3832	0.6550 ± 0.0310
Tanh	0.6327 ± 0.0055	21.2333 ± 2.1437	9.7929 ± 0.1194	0.6587 ± 0.0296
LeakyReLU	0.6713 ± 0.0108	21.5516 ± 2.4731	9.8702 ± 0.3882	0.6582 ± 0.0303
ELU	0.6612 ± 0.0053	21.4645 ± 2.4467	10.0442 ± 0.3001	0.6600 ± 0.0309
Softplus	0.5373 ± 0.0049	21.4253 ± 2.5427	13.3600 ± 0.5648	0.6504 ± 0.0313
GELU	0.6674 ± 0.0085	21.6471 ± 2.5226	10.6870 ± 0.4420	0.6580 ± 0.0313
SELU	0.6384 ± 0.0049	21.4751 ± 2.4207	10.9320 ± 0.3392	0.6418 ± 0.0307
SiLU	0.6718 ± 0.0030	21.6720 ± 2.5389	9.4336 ± 0.1641	0.6493 ± 0.0294
Mish	0.6767 ± 0.0086	21.6175 ± 2.7011	9.5342 ± 0.2662	0.6604 ± 0.0322
CoLU	0.6745 ± 0.0078	21.7710 ± 3.2995	10.8649 ± 0.4861	0.6527 ± 0.0305
Serf	0.6817 ± 0.0038	21.7869 ± 2.6890	9.7634 ± 0.2025	0.6461 ± 0.0290
Nipuna	0.6805 ± 0.0055	21.7396 ± 2.8740	10.4617 ± 0.1613	0.6587 ± 0.0322

6 Conclusions

In this study, we performed a comprehensive comparative analysis of fourteen activation functions applied to both dense and convolutional neural networks in computer vision tasks. Our evaluation was carried out on two benchmark datasets —MNIST and CIFAR-10— and three different network architectures with balanced parameter counts.

The results confirm that the choice of activation function significantly affects network behavior across multiple dimensions, including accuracy, convergence rate, and training/inference efficiency. Small differences in accuracy can make a substantial impact when it comes to self-driving cars, or technology to help the visually impaired navigate the streets. As such, when training a neural network, the main objective should be considered, depending on whether the goal is maximum accuracy, minimizing inference time, or ensuring fast training time and convergence for efficient and relevant prototy-ping. These results serve as a framework for future DNNs and CNNs designs, so people in the industry know what to expect of a certain activation function.

While some functions, such as GELU, Mish, and SiLU, demonstrated excellent accuracy and smooth learning dynamics, classical functions like Sigmoid and Perceptron consistently underperformed, particularly in deep architectures. In terms of training speed and implementation simplicity, ReLU remains a strong contender and is still a preferred default in production systems. However, more recent functions like Mish and GELU offer superior generalization and should be considered for tasks where accuracy is critical. We also introduced the Loss Area metric to quantitatively assess convergence behavior over training epochs, providing deeper insight into the learning efficiency of each activation function.

Building on these findings, future research can explore several directions:

- Scaling to complex datasets such as CIFAR-100 or ImageNet, where richer representations are required and differences between activation functions are likely to be amplified;
- Exploring hybrid activation functions or adaptive compositions of multiple functions within different layers of the same network;
- Resource-constrained deployment: analyzing the tradeoffs between inference time and accuracy for deployment on mobile or edge devices;
- Integration with normalization layers, e.g., BatchNorm and LayerNorm, to understand interaction effects between scaling and activation;
- Theoretical analysis of convergence properties and Lipschitz continuity in newer proposals such as CoLU, Serf, and Nipuna.

Therefore, this study aims to provide a comprehensive and standardized performance benchmark for activation functions under uniform experimental conditions, thereby serving as a practical reference for researchers and practitioners in deep learning when selecting suitable activation functions for their models.

Acknowledgments. This work was supported by the Scientific Research Funds of the University of Oradea.

Disclosure of Interests. The authors have no competing interests to declare that are relevant to the content of this article.

References

1. Cybenko, G.: Approximation by superpositions of a sigmoidal function. Math. Control Signals Syst. **2**(4), 303–314 (1989)
2. Hornik, K., Stinchcombe, M., White, H.: Multilayer feedforward networks are universal approximators. Neural Netw. **2**(5), 359–366 (1989)
3. Nair, V., Hinton, G.E.: Rectified linear units improve restricted Boltzmann machines. In: Proceedings of the 27th International Conference on Machine Learning (ICML-10), pp. 807–814 (2010)
4. Glorot, X., Bordes, A., Bengio, Y.: Deep sparse rectifier neural networks. In: Proceedings of the Fourteenth International Conference on Artificial Intelligence and Statistics (AISTATS), PMLR 15, pp. 315–323 (2011)
5. Maas, A.L., Hannun, A.Y., Ng, A.Y.: Rectifier nonlinearities improve neural network acoustic models. In: Proceedings of the 30th International Conference on Machine Learning (ICML 2013), JMLR W&CP, Atlanta, Georgia, USA, vol. 28, pp. 3–11 (2013)

6. He, K., Zhang, X., Ren, S., Sun, J.: Delving deep into rectifiers: surpassing human-level performance on ImageNet classification. In: Proceedings of the IEEE International Conference on Computer Vision (ICCV), pp. 1026–1034 (2015)

7. Clevert, D.-A., Unterthiner, T., Hochreiter, S.: Fast and accurate deep network learning by exponential linear units (ELUs). In: 4th International Conference on Learning Representations (ICLR), San Juan, Puerto Rico (2016)

8. Hendrycks, D., Gimpel, K.: Gaussian error linear units (GELUs). arXiv preprint, arXiv:1606.08415 (2016)

9. Ramachandran, P., Zoph, B., Le, Q.V.: Searching for activation functions. In: 6th International Conference on Learning Representations (ICLR), Vancouver, Canada (2018)

10. Misra, S.: Mish: A self regularized non-monotonic activation function. arXiv preprint, arXiv:1908.08681 (2019)

11. Elfwing, S., Uchibe, E., Doya, K.: Sigmoid-weighted linear units for neural network function approximation in reinforcement learning. Neural Netw. **107**, 3–11 (2018)

12. Barron, J.T.: Squareplus: a softplus-like algebraic rectifier. arXiv preprint, arXiv:2112.11687 (2021)

13. Klambauer, G., Unterthiner, T., Mayr, A., Hochreiter, S.: Self-normalizing neural networks. In: Guyon, I., et al. (eds.) Advances in Neural Information Processing Systems (NeurIPS), vol. 30, pp. 971–980 (2017)

14. Jin, X., Xu, A., Zhou, E.: Deep learning with S-shaped rectified linear activation units. In: International Conference on Learning Representations (ICLR) (2016)

15. Basirat, M., Roth, P.M.: The quest for the golden activation function. arXiv preprint arXiv:1808.00783 (2018)

16. Godfrey, L.B., Gashler, M.: A continuum among logarithmic, linear, and exponential functions and its potential to improve generalization in neural networks. In: Proceedings of the 7th International Conference on Knowledge Discovery and Information Retrieval (KDIR), pp. 34–45 (2015)

17. Kidger, P., Lyons, T.: Universal approximation with deep narrow networks. arXiv preprint arXiv:1905.08539 (2019)

18. Kratsios, A., Papon, L.: Universal approximation theorems for differentiable geometric deep learning. J. Mach. Learn. Res. (JMLR) **23**(1), 1–50 (2022)

19. Dubey, S.R., Singh, S.K., Chaudhuri, B.B.: Activation functions in deep learning: a comprehensive survey and benchmark. arXiv preprint arXiv:2109.14545 (2021)

20. Kidger, P.: Activation functions and their significance in deep learning. IEEE Trans. Neural Netw. Learn. Syst. **33**(4), 1456–1466 (2022)

21. Choi, M., Xiang, J., Wirth-Singh, A., Baek, S.-W., Shlizerman, E., Majumdar, A.: Transferable polychromatic optical encoder for neural networks. Nat. Commun. **16**, 5623 (2025)

22. Wong, K., Dornberger, R., Hanne, T.: An analysis of weight initialization methods in connection with different activation functions for feedforward neural networks. Evol. Intel. **17**, 2081–2089 (2024)

23. Shen, X., Espinoza, J.: Consistency and rate of convergence for deep ReLU neural networks. J. Stat. Theory Pract. **19**, 33 (2025)

Data Science

Cultivating Sustainable Data: Challenges and Directions in Digital Agriculture

Daniela Delinschi[1,3]([✉]) [iD], Rudolf Erdei[1,3] [iD], Emil Marian Pasca[1,3] [iD], Iulia Baraian[1,3] [iD], and Andrei Curos[2] [iD]

[1] Technical University of Cluj-Napoca, Cluj-Napoca, Romania
`daniela.delinschi@campus.utcluj.ro`
[2] University of Petrosani, Petrosani, Romania
[3] European University of Technology, European Union, Cluj-Napoca, Romania

Abstract. This paper addresses data sustainability in smart agriculture, examining challenges and best practices throughout the agricultural data lifecycle. It differentiates between efficiency, sustainability, and resilience in data systems while analyzing key sustainability challenges including infrastructure limitations, data quality issues, and ownership concerns. The research proposes and details a conceptual framework for quantifying data sustainability, based on dimensions such as data relevance, utility, criticality, age, and cost. A Romanian agriculture illustrative application illustrates sustainability challenges across different farm scales with targeted improvement strategies. The findings contribute to understanding how sustainable data management can enhance agricultural practices while balancing technical, environmental, economic, and social dimensions for long-term viability.

Keywords: Data Sustainability · Agricultural Data Lifecycle · Smart Agriculture · Data Management · Precision Agriculture · Digital Agriculture

1 Introduction

Data has become integral to technology, scientific discovery, and everyday activities. The creation and manipulation of data is a primary function of computerised information systems, helping beneficiaries make use of the data in various ways. As the volume and complexity of data continue to expand, the concept of data sustainability is gaining recognition even in sectors such as agriculture, which are not necessarily considered data-intensive. Consider the use of Machine Learning (ML) techniques to optimise irrigation schedules based on real-time weather data and soil moisture levels. The effectiveness of these approaches relies on the continued availability and reliability of the underlying datasets, the hardware on which the data is stored and processed, but also about the software components that are used to process said data.

C. Chira et al. (Eds.): InnoComp 2025, CCIS 2793, pp. 191–218, 2026.
https://doi.org/10.1007/978-3-032-12478-4_13

However, several benefits arise from large scale usage of data-driven agricultural practices or optimisations: domain accessibility for less experienced farmers, improved crop health, improved yields, limiting or even removing the need for dangerous chemicals (green agriculture [20]), resource efficiency (water, compost, fertilisation), timely decision making, enhanced traceability, precision agriculture. These benefits can be applied at scale, enabling smaller teams to manage larger farms, via using technology like 24/7 available sensors, drone imaging, satellite data and many others.

In this context, comprehensively defining Data Sustainability (DS) means understanding it as a holistic, long-term approach that seeks to integrate several well-known, well-established aspects like environmental protection, social equity, and economic viability [23,36], to ensure that both present and future generations can thrive within the limits of the planet's resources and ecosystems. In plainer words, we aim at raising productivity even in problematic areas (desertification), while making sure the process will not (irreversibly) harm the environment. It's a guiding principle for decision-making across all scales, from individual choices to global policy. In the specific context of data and agriculture, data sustainability aims to ensure that agricultural data resources are managed in a manner that supports long-term food security, environmental protection, and economic prosperity.

Efficiency, *Sustainability*, and *Resilience* represent key perspectives for designing robust data systems, notably within contexts like agriculture that present particular demands. Each dimension has a specific focus: *Efficiency* focuses on optimising resource allocation for performance outputs; *Sustainability* looks towards long-range environmental, social and financial viability; *Resilience* concentrates on a system's ability to endure disturbances and adapt. Despite these distinct angles, the three are intimately related, often interacting in various ways, as depicted in Fig. 1. For example, the drive for resource minimisation inherent in Efficiency (seen in reduced energy use, storage capacity, or processing load) clearly supports the environmental protection and economic goals associated with Sustainability. Similarly, the long-term outlook required by Sustainability and the need for preparedness in Resilience both rely on forward-thinking and adaptive strategies, overlapping considerably in design for adaptability and endurance. Operational effectiveness achieved through Efficiency can also enhance Resilience, potentially shortening recovery times or making resources available for building in tolerance to failures. The overall aim of bringing these facets together contributes to developing data systems that function effectively, are resource-conscious, economically viable, dependable, flexible, environmentally aware, and possess long-term persistence, although specific design decisions might occasionally require careful balancing between these potentially competing priorities.

While undoubtedly important, DS is difficult to analyse and quantify, thus the aim of this paper is to analyse the challenges and propose a conceptual framework for data sustainability quantification, which can guide future implementation and measurement. This difficulty is enforced by the fact that DS

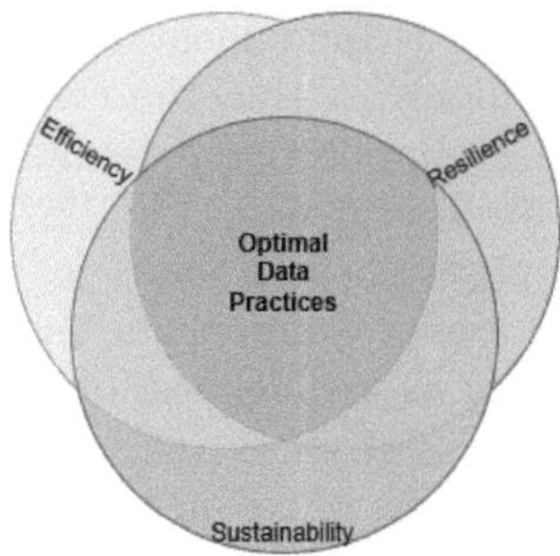

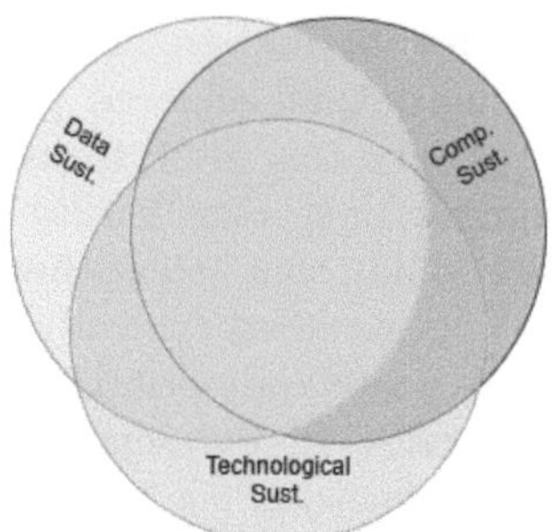

Fig. 1. Optimal data Practices regarding Efficiency, Resilience and Sustainability.

Fig. 2. Relations between Data, Computational and Technological Sustainability.

lies at the confluence between itself, Computational Sustainability and Technological Sustainability (see Fig. 2), with large overlaps and influences, which make clear definitions and quantifications difficult. Several attempts have been mane in other fields, like Industry [37], Public Health [14], and even Aquaculture [12], but specifics applied to Agriculture in its many forms (Smart Agriculture, Precision Agriculture, etc.) seem to be missing from the recent literature.

This paper makes the following key contributions: (a) A comprehensive analysis of the concept of DS within the context of smart agriculture, distinguishing it from efficiency and resilience. (b) An in-depth exploration of the agricultural data lifecycle, identifying key stages and associated challenges. (c) An identification and discussion of critical challenges hindering data sustainability in modern agricultural practices, drawing from recent literature. (d) The proposal of a conceptual, quantifiable framework for Data Sustainability, based on five measurable dimensions: relevance, utility, criticality, age, and cost. (e) A practical application illustrating the specific data sustainability challenges and tailored improvement strategies for different farm scales in Romania.

1.1 Efficiency vs. Sustainability

Considering the vital role data plays in supporting modern technology and driving advancements across various sectors, including agriculture, it's important to clarify the distinction between efficiency and sustainability. One of the more common confusions when discussing about optimisations is the distinction (or lack thereof) between the two aforementioned concepts. *Efficiency* is regarded as the simpler, more evident variant and often the first aspect which an engineer will focus on, as it is a property of the system itself. Every process from a system must be optimised for maximum output for the least input. This includes all consumed resources, like energy, time, storage, computation cycles, or network bandwidth. Efficiency can and should be measured, with lots of statistical tools and metrics being available, depending on the field which is being analysed.

In scientific and engineering discourse, efficiency is fundamentally a ratio: the useful output achieved per unit of resource input. Within the context of data systems, this often translates to maximizing computational results, data throughput, or storage density relative to inputs like energy (e.g., FLOPS/Watt, queries/Joule), time (algorithmic complexity), or physical space (bits/cm^2). Efficiency improvements typically target specific processes or components, aiming to minimize waste and optimize performance according to well-defined, often technical, metrics. It represents an optimization *within* a given system or process boundary, focusing on "doing the thing right" in terms of resource utilization for a predefined task.

While enhanced efficiency can contribute positively to environmental sustainability by reducing resource intensity per unit of activity, the two concepts are distinct and their relationship is complicated by systemic feedback mechanisms, most notably the Jevons Paradox. First described by William Stanley Jevons in relation to 19th-century coal consumption [17], the paradox observes that increasing the efficiency with which a resource is used tends to increase, rather than decrease, the overall rate of consumption of that resource. This counter-intuitive effect arises because efficiency gains lower the effective cost of utilizing the resource or service, thereby stimulating increased demand, new applications, and broader adoption, which can ultimately outweigh the per-unit savings [1,31]. In the context of data, dramatic improvements in computational, storage, and network efficiency have lowered the cost of data-intensive activities, potentially fuelling exponential growth in data generation (e.g., high-resolution video, complex Internet of Things (IoT) systems), complex Artificial Intelligence (AI) model training, and data transmission, leading to rising aggregate energy consumption in the Information and Communication Technology (ICT) sector despite per-unit efficiency gains [21,38].

Achieving data sustainability therefore requires a holistic framework that integrates lifecycle perspectives, considers multiple dimensions (environmental, social, economic), employs tools like renewable energy adoption and circular economy principles, addresses ethical considerations, and explicitly accounts for potential systemic interactions and demand-side responses [31]. In the context of agriculture, this could entail promoting the use of open data standards to facilitate data sharing and interoperability, while also addressing concerns about data privacy and security to ensure the responsible use of agricultural data.

1.2 Resilience of Data Practices

Resilience, while sometimes receiving less focus than Efficiency or Sustainability, represents a fundamental characteristic of any information system through which data passes. It describes the system's capacity to anticipate and react to disturbances. These disturbances may be internal, such as data corruption or software errors, or perhaps more importantly from a security standpoint, external, like power outages, network connectivity problems, or cyberattacks. Resilience affects both Efficiency and Sustainability, often in divergent ways; for example, a highly resilient system might consume more resources, thus being less efficient, yet potentially prove more sustainable due to its longevity. Key terms associated with resilient systems include dependability, fault tolerance, and adaptability in the face of change or failure.

The theoretical aim is a system achieving the most favourable balance across all coordinates depicted in Fig. 1. In such a scenario, data practices would be engineered to avoid causing any noteworthy decrease in performance along these axes, maintaining a system that is both responsive and secure. Attaining this state, however, presents considerable difficulty and often involves substantial costs, which can, in turn, severely constrain the long-term sustainability of the system itself. Nonetheless, improvements are attainable through a thorough grasp of Data Sustainability principles and the precise identification of system weaknesses. This identification process can follow from measuring relevant sustainability metrics across the system and evaluating which areas present opportunities for the greatest potential enhancement with comparatively modest effort.

1.3 Methodology and Approach

This paper adopts a conceptual and analytical approach to explore DS in smart agriculture. Our methodology comprises three main components:

(a) *Literature Synthesis and Conceptualization:* We conducted a comprehensive review of existing literature on data sustainability, digital agriculture, data lifecycle management, and related concepts (efficiency, resilience) to synthesize current understanding, identify key challenges, and inform the theoretical underpinnings of our proposed framework. This synthesis forms the basis for defining DS, outlining the agricultural data lifecycle (Sect. 2), and identifying key challenges (Sect. 3).

(b) *Framework Development:* Based on the literature synthesis, we propose a novel conceptual framework for quantifying Data Sustainability. This involves defining key dimensions (relevance, utility, criticality, age, and cost) and formulating a conceptual mathematical model (Eq. 1). The development of this framework is the primary original contribution of this paper.

(c) *Contextual Illustration and Qualitative Analysis:* To demonstrate the practical applicability and nuances of the proposed framework, we provide a qualitative analysis of data sustainability challenges and strategies within the context of Romanian agriculture (Sect. 5). This section is based on the authors' extensive collective experience and observations within the Romanian agricultural sector, complemented by general agricultural statistics and broader academic literature on agricultural modernization and technology adoption in similar contexts. It serves as an illustrative application to highlight how the conceptual framework's dimensions and challenges manifest in real-world agricultural settings, rather than an empirical study with a specific sample size or primary data collection.

This multi-faceted approach aims to bridge theoretical conceptualization with practical relevance, providing a structured basis for future empirical research and policy development in agricultural data sustainability.

The remaining article is structured as follows: Sect. 2 explores the agricultural data lifecycle from creation to destruction, Sect. 3 addresses key challenges in achieving data sustainability in agriculture, Sect. 4 presents the framework for measuring data sustainability and its quantifiable dimensions, Sect. 5 examines data sustainability challenges across Romanian farm types, and finally Sect. 6 provides conclusions and future directions.

2   Data Lifecycle in Agriculture

To appreciate the concept of DS within the agricultural sector, it is helpful first to outline the typical Data Lifecycle (DLc) relevant to this field [7,33]. The DLc describes the journey of data, covering key stages such as its creation, methods of storage, its application or usage, and eventual disposal or archival. For a general representation of these stages, which also pertains to agricultural systems, see Fig. 3. Note that progression through each stage shown is not always required, steps may be optional depending on the specific circumstances.

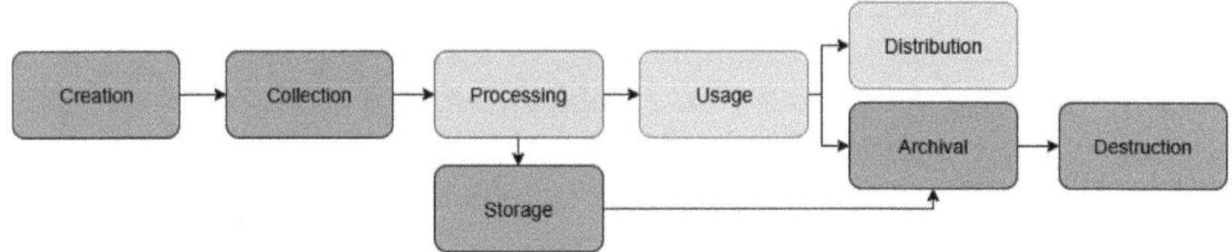

Fig. 3. General Data Lifecycle.

Data Creation: Creating, or generating useful data involves using various manual or automated strategies, which will potentially use various hardware components. In this stage we are interested in aspects like data precision/accuracy, noise levels, availability (the existence of missing values), all of these quantifiable via methodologies like the Data Quality Assessment Methodology (DQAM) proposed by Delinschi et al. [6].

The origins of such data are many. It can originate from in-field sensors monitoring environmental conditions like soil moisture or atmospheric temperature, or from wearable devices tracking livestock movement and well-being. Manual input remains a common source, encompassing handwritten logs, operator entries into machinery interfaces, or structured field scouting observations managed through spreadsheets.

Imaging technologies also contribute substantially, ranging from Unmanned Aerial Vehicles (UAVs) like drones capturing high-detail aerial views of specific plots, to satellites providing broader regional imagery, and fixed cameras observing livestock or monitoring farm areas. It is also important whether the data source is first-party, meaning directly managed and controlled by the user or organisation (like their own sensor network), or third-party, obtained from external entities such as weather service providers, satellite imaging companies, or other agricultural data brokers.

Data Collection: This stage concerns the consolidation of data gathered from disparate sources, which might be local to the farm, situated remotely, or provided by third parties. Each source type presents distinct benefits and drawbacks. Local data, once captured, can often be accessed promptly, with its primary cost related to the initial capital outlay for equipment decided by the farmer. Yet, the capabilities of the deployed hardware can substantially affect the resulting Data Quality (DQ). Limitations usually manifest as increased noise within the measurements, insufficient density of collection points, or data gaps caused by factors such as sensor malfunction or adverse environmental conditions.

Third-party data, when accessible, presents an alternative. It can often be procured (purchased or freely-available) according to specific quality benchmarks, or its quality can be evaluated prior to acquisition, generally offering access to datasets potentially meeting higher standards. However, these externally sourced datasets may include substantial amounts of irrelevant data or fail to capture certain necessary variables. Typically, datasets characterised by high quality and specificity command higher prices, a factor to weigh against the economics of collecting data using proprietary equipment.

Data collection logistics also present challenges, particularly concerning data transfer connectivity. Remote locations frequently suffer from poor or non-existent internet service, making data transmission difficult and potentially obliging the use of expensive satellite uplinks or even physical transportation of data storage media. These problems are amplified when dealing with numerous sites spread over a wide area, particularly when a large fraction of these sites are physically remote.

Data Processing: This step involves all transformations which are applied to data in order to make it useful or improve. It includes data cleaning, standardisation, aggregation, but also improvements like feature engineering. If data sources are diverse, this step will also include dedicated processes to ingest data, like Extract-Transform-Load/Extract-Load-Transform (ETL/ELT) pipelines. Ideally, the result of this step is a complete dataset, which is homogenous and useful for the topic it was gathered for.

Data Storage: This stage is responsible for persisting the data, both raw and processed, making it available for subsequent usage, distribution, and archival. There are several methods of data storage, including local storage (on the sensing device, local PC or local server), and cloud storage. Farm Management software components can be used to interface storage options with the rest of the system. Specific options exist, based on the type of data, including Relational Databases, Data Lakes, Data Warehouses, many of them supporting the addition of *metadata*, which aims to enrich the data source with information like source, units, time, location, or quality. The storage method is usually selected based on the scale of system (e.g.: local, regional, national).

Data Usage: Once transformed and persisted, data is used to extract information and actionable insights out of it. Dashboards are common overviews which require few resources and offer a rich overview of the farm, with the limitation that interpretation is left to the user. ML algorithms are used to forecast or predict future trends, helping the farmer make informed decisions and maintaining optimal conditions. More recently, Large Language Model (LLM) technologies enable extensive interpretation of farm conditions, complete with in depth suggestions of possible actions or mitigations measures. Data Usage is the step which shows the value of data, through the wealth of information it is able to extract and present out of it.

Data Distribution: This is an optional step, which may involve sharing raw data, processed data, or insights derived from data. Modern systems, for instance, leverage Federated Learning [9,10] to disseminate insights or models trained on a farm's data, without directly sharing the sensitive raw data itself. Other centralized systems might privately aggregate data to build ML models and generate statistics. It is important to distinguish between the distribution of raw or processed *data* and the dissemination of *knowledge* or insights derived from that data, as they involve different privacy and utility considerations.

Data Archival: Data archival can occur directly after processing for long-term historical records, or after active usage when data is no longer immediately needed for operational purposes. The long-term preservation of data not currently in active use, known as archival, serves several functions including enabling historical trend analysis, fulfilling regulatory compliance requirements, and supporting extended monitoring activities. It is often acceptable for archived data to possess lower granularity than active data. Consequently, techniques like lossy compression or data aggregation might be employed, summarising raw data into key statistical indicators, for example, *min*, *max*, *std_dev*, and *quartiles*. A critical aspect of this process is the judicious choice of aggregation intervals, balancing the goal of substantial storage space reduction against the need to retain meaningful information. Planning for data privacy is also important during archival to ensure data protection in case of unauthorised access attempts. This stage is shaped by the objectives of minimising storage expenses, strengthening data privacy, and preserving the informational value potentially required for future use.

Data Destruction: While the terms data *destruction* and data *deletion* are sometimes used with overlapping meanings, this paper specifically uses *destruction* to refer to the secure and irreversible removal of information that is either no longer required or cannot be legally retained (for instance, upon a user's request for data removal). The primary objective is the secure elimination of data from the system, especially if that data is considered sensitive or vulnerable [11]. For data assessed as having lower vulnerability, perhaps certain types of sensor readings in agricultural systems where subsequent overwriting is common and recovery poses minimal privacy risk, a standard operating system deletion process might be considered sufficient. However, for information demanding greater protection, more robust methods are necessary. These can range from employing secure deletion software protocols and cryptographic erasure techniques to the physical destruction of the storage media itself.

3 Key Challenges in Achieving Data Sustainability in Agriculture

Data sustainability in agriculture faces numerous challenges that span technical, organizational, environmental, and socioeconomic dimensions (see Fig. 4). Based on recent literature and industry practices, this section outlines the primary obstacles to achieving sustainable data management in smart and precision agriculture, drawing on examples and findings that are broadly applicable across various regions, including Europe and other global agricultural contexts. While specific nuances may exist between different countries or European Union (EU) states, the identified challenges represent fundamental issues in digital agriculture.

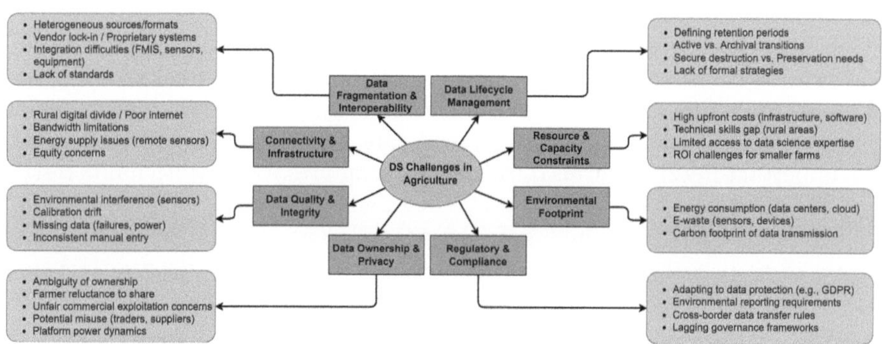

Fig. 4. Data Sustainability Challenges in Agriculture.

Data Fragmentation and Interoperability Issues: The agricultural data ecosystem is defined by **data sources** and formats, leading to integration challenges. As noted by Wolfert et al. [42], farm operations involve vendors, legacy systems, and formats, creating barriers. This prevents data between farm management systems and equipment, restricts of sensor networks with decision support tools, and data collected at scales. For instance, a farm might use soil moisture sensors, weather stations, and equipment telemetry, all of which produce data in forms that require integration.

Kernecker et al. [18] noticed that 73% of questioned farmers in Bavaria, Germany stated technical challenges as a key obstacle for adoption of digital agricultural technology, emphasising the importance of interoperability in precision agriculture. These difficulties included problems with data transfer between different platforms, software compatibility issues, and integration challenges between legacy equipment and newer digital systems.

A practical example is variable-rate application maps developed in one software system that cannot be correctly translated by fertiliser spreaders or irrigation systems, which limits the potential benefits of precision techniques. According to Janssen et al. [16], data structures and sharing methods is to overcome these constraints.

Promising approaches include development of open data standards and interoperability protocols specific to agriculture, as demonstrated by initiatives like ISOBUS for farm equipment data exchange and AgGateway's ADAPT framework for field operations data Harper et al. [15]. Edge computing solutions show potential to mitigate connectivity constraints by enabling local processing of time-sensitive data while optimizing transmission of aggregated insights Wolfert et al. [42]. This approach has proven effective in livestock monitoring systems that process behavioural data locally while sending only health alerts and summaries to cloud platforms.

Connectivity and Infrastructure Limitations: Rural digital infrastructure remains an important limitation for data sustainability in agriculture. Salemink et al. [30] identified the remaining digital gaps in agricultural regions worldwide, with significant implications. Unstable internet access restricts real-time data gathering and cloud-based processing, as seen in vineyard management systems where soil moisture and microclimate monitoring stations fail to provide important freezing alarms during network failures. Bandwidth limits restrict the transmission of high-resolution imaging and sensor data, enforcing compromises in data quality or timeliness. For example, drone-captured multispectral images of crop health sometimes requires downsampling or delayed transmission in places with weak connection, minimising its immediate value for decision making.

Community based digital infrastructure deployment models have successfully addressed connectivity challenges in several agricultural regions Rojo et al. [28], with farmer cooperatives in rural Brazil establishing shared wireless networks specifically dimensioned for agricultural data needs.

Energy supply constraints have an impact on the performance of remote sensing equipment, especially in expanding agricultural regions. Hackfort et al. [13] reported that solar-powered IoT sensor networks in rural agricultural regions have data gaps during cloudy times, affecting environmental monitoring continuity. These connection challenges particularly impact smallholder farmers and those in economically disadvantaged areas, raising fairness concerns about data-driven agricultural adoption.

Data Quality and Integrity Challenges: Agricultural data gathering under hazardous and dynamic circumstances presents particular obstacles to data integrity. According to DQAM proposed by Delinschi et al. [6], agricultural data dependability is impacted by various aspects. Temperature changes, humidity, and dust all cause interference with sensor readings in field deployments, as demonstrated with soil nutrient sensors, whose accuracy reduces under excessive wet conditions. Calibration drift with agricultural-deployed equipment is another difficulty, with studies indicating that weather stations in agricultural settings require recalibration 2–3 times more frequently compared to urban alternatives due to pesticide exposure and field operations.

Missing data from equipment failures or power failures creates discontinuities in time-series data, which is crucial for crop prediction and decision support. Weersink et al. [39] conducted a case study of irrigation management systems in California's Central Valley, showing data completeness rates as low as 78% during peak growing seasons due to equipment failures and maintenance issues. Inconsistent manual data input techniques increase quality difficulties, especially in processes like as pest monitoring and harvesting data collection when automated technologies are not completely integrated. These quality challenges affect the dependability of analytics and decision support systems, with poor data quality stated as a main cause for precision agricultural technology desertion among early users.

Data Ownership and Privacy Concerns: The question of **who owns agricultural data** remains contentious and poorly resolved. Wiseman et al. [41] documented growing farmer concerns regarding data ownership with multifaceted implications. Willingness to share data with third-party service providers has declined as awareness of potential data monetization has increased. For example, surveys of corn and soybean producers in the US Midwest revealed that 63% expressed concerns about sharing yield data with input suppliers, fearing it would be used to calibrate pricing algorithms. Commercial exploitation of farm data without fair compensation has emerged as farmers recognize that aggregated field performance data has significant market value when combined across regions.

Potential misuse of data by commodity traders, input suppliers, and landowners represents another concern, with documented cases where rental rates increased following adoption of precision agriculture systems that made yield potential more transparent. Bronson et al. [2] analysed the power dynamics in agricultural data ecosystems, finding asymmetric relationships between farmers and agribusiness corporations that control data platforms. In response, frameworks like that proposed by Chereja et al. [4, 5] offer approaches to privacy assessment in multi-layered ecosystem architectures, but implementation remains challenging in the complex web of agricultural data systems. **Federated learning** approaches that preserve data privacy while enabling collective intelligence offer particular promise for agricultural applications Erdei et al. [9]. Early implementations in pest and disease forecasting demonstrate the ability to develop robust predictive models without centralizing sensitive farm data.

Data Lifecycle Management: Effective management across the entire **data lifecycle** presents significant challenges in agricultural contexts. According to Demestichas et al. [7], determining appropriate data retention periods for different types of agricultural data proves difficult when balancing immediate operational needs against long-term value. For instance, high-resolution imagery for crop scouting may have short-term tactical value but also contribute to multi-year analyses of field productivity patterns. Managing the transition between active use and archival storage creates practical difficulties, particularly for seasonal data that may be dormant for months but critical during specific periods.

Ensuring appropriate data destruction while preserving historical records needed for long-term analysis requires nuanced governance approaches rarely implemented in agricultural settings. Thompson et al. [34] found that most US commercial farms lack formal data lifecycle management strategies, resulting in accumulation of redundant data and potential loss of valuable historical information. A case study of dairy operations revealed that while daily milk production data was meticulously preserved, corresponding feed composition data was routinely discarded after regulatory retention periods expired, compromising the value of longitudinal performance analysis.

Resource and Capacity Constraints: Implementing sustainable data practices requires resources that may be unavailable to many agricultural operations. Klerkx et al. [19] identified several **resource-related barriers** with practical consequences. High upfront costs for data collection infrastructure and software create adoption thresholds, as exemplified by precision irrigation systems that require substantial investment in soil moisture sensors, weather stations, control systems, and analytics platforms before delivering return on investment. A technical skills gap in rural communities limits effective utilization of available technologies, with studies showing that many farmers utilize less than 30% of the functionality in farm management information systems due to training limitations.

Limited access to specialized expertise in data science and analytics prevents many agricultural operations from extracting full value from collected data. The shortage of agronomists with strong data analytics skills has created bottlenecks in service delivery, with wait times for specialized precision agriculture consulting exceeding six months in some regions during critical planning periods. For smallholder and medium-sized farms, Eastwood et al. [8] found that the economic benefits of advanced data systems often fail to justify their costs, creating adoption barriers to sustainable data practices.

Environmental Footprint of Data Systems: The energy consumption and environmental impact of agricultural data systems present a paradox for sustainability efforts. Vanloqueren et al. [35] documented that **data centers and**

cloud computing services consume significant energy, with implications for agriculture's carbon footprint. A lifecycle assessment of precision viticulture systems found that server infrastructure and data transmission accounted for 18% of the total carbon footprint, partially offsetting the environmental benefits achieved through more precise resource application. Electronic sensors and devices contribute to e-waste, with the growing deployment of IoT devices in fields creating end-of-life disposal challenges, particularly for sensors containing heavy metals or batteries.

The carbon footprint of data transmission networks can offset environmental gains from optimized field operations. For example, continuous transmission of high-definition video feeds from autonomous agricultural equipment can generate more emissions through network infrastructure than are saved through optimized equipment routing. This creates tension between data-intensive precision agriculture approaches and environmental sustainability goals. Rose et al. [29] argue that lifecycle assessment of agricultural data systems must account for these environmental costs alongside benefits to ensure genuine sustainability improvements.

Regulatory and Compliance Challenges: The evolving regulatory landscape creates additional complexities for agricultural data management. Wiseman et al. [41] highlighted emerging **compliance challenges** affecting the sector. Adapting to data protection regulations such as GDPR in Europe has proven particularly difficult for agricultural data platforms that were not designed with privacy-by-design principles. For instance, farm management systems that integrate worker productivity tracking have faced legal challenges in jurisdictions with strong labor data protection provisions. Meeting increasingly stringent requirements for environmental reporting demands more sophisticated data management practices, as evidenced by the expanded data collection requirements for carbon footprint certification in agricultural export markets.

Navigating cross-border data transfer restrictions introduces further complexity for multinational agricultural operations and supply chains. A case study of seed production operations spanning North America and South America revealed significant data integration challenges arising from divergent national regulations on agronomic data transfer and storage. Reins et al. [27] noted that agricultural data governance frameworks often lag behind technological developments, creating uncertainty for stakeholders seeking to implement sustainable data practices.

Collaborative governance frameworks balancing innovation with equity and ethics continue to evolve Bronson et al. [2], with several agricultural technology providers adopting transparent data usage policies in response to farmer concerns.

4   Sustainability Dimensions in Agriculture. Making Data Sustainability Measurable

The combination of agricultural techniques and data management brings unique problems and opportunities for achieving sustainability. As agriculture grows more dependent on data-driven decision-making, understanding how to analyse and evaluate the sustainability of these data systems is crucial. This section presents an approach for conceptualising and proposing a quantifiable framework for assessing data sustainability in agricultural environments, defining characteristics that enable systematic assessment and development. Making data sustainability measurable allows stakeholders to better match technical implementations with larger ecological, economic, and social sustainability goals. The next subsections investigate both generic data dimensions applicable across domains and agriculture-specific aspects, which combined offer a complete approach to evaluating data sustainability in agricultural systems (Fig. 5).

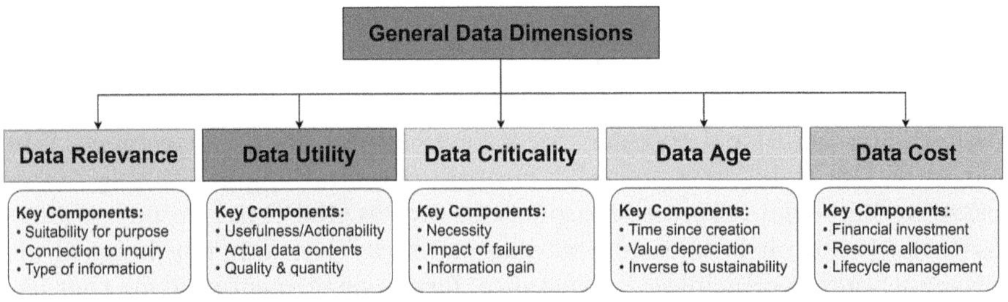

Fig. 5. Hierarchical Framework of General Data Dimensions.

4.1   General Data Dimensions

In this section, we will explore the generalized metrics, referred to as *dimensions* (encompassing various data qualities or characteristics), which are important in the process of quantifying a DS Index metric. Being generalized, no specific or exact metric is proposed, these being chosen at the moment of system design, however, we propose some guidelines, which will aid in choosing the right metrics for that particular system.

Data Relevance relates to how suitable or applicable data is to a defined purpose, such as addressing a specific question, completing a task, or informing a decision. Data possesses relevance if it has a direct connection to the matter being investigated. It is important to clarify that, in this framework, relevance centres on the suitability of the *type* of information collected (what is being measured), as opposed to analysing the pertinence of the specific data *contents* (the actual values obtained).

Key questions: *What is the alignment score of the dataset with farm objectives? What percent of data fields are used in key decision models?*

Data Utility refers to the *usefulness* or *actionability* of data. It describes whether the data, in its current form or after processing, can be effectively used to achieve a goal, improve a process, or enable a function. This concept complements Data Relevance; whereas relevance pertains to the type of information collected (what is measured), utility is determined by the actual data *contents* (the specific values). Both Data Quality and Data Quantity bear considerable influence on the attainable utility of a dataset.

Key questions: *What is the correlation of that data field with yield improvements? What is the reduction in resource use (water/fertilizer) of said data field? What is the actionability score based on farmer feedback?*

Data Criticality relates to two key aspects: firstly, the *necessity* of particular data for a specific function, system, or decision-making process, and secondly, the *severity of impact* should that data become unavailable, inaccurate, or compromised. This implies that the data is highly correlated with the end-result, or that the Information Gain is sufficiently high. Within agriculture, data frequently exhibits high criticality. Failures affecting data integrity or availability, especially within highly automated and mechanised systems, possess the potential to cause substantial negative consequences, possibly extending to crop loss.

In agriculture, depending on the time, data criticality might increase, decrease, or even spike. For example, in the winter, data might become almost non-critical, so it will loose its value, while in summer or near harvest, data criticality will be at its highest. This temporal dependency will result into a conceptualisation of a *real-time* or *context-aware* DS score (see Sect. 4.2).

Key questions: *What is the estimated cost of system downtime due to data unavailability (e.g.: cost of lost crops)? What is the dependency level of automated systems on this data?*

Data Age denotes the period elapsed from the moment a particular data entry was created. A core assumption within this framework is that data typically becomes less valuable as its age increases. Consequently, an inverse proportionality is understood to exist between Data Age and Data Sustainability.

Despite its apparent simplicity, each data type is subject to a concept of degradation, analogous to *half-life* in physics (symbol $t_{1/2}$) [22], in our case representing the time taken for data to lose half its value. This degradation, visually represented in Figure 6, directly informs the inverse proportionality of Data Age (D_{Age}) within the conceptual sustainability framework: as data ages, its value and contribution to overall sustainability diminish. The speed at which data degrades is variable, some become obsolete more rapidly than others. This allows for the following possible classification:

- *Very Short Half-Life Data (Seconds/Minutes)*: Data from real-time sensors used for immediate control, e.g., GPS corrections for auto-steer systems, valve statuses in irrigation systems, and current drone altitude. Such data is highly valuable at the point of collection, but its usefulness diminishes extremely rapidly;

- *Short Half-Life Data (Hours/Days)*: This includes daily weather forecasts used for operational planning, up-to-date soil moisture readings to guide irrigation, and alerts from pest traps checked on a daily basis. This data is useful for making decisions either immediately or in the very short term, but it is soon superseded by more recent information;
- *Medium Half-Life Data (Weeks/Months/Season)*: Examples include weekly growth stage reports, cumulative rainfall totals across the season, the results of soil nutrient tests used to plan seasonal fertiliser application, and imagery from drones used to spot issues mid-season. Data of this kind is helpful for tactical decisions made during a specific growing season or cycle, but unless it is collated or used for comparison, it is of limited use for planning future cycles;
- *Long Half-Life Data (Years/Decades)*: Examples in this category are maps of historical yields, baseline soil composition analyses, climate data covering long periods that can be used for trend analysis, genetic details of crop types and livestock, records showing farm boundaries, and land purchase documents. This retains its value for strategic planning, long-term trend analysis, benchmarking, for research, or to meet regulatory needs over considerable periods.

It should be noted that the degradation of the same data will differ across different applications, which can prove to be useful although old. Data storage policies can adjust for this, taking into account the potential for the data to be used in the future.

Key question: *Should the data field be weighted by data degradation/half-life? At what factor?*

Data Cost signifies all the cumulative financial and resource investments related to managing data throughout its lifecycle. These costs span various activities, such as acquiring or generating the data (e.g., purchasing datasets, deploying hardware), collecting and transmitting it (e.g., network charges), storing it (e.g., server or cloud fees, electricity), processing and analysing it (e.g., computation time, software licenses), ensuring its quality and security, meeting regulatory requirements, and performing secure deletion or long-term archiving.

Key questions: *What is the total lifecycle cost estimation (hardware amortisation + energy + licenses + personnel time + any recurrent costs like communication or cloud storage)?*

4.2   Relation Between Dimensions

The dimensions of Data Sustainability, as outlined previously, are intended to provide a comprehensive framework for analysis. While acknowledging that a complete enumeration of all relevant factors may be unattainable, these dimensions offer a structured approach to assessing Data Sustainability, particularly within the context of agricultural applications. The derived metrics are significant because their values directly inform decisions concerning data management practices. For instance, a low score on the DS metric might trigger the development of improved policies aimed at improving data discoverability and retrieval. A negative score might trigger a complete re-design of the platform, or abandoning of the current project or application. Such policies and standards are essential components of effective Data Governance [40].

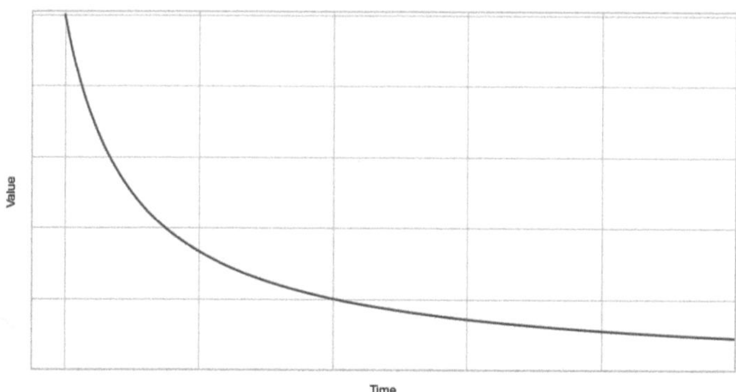

Fig. 6. Data intrinsic value decay over time (Data half-life, $t_{1/2}$).

Equation 1 formalizes a conceptual model for quantifying Data Sustainability (D_{Sust}). This model proposes that data sustainability is directly proportional to its inherent value and importance, and inversely proportional to the resources required to manage it and its temporal relevance. Specifically:

- Data Relevance (D_{Rel}), Data Utility (D_{Util}), and Data Criticality (D_{Crit}) are placed in the numerator because they represent the positive contributions of data. Higher relevance, utility, and criticality imply greater value derived from the data, thus enhancing its sustainability. For instance, data that directly supports critical farm decisions (high D_{Crit}) or significantly improves yields (high D_{Crit}) is inherently more sustainable as it delivers greater impact for its existence.
- Data Cost (D_{Cost}) and Data Age (D_{Age}) are placed in the denominator because they represent diminishing factors. Higher costs (financial, energy, resources) reduce sustainability, as more investment is required per unit of value. Similarly, as data ages, its value typically degrades (as illustrated in Figure 6), making older data less sustainable unless it retains specific long-term value.

$$D_{Sust} = \frac{D_{Rel} \cdot D_{Util} \cdot D_{Crit}}{D_{Cost} \cdot D_{Age}} \tag{1}$$

This inverse proportionality for D_{Age} is also highlighted in Fig. 6, where the non-linear relation between the value of data and time is apparent. The current model serves as a heuristic framework, highlighting inherent trade-offs which are present in any data source. The equation underscores that sustainable data practices involve not just maximizing the utility and value derived from data, but also actively managing and minimizing the associated costs and critically evaluating the continued need for ageing data.

It is important to acknowledge that this is a conceptual simplification. Precisely quantifying each dimension poses significant challenges and would likely depend heavily on the specific context and application. Context-specific weights (w_i) can be added to each of the members of the equation, enhancing the importance of any item. This could prove useful for situations where one metric is more important than others, for example farms might focus on *Criticality*, while research entities might favour *Utility*. Furthermore, the model assumes a certain level of independence between factors and may not fully capture pertinent aspects like data quality, completeness, ethical considerations, or compliance mandates, although these could be integrated by refining the definitions of the primary dimensions (e.g., low quality reduces D_{Util}, high ethical risk increases effective D_{Cost}). Despite these limitations, the model provides a valuable structure for discussing, comparing, and designing more sustainable data management strategies in agriculture.

5   Data Sustainability in Agriculture, an Illustrative Application for Romanian Farmers

Data sustainability is a crucial issue for Romanian agriculture as it deals with the challenges of technological advancement, climate change, and market integration. This section focusses at practical techniques for enhancing data sustainability through Romania's diverse farming setting, recognising that solutions must be adapted to the unique problems and resources of each farm type [24]. By addressing the particular challenges that family farms, agricultural SMEs, and large-scale enterprises encounter, we can propose practical approaches to more sustainable and successful data practices. These enhancements not only promise increased operational efficiency and economic sustainability, but also contribute to larger aims such as environmental stewardship and rural development.

5.1   Farm Types in Romania and Their Data Sustainability Related Challenges

Romania's agricultural sector presents a complex tapestry [3], characterized by its significant potential, diverse farm structures, and the ongoing transition towards modernization within the European Union framework [25]. Examining data sustainability within this context reveals distinct challenges and opportunities tied to the specific operational realities of different farming archetypes. For this analysis, we categorize Romanian farms into three primary groups: small family farms, local agricultural SMEs, and large-scale commercial operations [32]. The characterizations and associated data sustainability challenges described in the following subsections are derived from the authors' experience and observations within the Romanian agricultural landscape, corroborated by general agricultural trends, national statistical data (as referenced in Table 1), and broader literature on agricultural technology adoption in similar European contexts.

Small Family Farms: These farms typically represent the backbone of rural Romania, often operated by 2–4 family members on small or fragmented plots, generally under 10 hectares. Many operate at a subsistence or semi-subsistence level, contributing significantly to local food security but often facing economic precarity [24,26]. These farms are generally facing some of the following DS challenges:

Minimal Data Generation: Farming practices are heavily reliant on traditional knowledge passed down orally. Decision-making is experience-based rather than data-driven. Manual tools and older, small-scale motorized equipment lack automation or data-logging capabilities. Consequently, critical operational data (soil conditions, precise input usage, yield variations within plots) is rarely known or captured systematically.

Severe Access Barriers: Financial constraints are a primary obstacle, preventing investment in even basic sensors or digital tools. Coupled with limited access to modern know-how, often due to age, educational background, and a significant language barrier (most research and tech documentation is in English), farmers struggle to access or understand relevant external data sources (e.g., weather forecasts beyond basic media reports, market prices, research findings). Reliable internet connectivity can also be a challenge in more remote rural areas.

Usability and Literacy Gap: Even if data were accessible, a lack of digital literacy and familiarity with data interpretation tools would hinder its effective use. Concepts like precision agriculture (albeit somewhat intuitively applied sometimes) are largely alien. Data required for accessing certain support schemes might be perceived as a bureaucratic burden rather than a management tool.

Environmental Data Void: Dependence on weather patterns is profound, yet there's little capacity to leverage predictive weather data or monitor microclimatic conditions. For small greenhouses, low-cost IoT solutions for climate control and monitoring are often unknown or perceived as too complex or expensive, leading to inefficient resource use (water, energy). This lack of environmental data hampers adaptation to climate change and efforts towards demonstrable sustainable practices.

Long-term Viability: Without data capture and utilization, these farms struggle to improve efficiency, document sustainable practices (often inherent in traditional methods but not formally recorded), or demonstrate creditworthiness, threatening their long-term economic and social sustainability within the evolving agricultural landscape.

These aspects will adversely affect all presented metrics, with low values for D_{Rel}, D_{Util} and D_{Crit}, resulting in very low *perceived* values for DS, which translates in the extremely low adoption rates (almost non-existent). Of course, this aspect negatively affects yields and any growth potential which might exist.

Local Agricultural SMEs. These farms operate on a moderate scale (typically 10–100 hectares), employing 10–20 local staff. They represent a crucial segment bridging traditional practices and commercial agriculture, often run by entrepreneurs or families with a more business-oriented approach. For these farms, DS challenges include:

Inconsistent Data Generation and Integration: While potentially using more modern machinery than family farms, data generation might be inconsistent. Some tractors or combines may have basic sensors, but the data isn't always collected, stored centrally, or integrated with other farm management information (if present). Investment in a comprehensive Farm Management Information System (FMIS) is often constrained by budget.

Constrained Access to Expertise & Advanced Data: Access to modern know-how and data analysis capabilities often depends on hiring specialized personnel (e.g., an agronomist with digital skills), which significantly increases operational costs. Access to advanced precision agriculture services (e.g., satellite imagery analysis, variable rate application maps) is often limited by cost.

Subsidy-Driven Data Practices: Reliance on national and EU subsidies can shape data practices. Data collection might be geared primarily towards fulfilling reporting requirements for subsidies (e.g., plot declarations via APIA - Agency for Payments and Intervention in Agriculture), rather than optimizing operational decisions or long-term strategic planning.

Reactive Data Use: While awareness of weather dependency exists, the use of predictive analytics or integrated weather data for proactive decision-making (e.g., optimizing irrigation schedules, pest control timing) might be limited by the cost and complexity of relevant tools and data services.

Data for Market Access: As these farms often aim for broader markets, the lack of robust data capture for traceability or quality assurance can be a barrier to accessing more lucrative contracts or certifications.

In their case, perceived advantages might be insufficient when compared to costs (D_{Cost}), farmers being reluctant to invest heavily in data-driven platforms. This also leads to limited growth when compared to potential, as optimised and sustainable practices are not used to the highest potential.

Large Regional/National Farms. Operating on vast tracts of land (often hundreds or thousands of hectares), these entities employ significant workforces (>50–100 people) and utilize advanced, large-scale machinery. They are key players in national commodity production and exports. Their DS challenges include:

Data Overload and Integration Complexity: These farms generate vast amounts of data from diverse sources (telematics from machinery fleets, IoT sensors, drones, weather stations, ERP systems). The primary challenge shifts from data generation to effective management, integration, and analysis. Data silos between different systems and equipment brands are common, hindering a holistic view of operations.

Skilled Personnel Gap: While possessing advanced technology, finding and retaining personnel skilled in operating and interpreting the data from this technology is a major issue. There's also a need for data analysts and IT specialists familiar with agricultural contexts, who are scarce and expensive. The difficulty in finding reliable seasonal manual labour also indirectly impacts data quality if certain tasks (like selective harvesting or manual data collection points) cannot be executed properly.

High Operational Costs and Data Return on Investment (ROI): Significant investments in technology and high running costs necessitate demonstrating clear ROI from data-driven practices. Optimizing logistics, input application, and predictive maintenance using data is crucial but complex to implement effectively across large areas.

Data for Efficiency and Loss Reduction: Significant post-harvest losses can occur due to storage limitations or logistical bottlenecks. Real-time data on yields, storage conditions, and market demand could optimize harvesting schedules and logistics, but requires sophisticated data infrastructure and analytics.

Data for Compliance and Market Differentiation: The need to use crop protection products often conflicts with the growing market demand for organic (BIO) or sustainably certified produce. Precise, verifiable data on input usage, soil health, and biodiversity indicators becomes critical for minimizing chemical use, justifying practices to regulators, and potentially achieving premium certifications. This requires robust data tracking and traceability systems.

These farmers fully understand the potential of data-driven agriculture and invest in extracting the full potential from their available land and conditions. They actively improve relevance and utility, by researching new ways to use available data and, consequently, improving its DS.

5.2   Overview of Farm Scales in Romania

Table 1. Romanian farm statistics in 2023 (Source: INS [32])

Farm Type	Avg. Surface	Avg. Workers	Percent from Total Count	Percent from Total Surface Area
Family Farms	4–7 ha	2–4	90–92%	20–25%
Local Agricultural SMEs	10–100 ha	10–20	6–7%	25–30%
Large Regional/ National Farms	>100 ha	>30	>1%	~50%

Table 1 highlights the highly fragmented structure of Romanian agriculture, with a vast majority of small family farms operating on a limited share of the total surface area. This structural disparity implies a significant digital divide across farm types in Romania, challenging the homogenous application of data sustainability principles. Fundamental issues relating to access (cost, infrastructure, language), digital literacy, and the availability of context-specific affordable solutions persist. While larger farms contend with managing and utilizing complex data streams, smaller farms often struggle with basic data generation and access. This disparity limits their capacity to adapt, compete, and demonstrate sustainable practices effectively.

5.3 Improving Data Sustainability Practices in Romanian Farms

Small Family Farms: Often the most persistent issue here is that the farmers will refuse to acknowledge the needs, also refusing to take improving action. However, if this barrier is overcome, some further actions should focus on immediate, practical problems and observable gaps. Consultants might be employed to assist and help plan for farm digitalisation. Simple record keeping strategies, like free apps, spreadsheets, can be fast to setup and relatively easy to use, also with low to non-existent costs. Geo-tagging and photo documentation can also be relatively easy to do, since the smartphones today have become ubiquitous.

Consultants might suggest some training and accessing relevant information sources: digital literacy courses, improved weather forecasting, accessing pest and disease forecasts/announcements. Potential funding opportunities might also be suggested. All these practices will increase DS and ultimately the entire farm sustainability and competitiveness.

These actions aim to improve the three positive metrics (D_{Rel}, D_{Util} and D_{Crit}), while maintaining D_{Cost} as low as possible. These actions ensure useful exploitation of farm potential and plan for future growth.

Local Agricultural SMEs: These farmers usually have a basic understanding of data importance and also have some basic data collection processes. Improvements might imply overviewing existing strategies, performance analysis of the processes, evaluating technology use, and setting up an improvement plan and strategy. These improvements might include the adoption of dedicated FMIS software and basic optimisations regarding their work. Sustainability of these practices is usually secondary to optimising profit generation, which will aid in the farm's survival on the market.

From the DS standpoint, improvements would include setting up clear policies for Data Management and Usage, implementing already-known sustainable practices. Satellite and drone imagery might be useful, if any of the farm personnel understands them, or a dedicated person can be found nearby. To conclude, these types of farms would benefit the most from understanding what they can currently do with the available resources and fully utilise those. Depending on the revenue stream, some improvements could be done which would increase the amount of data which is processed and help further understand possible actions to improve the farm.

At this scale, improvements in the metrics are easy to observe, with a positive impact being visible at any stage, with most modifications. Farm practices tend to align to those of larger scale farms, diversifying farm products (e.g.: garbage turns into composts which can also be sold), resulting in more sustainable and greener farms.

Large Regional/National Farms: these types of farms usually have well defined Data Management and Data Governance policies and practices, and possibly even a well-documented FMIS system installed. They fully understand the importance of data in their operations, managing to fully integrate data analytics in the decision-making process. This process is optimised and tailored to their specifics, leaving little room for further improvement. They access markets at a national level, establishing themselves also in the international landscape, with products which have predictable high quality.

In terms of improvements, these farms can align to EU's Common Agricultural Policy (CAP) and the Farm Sustainability Data Network (FSDN). Their options are aligning with international formats, enabling them access and integration with the latest cutting-edge research results. The CAP has been designed with sustainability in mind, and is fully aligned with the EU's Green Deal. It aims to create a balanced system that supports farmers economically, ensures food availability for consumers, fosters vibrant rural communities, and promotes environmentally sustainable agricultural practices across the EU, with significant financial resources directed towards achieving these goals. Any improvements will directly translate in lower costs and higher payouts to employees, raising the quality of life in the region.

These actions will establish these large-scale farms on international markets, capable of offering high-quality products at competitive prices, all this while maintaining a green farm, with sustainable practices. Most of these farms will also be involved in active research, further raising the sustainability of their data. Some also choose to align their data to FAIR principles, making the data publicly available, in an anonymised format. This will offer the possibility to lower scale farms to use said data for their own optimisations.

6   Conclusions

This paper has explored the critical concept of DS within the agricultural sector, highlighting it as a vital practice for balancing the immense value derived from data with its associated lifecycle costs encompassing environmental, economic, and social dimensions. Sustainable data practices in agriculture require a holistic approach, integrating considerations across technology, governance, and economics, from initial data conception through to archival or deletion.

We have argued that achieving Data Sustainability necessitates moving beyond simply collecting vast amounts of data. It requires critical assessment, potentially guided by frameworks like the conceptual DS index discussed, weighing data's relevance, utility, and criticality against its cost (including energy footprint and hardware resources) and age-related decay. Implementing clear governance policies from the outset defining data ownership, privacy, access controls, and usage rights is foundational. These policies guide system engineers in designing and optimising agricultural data systems that are not only effective but also secure, compliant, and aligned with sustainability goals.

Furthermore, designing for sustainability encourages integrated and standardised data systems, rather than purely monolithic approaches, to manage complexity, ensure interoperability, and reduce long-term maintenance costs. Technical optimisations, particularly in energy consumption and hardware lifecycle management, are crucial levers for improving sustainability metrics. Such improvements directly contribute to the economic viability of data-driven agriculture, enhancing the ROI for farms of all sizes.

Ultimately, embracing DS in agriculture offers benefits beyond environmental stewardship and efficient resource management. By fostering cost-effective, robust, and user-friendly data systems, sustainable practices can significantly boost farm profitability and resilience. This enhanced economic footing can, in turn, contribute to broader socio-economic sustainability in rural areas through job creation and improved market access. Achieving truly sustainable data ecosystems requires ongoing innovation and commitment, but it holds the key to unlocking the full potential of data in agriculture responsibly and equitably for the future.

6.1 Future Directions

While this paper outlines key principles and challenges for DS in agriculture, significant opportunities remain for future research and development to advance this critical field. Several promising directions warrant further investigation:

Refinement of DS Metrics and Quantification into a full-fledged methodology: There is a pressing need for standardised methodologies and practical tools to accurately quantify DS in diverse agricultural contexts. Future work should focus on operationalizing conceptual frameworks, potentially refining the DS index proposed herein, to incorporate measurable environmental impacts (e.g., full lifecycle carbon accounting for data operations), economic costs, and social factors. Benchmarking the sustainability performance of different agricultural data pipelines (e.g., remote sensing vs. edge IoT networks) would also provide valuable insights for optimisation.

Development of Sustainable-by-Design Technologies and Practices: Innovation is required in hardware and software tailored for agricultural DS. This includes research into ultra-low-power sensors, energy-efficient edge computing hardware suitable for harsh farm environments, and novel data compression and transmission techniques optimised for agricultural data types. Furthermore, advancing energy-aware AI/ML algorithms (including Federated Learning and TinyML approaches) for common agricultural analytics tasks is crucial. Investigating next-generation, low-energy archival solutions for preserving valuable longitudinal agricultural data long-term is also essential.

Advanced Data Lifecycle Management Strategies: Research should explore more sophisticated and automated methods for managing the data lifecycle sustainably. This includes developing intelligent systems capable of assessing data relevance, utility, and predicting obsolescence to inform dynamic retention and deletion policies. Context-aware data minimization strategies, which adapt data collection intensity based on real-time needs and predicted value, represent another key research area. Developing adaptable frameworks for sustainable lifecycle management tailored to different scales of farming operations (from smallholders to large enterprises) is also needed.

Governance, Socio-Economic, and Ethical Dimensions: Further exploration is needed into effective and equitable data governance models for agricultural data sharing that foster collaboration while upholding sustainability principles. Investigating effective incentive structures (policy-based, economic) to promote the adoption of sustainable data practices among farmers and technology providers is vital. Moreover, deeper investigation into the socio-economic impacts, ethical considerations (including data sovereignty and farmer empowerment), and practical barriers to adopting sustainable data technologies on farms is required.

Integration and Co-optimization: Future studies should continue to explore the complex interplay and potential trade-offs between Data Sustainability, Efficiency, and Resilience within agricultural systems. Developing methodologies and frameworks for co-optimizing these dimensions, ensuring that efforts to enhance one aspect do not inadvertently compromise another, will be critical for building truly holistic and robust systems.

Other aspects, which can be addressed, could include research into the interplay between DS and data/algorithmic bias in agriculture, developing a practical DS assessment toolkits for farmers and consultants, investigating the role of open-source hardware/software and open data standards in promoting DS, particularly for smaller farms, and exploring the concept of *"data half-life"* specific to different agricultural data types to better inform the D_{Age} factor. This last aspect might be turned into a complete methodology for archival or distribution, which might drastically increase the *Utility* and *Relevance* of data.

Addressing these future directions will require a multi-disciplinary effort, bringing together expertise from computer science, agricultural engineering, environmental science, economics, and social sciences. Continued focus in these areas is paramount to ensuring that the data revolution in agriculture proceeds in a manner that is both innovative and fundamentally sustainable.

Acknowledgments. This work was supported by a grant of the Romanian National Authority for Scientific Research and Innovation, CCCDI - UEFISCDI, project number ERANET-M-3-ERANET-REPLACER 17/2024, within PNCDI IV. This work is also supported by the project "Collaborative Framework for Smart Agriculture" COSA that received funding from Romania's National Recovery and Resilience Plan PNRR-III-C9-2022-I8, under grant agreement 760070. This research has been supported by the CLOUDUT Project, cofunded by the European Fund of Regional Development through the Competitiveness Operational Programme 2014–2020, contract no. 235/2020.

References

1. Alcott, B.: Jevons' paradox. Ecol. Econ. **54**(1), 9–21 (2005). https://doi.org/10.1016/j.ecolecon.2005.03.020
2. Bronson, K.: Smart farming: including rights holders for responsible agricultural innovation. Technol. Innov. Manag. Rev. **8**(2), 7–14 (2018)
3. Burja, C., Burja, V.: Farms size and efficiency of the production factors in Romanian agriculture. Ekonomika poljoprivrede **63**(2), 361–374 (2016)
4. Chereja, I., Erdei, R., Delinschi, D., Pasca, E., Avram, A., Matei, O.: Privacy-conducive data ecosystem architecture: by-design vulnerability assessment using privacy risk expansion factor and privacy exposure index. Sensors **25**(11), 3554 (2025)
5. Chereja, I., Erdei, R., Pasca, E.M., Delinschi, D., Avram, A., Matei, O.: A privacy assessment framework for data tiers in multilayered ecosystem architectures. Mathematics **13**(7), 1116 (2025)
6. Delinschi, D., Erdei, R., Pasca, E.M., Matei, O.: Data quality assessment methodology. In: International Conference on Soft Computing Models in Industrial and Environmental Applications, pp. 199–209. Springer, Cham (2024)
7. Demestichas, K.P., Daskalakis, E.: Data lifecycle management in precision agriculture supported by information and communication technology. Agronomy (2020). https://api.semanticscholar.org/CorpusID:228998633
8. Eastwood, C., Ayre, M., Nettle, R., Rue, B.D.: Making sense in the cloud: farm advisory services in a smart farming future. NJAS-Wageningen J. Life Sci. **90**, 100298 (2019)
9. Erdei, R., Delinschi, D., Bărăian, I., Matei, O.: Aggregation strategy for federated machine learning algorithm. In: International Conference on Soft Computing Models in Industrial and Environmental Applications, pp. 157–167. Springer, Cham (2024)
10. Erdei, R., Delinschi, D., Matei, O.: Security centric scalable architecture for distributed learning and knowledge preservation. In: International Workshop on Soft Computing Models in Industrial and Environmental Applications, pp. 655–665. Springer, Cham (2022)
11. Erdei, R., Pasca, E., Delinschi, D., Avram, A., Chereja, I.: Privacy assessment methodology. In: The 19th International Conference on Soft Computing Models in Industrial and Environmental Applications SOCO 2024: Salamanca, Spain, 9–11 October 2024 Proceedings, vol. 2, p. 210. Springer, Cham (2024)
12. Garlock, T.M., et al.: Environmental, economic, and social sustainability in aquaculture: the aquaculture performance indicators. Nat. Commun. **15** (2024). https://api.semanticscholar.org/CorpusID:270637255
13. Hackfort, S.: Patterns of inequalities in digital agriculture: a systematic literature review. Sustainability **13**(22), 12345 (2021)
14. Hall, A., et al.: Evaluation of measures of sustainability and sustainability determinants for use in community, public health, and clinical settings: a systematic review. Implementation Sci.: IS **17** (2022). https://api.semanticscholar.org/CorpusID:254633835
15. Harper, A., Mishra, U., Shantharam, J., Zhu, C.: The internet of things and agriculture: how IoT can improve efficiency and reduce waste in the agriculture industry. J. Clean. Prod. **199**, 869–878 (2018)
16. Janssen, S.J., et al.: Towards a new generation of agricultural system data, models and knowledge products: information and communication technology. Agric. Syst. **155**, 200–212 (2017)

17. Jevons, W.S.: The Coal Question; An Inquiry Concerning the Progress of the Nation, and the Probable Exhaustion of Our Coal-Mines. Macmillan and Co., London and Cambridge (1865)
18. Kernecker, M., Knierim, A., Wurbs, A., Kraus, T., Borges, F.: Farmers' perspectives on field crop robots-evidence from Bavaria, Germany. Agriculture **10**(10), 441 (2020)
19. Klerkx, L., Jakku, E., Labarthe, P.: A review of social science on digital agriculture, smart farming and agriculture 4.0: new contributions and a future research agenda. NJAS-Wageningen J. Life Sci. **90**, 100315 (2019)
20. Martínez-Falcó, J., Sánchez-García, E., Millán-Tudela, L.A., Marco-Lajara, B.: The role of green agriculture and green supply chain management in the green intellectual capital–sustainable performance relationship: a structural equation modeling analysis applied to the Spanish wine industry. Agriculture (2023). https://api.semanticscholar.org/CorpusID:256860114
21. Masanet, E., Shehabi, A., Lei, N., Smith, S.J., Koomey, J.: Recalibrating global data center energy-use estimates. Science **367**, 984 – 986 (2020). https://api.semanticscholar.org/CorpusID:211554280
22. Muller, R.A.: Physics and Technology for Future Presidents: An Introduction to the Essential Physics Every World Leader Needs to Know. Princeton University Press (2010)
23. Nyfeler, J.K.: The three pillars of sustainability (2013). https://api.semanticscholar.org/CorpusID:211107069
24. Ogrezeanu, M., Mihai, M., Dumitraş, D.E., Stanca, L., Pocol, C.B.: Digital technologies on the farm: from the improvement of management practices and human resources to sustainability. ProEnvironment **18**(61), 1–11 (2025)
25. Popescu, A., Alecu, I.N., Dinu, T.A., Stoian, E., Condei, R., Ciocan, H.: Farm structure and land concentration in Romania and the European union's agriculture. Agric. Agric. Sci. Procedia **10**, 566–577 (2016)
26. Radulescuu, C.V., Popescu, M.L., Negescu, M.D.O., Bodislav, D.A.: Digital technologies applied in agriculture for sustainable development. Eur. J. Sustain. Dev. **8**(5), 75 (2019)
27. Reins, L.: Regulating New Technologies in Uncertain Times. Springer, Cham (2019)
28. Rojo, G., Salas-Olmedo, M.H., García-Palomares, J.C.: Community-based connectivity solutions for rural digital development: addressing last-mile challenges in remote agricultural regions. Land Use Policy **99**, 104868 (2020)
29. Rose, D.C., Wheeler, R., Winter, M., Lobley, M., Chivers, C.A.: Agriculture 4.0: making it work for people, production, and the planet. Land Use Policy **100**, 104933 (2021)
30. Salemink, K., Strijker, D., Bosworth, G.: Rural development in the digital age: a systematic literature review on unequal ICT availability, adoption, and use in rural areas. J. Rural. Stud. **54**, 360–371 (2017)
31. Sorrell, S.: Jevons' paradox revisited: the evidence for backfire from energy efficiency improvements. Energy Policy **37**(4), 1456–1469 (2009). https://doi.org/10.1016/j.enpol.2008.12.003
32. de Statistic, I.N.: Anchet Structural în Agricultur (2023). https://insse.ro/cms/ro/content/ancheta-structural-in-agricultur. Accessed 18 July 2025
33. Thompson, N.M., DeLay, N.D., Mintert, J.: Understanding the farm data lifecycle: collection, use, and impact of farm data on U.S. commercial corn and soybean farms. Precis. Agric. **22**, 1685–1710 (2021). https://api.semanticscholar.org/CorpusID:235508726

34. Thompson, N.M., Delay, N.D., Mintert, J.: Understanding the farm data lifecycle: collection, use, and impact of farm data on us commercial corn and soybean farms. Precision Agric. **22**, 1685–1710 (2021)
35. Vanloqueren, G., Baret, P.V.: How agricultural research systems shape a technological regime that develops genetic engineering but locks out agroecological innovations 1. In: Food Sovereignty, Agroecology and Biocultural Diversity, pp. 57–92. Routledge (2017)
36. Vasto, P.H.D., Moreno, V.M., Gálvez-Sánchez, F.J.: The three pillars of sustainability trends: a bibliometric analysis. Eur. Public Soc. Innov. Rev. (2024). https://api.semanticscholar.org/CorpusID:274335660
37. Vatin, N.I., Negi, G.S., Yellanki, S., Mohan, C., Singla, N.: Sustainability measures: an experimental analysis of ai and big data insights in industry 5.0. In: BIO Web of Conferences (2024). https://api.semanticscholar.org/CorpusID:267171886
38. Wang, H., Shi, D., Zhang, C., Ding, N., Cheng, C.: Digital transformation and visual knowledge map analysis of intelligent factory for sensor information of internet of things. Intell. Decis. Technol. **18**, 3437–3451 (2024). https://api.semanticscholar.org/CorpusID:275545090
39. Weersink, A., Fraser, E., Pannell, D., Duncan, E., Rotz, S.: Opportunities and challenges for big data in agricultural and environmental analysis. Ann. Rev. Resour. Econ. **10**(1), 19–37 (2018)
40. Wiseman, L., Jouanjean, M.A., Casalini, F., Gray, E.: Issues around data governance in the digital transformation of agriculture. Nineteenth-Century Literature (2020). https://api.semanticscholar.org/CorpusID:229209381
41. Wiseman, L., Sanderson, J., Zhang, A., Jakku, E.: Farmers and their data: an examination of farmers' reluctance to share their data through the lens of the laws impacting smart farming. NJAS-Wageningen J. Life Sci. **90**, 100301 (2019)
42. Wolfert, S., Ge, L., Verdouw, C., Bogaardt, M.J.: Big data in smart farming-a review. Agric. Syst. **153**, 69–80 (2017)

Automated Classification of Romanian Mammography Reports

Cristiana Moroz-Dubenco(✉) and Anca Andreica

Faculty of Mathematics and Computer Science, Babeş-Bolyai University, Mihail Kogalniceanu 1, 400084 Cluj Napoca, Romania
`cristiana.moroz@cbbcluj.ro`

Abstract. This study presents an automated system for classifying breast cancer diagnosis and BI-RADS scores based on free-text mammography reports written in Romanian. Utilizing a newly collected and publicly available Romanian Dense Breast Mammography Collection (RDBMC) dataset, we define two classification tasks: (1) assessing the patient's overall condition as healthy, benign, or malignant, and (2) predicting the severity of findings using the BI-RADS scoring system. To address the scarcity of research on Romanian medical reports, we experiment with both the original Romanian texts and their English translations. We evaluate four classifiers (Random Forest, Logistic Regression, Decision Trees, and Naive Bayes) combined with two text embedding methods (TF-IDF and Latent Semantic Indexing), incorporating patient age as an additional feature. Our results show that the Random Forest classifier with TF-IDF embeddings achieves the best performance for Romanian reports, with an accuracy of 80% in diagnosis classification and 73% in BI-RADS scoring, demonstrating robustness and potential clinical applicability. Although English translations yield slightly higher scores, differences are small, indicating that the system is effective across languages.

Keywords: Mammography Reports · Medical Text Classification · Romanian Language · Breast Cancer Diagnosis · BI-RADS Scoring

1   Introduction

According to the *Cancer Over Time* project [14], in 2020, breast cancer became the most prevalent type of cancer for the first time in history, surpassing lung cancer. Among women, it is not only the most frequently diagnosed form of cancer, but also the leading cause of cancer-related deaths, accounting for 15.5% of all cancer fatalities.

Although the incidence of breast cancer has increased over the last decade, the number of deaths has decreased due to progress in early diagnosis and effective treatment [3]. If cancer is detected early and remains confined to the breast, the 5-year relative survival rate reaches as high as 99%, based on data from the American Cancer Society [4].

© The Author(s), under exclusive license to Springer Nature Switzerland AG 2026
C. Chira et al. (Eds.): InnoComp 2025, CCIS 2793, pp. 219–242, 2026.
https://doi.org/10.1007/978-3-032-12478-4_14

One of the most important tools in the early detection of breast cancer is screening mammography, which involves capturing X-ray images of the breast to identify potential signs of malignancy. These images are typically accompanied by textual reports written by radiologists, which contain critical observations and diagnostic assessments. Analyzing these mammography reports plays a vital role in interpreting findings, tracking changes over time, and supporting decisions about further investigations such as biopsies [1].

Notably, a malignant tumor can appear on a mammogram up to three years before it becomes palpable [2], emphasizing the importance of regular screenings. Despite their clinical value, mammograms are not infallible – false positives and false negatives remain significant concerns, with reported sensitivity and specificity values of 86.9% and 88.9% respectively in the United States [18]. The diagnostic accuracy depends heavily on the radiologist's experience and the quality of the imaging equipment. To improve consistency and reduce the workload, radiologists increasingly rely on computer-aided detection (CADe) and diagnosis (CADx) systems.

In this context, automated analysis of mammography reports emerges as a complementary approach, helping to extract relevant historical data, highlight suspicious patterns, and support longitudinal evaluations – ultimately contributing to more informed clinical decisions. Mammography reports, which contain detailed textual descriptions of imaging findings, can be processed using document-level text analysis (DLA) techniques to automatically classify them. In a machine learning context, the DLA task is framed as a classic text classification problem, where the system assigns a single label to an entire document. The simplicity of this approach has encouraged significant research, especially in the early stages of automated medical text analysis.

However, a main limitation of document-level analysis is the assumption that a document, regardless of its length or content complexity, belongs to a single class. In mammography reports, this assumption does not always hold. A report often describes multiple abnormalities, potentially including both benign and malignant lesions, or findings from both breasts in the same document. This poses challenges for accurate classification and clinical interpretation.

To address this, we propose an approach for classifying mammographies from medical reports, based on the methodology presented in our previous work [7]. Considering that mammograms are typically performed annually, a patient may accumulate a substantial history of exams. When analyzing a new mammogram, comparison with previous exams is crucial to detect subtle changes. As a means of accelerating this process, our method explores the potential of classifying previous mammographies as benign or malignant based solely on the textual reports, without requiring access to the original images.

We define two types of classification tasks. The first aims to assess the patient's overall condition (or diagnosis) as *healthy*, *benign*, or *malignant*. The second targets the severity of the findings based on the BI-RADS (Breast Imaging Reporting and Data System) scoring. Our system diagnoses breast cancer and predicts BI-RADS scores from free-text mammography reports written in Romanian. To the best of our knowledge, no such study on Romanian reports exists in literature.

More broadly, there is a lack of research involving Romanian medical free-text data for clinical decision support, making our study one of the first to explore this direction. To assess the influence of language on classification performance, we also experiment with their English translations. Furthermore, we incorporate patient age into the models to evaluate the effect of demographic factors.

We compare the performance of two different embedding methods and four classification algorithms for predicting both the diagnosis and the BI-RADS score. Additionally, we analyze the impact of stopwords, patient age, and report language on classification accuracy. The proposed approaches achieve over 70% accuracy on the Romanian Dense Breast Mammography Collection (RDBMC)[1] dataset, demonstrating the feasibility and robustness of automatic mammography report analysis in low-resource language contexts. Our model is designed to handle medical terminology and variations in reporting style, while also accounting for linguistic and demographic nuances. In doing so, this work addresses the under-representation of non-English medical datasets and highlights domain-specific challenges in building robust automated diagnostic tools from free-text reports.

While our methodology does not introduce novel machine learning algorithms, our contribution is centered on establishing the first baseline for Romanian-language mammography report classification, a task that has been overlooked in previous research. Given the lack of existing resources and studies in this area, we intentionally employ classical, well-established machine learning methods, which offer the advantages of interpretability, simplicity, and low computational cost. These methods are more likely to be adopted in low-resource environments, such as smaller medical centers in Romania, where computational resources and large-scale annotated datasets are often unavailable. We believe this work provides a robust, transparent, and reproducible benchmark that will serve as a reference point for future research on Romanian medical free-text data.

2   Related Work

Recent advances in natural language processing (NLP) and machine learning have enabled the automated analysis of unstructured medical texts, including radiology and mammography reports. Several studies have focused on the classification of such reports to support early breast cancer detection, improve clinical workflows, and facilitate large-scale data analysis. This section reviews existing approaches to medical text classification, with a focus on applications in mammography, highlighting the challenges and methodologies relevant to our work on Romanian-language clinical data.

In [5], Boroumandzadeh and Parvinnia propose a novel method for automated BI-RADS scoring from mammography reports in English. The methodology consists of four steps: 1. *pre-processing* – removal of punctuation marks and stopwords, conversion to lowercase, and reduction to words' radicals; 2. *feature extraction* – employment of a combination between word2vec and TF-IDF techniques; 3. *feature engineering* – extraction of 20 features from the Hospital

---

Information System (HIS), such as lactation or pregnancy history, marital status, age, and selection of the most important 7 features, based on medical specialists' ranking; 4. *classification* – prediction of BI-RADS score using Extreme Gradient Boosting (XGBoost), Multi-Class Support Vector Machine (SVM), Multi-Level Fuzzy Min-Max Neural Network (MLF) and Naive Bayes (NB) algorithms.

The experiments, performed on a dataset collected from the Namazi Hospital and Saadi Hospital in Fars province, Iran and consisting of 5076 patients, have shown that MLF performs best, obtaining an accuracy score of 89%. Moreover, the experiments were repeated without using the HIS features, leading to lower performance, thus concluding that the usage of these features has a positive effect on the final classification.

Lopez et al. [19] advance a system with two main goals: 1. BI-RADS scoring, and 2. classification of high- and low-priority cases, to be used for both mammography and breast ultrasound reports in Spanish. First, the text is pre-processed by converting it to lowercase and removing all references to the BI-RADS classification so as not to bias the algorithms. Following, three BERT-based models that support Spanish were tested: XLM-RoBERTa [12], BETO [9] and Bio-BERT-Spanish on three datasets – one containing only mammography reports (27,683), one containing only breast ultrasound reports (53,733 records), and one containing both types (43,987). In terms of BI-RADS scoring, BETO achieved the best performance, between 76.05% and 77.59% for each of the datasets, while for the prioritization task, the Bio-BERT-Spanish model achieved a maximum accuracy of 91.02% for the combined dataset (mammography and ultrasound reports).

Castro et al. [10] also focus on two tasks: BI-RADS scoring and laterality classification of each BI-RADS category value from 2159 mammography, breast ultrasound, computed tomography and magnetic resonance reports in English. The study first evaluated a rule-based system (BI-RADS Observation Kit [23]) before building a trained classifier to improve performance. For BI-RADS annotation, a linear chain Conditional Random Fields model [17] is developed, while for lateratility classification, the Rules from Partial Decision Trees algorithm [15] is applied on the results yielded from the scoring task, reaching a weighted F1-score of 0.91.

Although recent studies on medical report classification for English and Spanish datasets have explored complex architectures such as BERT variants or domain-specific neural models, no prior work has addressed the automatic classification of mammography reports written in Romanian. Our study explicitly focuses on addressing this gap by establishing a reproducible baseline for this under-explored language using classical, interpretable machine learning techniques. This work provides a starting point for future advancements in Romanian-language medical NLP, where annotated resources and computational capacities remain significantly more limited than for high-resource languages.

3   Proposed Approach

In this section, we describe the methodology developed for the automated classification of Romanian mammography reports. Our approach is structured in three main stages: data preparation, data representation, and classification. First, we apply a series of preprocessing steps to normalize the reports and handle linguistic variations. Then, we represent the textual data using two different embedding techniques, capturing both semantic and syntactic information. Finally, we evaluate the performance of four classification algorithms on two tasks: diagnosis prediction and BI-RADS score estimation. This pipeline is designed to handle domain-specific medical language and adapt to variations in style, structure, and patient demographics, while remaining applicable in low-resource language settings.

The primary research question addressed in this study is whether commonly available text representations combined with classical classifiers can provide a solid and interpretable baseline for the automatic classification of Romanian-language mammography reports. This work contributes to the field by offering the first structured baseline results and dataset for this specific low-resource medical language setting, highlighting both the opportunities and current limitations.

3.1 Data Preparation

To enable a comparative analysis of language influence on classification performance, as previously stated, we evaluate our system on both the original Romanian reports and their English translations. The same pre-processing methodology, detailed below, is applied to both language versions.

In order to incorporate patient demographics, the age associated with each report is appended at the end of the text, allowing us to explore its potential impact on diagnostic outcomes.

We apply a standard text pre-processing pipeline. First, all reports are converted to lowercase to ensure consistency and reduce sparsity in the textual data. We also examine the effect of stopwords removal (i.e. filtering out commonly used words with little or no significance, such as *the, a, is* etc.) by using language-specific lists from the *advertools* library[2]. Experiments are conducted both with and without stopwords filtering to assess its influence on model accuracy.

3.2 Data Representation

Once the pre-processing step is complete, the textual data must be converted into a numerical format suitable for machine learning models. This transformation is achieved through embedding techniques, which represent each document as a fixed-length vector of numerical features. In our study, we evaluate two such methods: *Term Frequency-Inverse Document Frequency (TF-IDF)* and *Latent Semantic Indexing (LSI)*.

---

[2] https://github.com/eliasdabbas/advertools.

Term Frequency-Inverse Document Frequency is a widely adopted technique for transforming textual data into numerical vectors that can be utilized in machine learning models. As the name implies, it merges two core components: Term Frequency (TF), which reflects how often a specific word w appears within a document *doc*, and Inverse Document Frequency (IDF), which quantifies how rare that term is across the entire document collection.

TF is defined as the total count of a term within a document and is computed using the following formula:

$$TF(t, st) = \sum_{x \in S} fr(x, t), \tag{1}$$

where $fr(x, t)$ is the frequency function:

$$fr(x, t) = \begin{cases} 1 & \text{if x=t} \\ 0 & \text{otherwise} \end{cases} \tag{2}$$

The IDF component evaluates the uniqueness of a term by comparing its presence across all documents in the corpus. Its computation is given by:

$$IDF(t, S) = log(\frac{N}{st \in S : t \in st}) \tag{3}$$

This weighting scheme ensures that terms appearing in fewer documents receive a higher importance score, thereby reducing the influence of ubiquitous but semantically weak terms. In effect, TF captures term relevance within individual documents, while IDF adjusts for its significance across the corpus.

The final TF-IDF score is obtained by multiplying these two components:

$$TF - IDF(w, doc, S) = TF(w, doc) \cdot IDF(w, S) \tag{4}$$

Latent Semantic Indexing is a dimensionality reduction technique used to convert documents into fixed-length vector representations by uncovering latent patterns in word usage. It begins by constructing a document-term matrix, capturing word occurrences across the corpus, and then applies Singular Value Decomposition (SVD) to reduce this high-dimensional matrix into a lower-dimensional semantic space.

This reduction allows LSI to capture the underlying relationships between terms and documents, even when synonymous terms are used differently across reports. In our context, each mammography report–whether in Romanian or English–is represented as a numeric vector over a reduced set of latent features.

Formally, every document *doc* is embedded as:

$$\mathcal{F} = \{ft_1, ft_2, \ldots, ft_{size}\}$$

where $\mathcal{F}$ is the set of *size* features extracted from the text directly using latent features extracted from the original term-document matrix. Thus, the LSI-based representation of a document is:

$$doc^{LSI} = (doc_1^{LSI}, \cdots , doc_{size}^{LSI})$$

where doc_i^{LSI} ($\forall 1 \leq i \leq size$) corresponds to the i-th feature derived through the LSI process.

3.3   Classification

To evaluate the effectiveness of the TF-IDF and LSI embeddings for automatic classification of mammography reports in both Romanian and English, we train and test four commonly used machine learning classifiers: Decision Tree (DT), Random Forest (RF), Naive Bayes (NB), and Logistic Regression (LR). These models were chosen for their simplicity, interpretability, and proven efficiency in text classification tasks, especially in low-resource language settings. All four classifiers have previously been employed with promising results in Romanian-language document-level classification tasks [8,11,20,22].

Decision Tree (DT) [21] is a simple classifier, building/determining a set of if-then-else statements to categorize the samples into the specified classes. The tree is constructed by recursively dividing the dataset into smaller subsets according to the value of a specific feature or attribute. The dataset is partitioned to ensure that the resulting subsets are as homogeneous as possible in terms of the target variable. The process continues until each subset contains only a single class of the target variable or until a stopping condition is satisfied.

Random Forest (RF) [6] is an ensemble learning method for classification and regression tasks in machine learning. The random forest algorithm combines multiple decision trees that are trained on different subsets of the input features and training data, and aggregates their predictions to improve the overall classification accuracy and robustness. It uses bagging (bootstrap aggregating) to generate different subsets of the training data with replacement, and random feature selection to generate different subsets of the input features for each tree. This randomness helps to reduce the overfitting of the individual decision trees and improve their diversity and generalization performance.

Gaussian Naive Bayes (GNB) algorithm [16] is based on the Bayes theorem, with the assumption of strong independence between features (thus, the naivety) and a Gaussian distribution. It provides a simple, yet highly functional, probabilistic classification method. It involves computing the posterior probability of each class using Bayes' theorem, which indicates that the posterior probability of a class, conditioned on the features, is proportional to the product of the class's prior probability and the likelihood of the features for that class.

Despite its name, Logistic Regression (LR) [13] is commonly used for binary classification tasks in machine learning and data science. The algorithm models the relationship between a binary target variable and one or more independent variables (features) by estimating the probability of the target variable based on the input features. The logistic regression model is based on the logistic function, which converts any real-valued input to a range between 0 and 1, representing the probability that the target variable is 1 given the input features.

4   Experimental Results

4.1   Data

To assess the performance of the proposed classification approach, we use the Romanian Dense Breast Mammography Collection (RDBMC), a dataset collected by us and publicly available. The dataset contains 251 cases, all originating from patients with dense breasts. Among these, 199 cases contain data for both breasts, while 52 include only one breast, summing up to a total of 450 breasts. Each case is accompanied by both the original mammography report written in Romanian and an English translation. For each case, detailed annotations were extracted from the mammography reports, histopathology results, ultrasound reports (where available), and the DICOM metadata of the mammographic images.

The dataset includes classification labels for each breast, according to two criteria: a diagnostic label (healthy, benign, or malignant) and a BI-RADS assessment score (1 to 6). To generate a single classification label per patient case, we consider the most serious condition present across both breasts. For example, if one breast is healthy and the other contains a benign lesion, the case is labeled as benign; similarly, if any malignant lesion is present, the case is labeled as malignant. In the absence of any abnormalities, the case is labeled as healthy. The same rule is applied to the BI-RADS score: the score corresponding to the breast with the most serious findings is retained as the label for the case.

The resulting dataset contains 35 healthy, 101 benign, and 115 malignant cases, as summarized in Table 1. For BI-RADS classification, one class (BI-RADS 3) is not represented and is therefore excluded, leading to a five-class classification task. This label distribution is notably imbalanced, with BI-RADS 5 being over-represented and BI-RADS 6 severely under-represented. However, we do not apply any class balancing strategies (e.g., weighting, oversampling, hierarchical modeling), so as to align with our goal of providing a transparent and reproducible baseline on the naturally imbalanced dataset. We acknowledge that addressing class imbalance through dedicated techniques could improve robustness and fairness, and this remains a direction for further research.

Table 1. Data distribution within the RDBMC dataset.

	Class	Cases
Diagnosis	Healthy	35
	Benign	101
	Malignant	115
BI-RADS	1	35
	2	65
	3	0
	4	51
	5	96
	6	4

Both the Romanian and the translated English versions of the reports are used in our experiments to evaluate the robustness and language sensitivity of the proposed approach. In addition to the reports themselves, the dataset provides each patient's age at the time of the mammogram. To incorporate this potentially important feature into our models, we append the age value to the end of each report. While this strategy is unconventional, we adopt this approach to preserve the simplicity of a text-only classification pipeline, without relying on structured data fields. Our goal is to explore whether even this minimal integration of demographic information could yield observable benefits in classification performance, particularly in low-resource medical contexts where structured data might not always be standardized or available.

To provide a better understanding of the dataset's content and structure, Table 2 presents a few representative examples of original mammography reports, their English translations, and patient's age, along with their corresponding diagnostic labels and BI-RADS scores.

To evaluate the performance of the proposed models, the dataset was divided into 80% for training and 20% for testing. The split was performed at the patient level to avoid any data leakage between training and testing sets. To ensure reproducibility and consistency of results, all experiments were conducted in a deterministic manner, with identical training and evaluation conditions across all models.

4.2   Evaluation Metrics

To evaluate the performance of the proposed approach, we rely on four commonly used classification metrics:

1. **Accuracy** – the overall percentage of correctly classified cases:

$$\text{Accuracy} = \frac{TP + TN}{TP + TN + FP + FN} \tag{5}$$

2. **Precision** – the proportion of correctly predicted positive instances among all instances predicted as positive:

$$\text{Precision} = \frac{TP}{TP + FP} \tag{6}$$

3. **Recall** – the proportion of actual positive instances that were correctly identified:

$$\text{Recall} = \frac{TP}{TP + FN} \tag{7}$$

4. **F1-score** – the harmonic mean of precision and recall, providing a balance between the two:

$$\text{F1-score} = 2 \cdot \frac{\text{Precision} \cdot \text{Recall}}{\text{Precision} + \text{Recall}} \tag{8}$$

Table 2. Examples of mammographic reports.

Original report	English translation	Age	Diagnostic label	BI-RADS score
San drept dens heterogen. Fara opacitati sau microcalcificari suspecte. Aspect mamografic stationar comparativ cu examinarea anterioara. Fara aspecte suspecte de bilateralizare	Right breast: heterogeneously dense. No suspicious opacities or microcalcifications. Mammographic appearance remains stable compared to the previous examination. No suspicious signs of bilateral involvement	55	healthy	1
Sani cu structura glandular heterogena.Leziune cu calcifieri "in popcorn" la unirea cadranelor superioare, cu dimensiuni de 15/12 mm- aspect de fibroadenom vechi.Opacitate ovalara circumscrisa de 10 mm in CSE san drept - stationara.Fara opacitati sau microcalcifieri suspecte	Breasts with a heterogeneous glandular structure. A lesion with "popcorn" calcifications is observed at the junction of the upper quadrants, measuring 15×12 mm, with the appearance of an old fibroadenoma. In the upper outer quadrant (UOQ) of the right breast, there is a well-circumscribed oval opacity, 10 mm in size, which remains unchanged. No suspicious opacities or microcalcifications are detected	43	benign	4
Sani cu structura glandular heterogena.Sanul stang crescut in volum, cu edem interstitial difuz, cu eevidentierea unei opacitati de 4.5/3.5 cm in cadrane centrale, extinsa in cadranele superioare, cu retractie mamelonara si caractere tu (intensitate crescuta, partial bine delimitata, partial fluu, fara microcalcifieri incluse). Fara modificari in sanul drept	Breasts with heterogeneous glandular structure. The left breast is enlarged, showing diffuse interstitial edema, with a 4.5/3.5 cm opacity observed in the central quadrants, extending into the upper quadrants. Nipple retraction is present, and the mass has tumor-like characteristics (increased intensity, partially well-defined, partially fluid, without included microcalcifications). No abnormalities detected in the right breast	77	malignant	5

In the formulas above, the terms are defined as follows:

- **TP (True Positives)** – the number of samples correctly predicted as belonging to a given class
- **TN (True Negatives)** – the number of samples correctly predicted as not belonging to a given class
- **FP (False Positives)** – the number of samples incorrectly predicted as belonging to a given class
- **FN (False Negatives)** – the number of samples incorrectly predicted as not belonging to a given class.

The performance metrics are computed for each individual class and then averaged in a weighted (weighted arithmetic mean) manner, according to the number of samples in each class. This ensures that the evaluation is not biased by class imbalance.

4.3 Diagnosis

This subsection presents the results obtained for the diagnosis task, i.e., the classification of the patient's overall condition into one of three categories: *healthy*, *benign*, or *malignant*.

Tables 3 and 4 hold the results obtained using TF-IDF and LSI embeddings of mammography reports, respectively. Tables 5 and 6 contain the results of similar experiments applied on the reports to which the age was concatenated. For all the classifiers, we provide the results obtained with (i.e. *Exclude stopwords = FALSE*) and without (i.e. *Exclude stopwords = TRUE*) stopwords. The highest value for each metric is marked with bold for Romanian and italic for English.

Stopwords. In order to assess whether stopwords removal influences the classification performance, we analyze the results from each table individually. For TF-IDF for Romanian, NB obtains the same results with and without stopwords, while LR works best when stopwords are not excluded, regardless of the representation. For DT, the same case yields higher performance for both representations. However, when it comes to RF, if applied on TF-IDF representations, the removal of stopwords seems beneficial, whilst for LSI embeddings, better results are obtained when stopwords are not excluded. Trying to draw a conclusion, we can state that, except for the combination between LSI and RF, the removal of stopwords negatively influences the classification results.

The same conclusions hold for English as well, for both TF-IDF and LSI, with the exception of DT. For this classifier, better performance is achieved when stopwords are excluded, as opposed to the results obtained for Romanian. This observation suggests that DT is the least robust algorithm when it comes to language adaptation.

For the variant where a patient's age is concatenated to their report, LR reaches better performance when stopwords are not excluded for both representations and languages, thus being the only consistent algorithm across all variants. For Romanian language, TF-IDF leads to the same performance with and without stopwords for DT, RF and NB, while DT and NB work better when stopwords are kept and RF, when they are removed, for English. For LSI embeddings, NB reaches better performance when stopwords are excluded, and DT with stopwords for Romanian and without for English, and vice-versa for RF.

Across all four variants, DT works better when stopwords are not excluded for Romanian and the opposite for English. As already noticed, LR benefits from stopwords regardless of the language. For RF and NB, no conclusion can be drawn either on language, representation or type of data (only reports or reports concatenated with age).

Language. When it comes to the influence the language has on the classification results, the results from Tables 3, 4, 5 and 6 show that the language does not play an important role, as some algorithms obtain better performance on English, while others, on Romanian. The only classifier obtaining consistently better results on English is NB, with differences from 2% in accuracy for TF-IDF, to 11% for LSI.

Age. As for the age, it has a negative effect when TF-IDF representation is used, while LSI benefits from the addition of it to the data, for all algorithms and both languages. Thus, whether or not age should be taken into consideration fully depends on the representation used.

Representation. Determining the most suitable representation for the reports in the RDBMC dataset proves to be particularly challenging. For the Romanian reports, all classifiers – except Random Forest – achieve better performance when using the LSI representation. However, when age is concatenated with the report, the Naive Bayes classifier performs better with TF-IDF, while the optimal representation for the other classifiers remains unchanged. For the English reports, Decision Tree, Random Forest, and Naive Bayes yield superior results with TF-IDF, whereas Logistic Regression achieves better performance with LSI. When age is added to the report content, NB once again shifts in preference, showing improved results with the LSI representation.

Classification. The only clear conclusion is with regards to the best-suited combination of algorithms (for embedding and classification) for our data. For Romanian, the best results are obtained with Random Forest classifier and TF-IDF embeddings on the reports alone, excluding stopwords: 80% accuracy, precision, recall and F1-score. The same configuration yields the best performance for English also, with an increase of 2% in accuracy in recall (82%), 4% in precision (84%) and 1% in F1-score (81%).

Most classification errors in the diagnosis task were observed between the benign and malignant classes, which is consistent across both the Romanian and English datasets. This confusion typically arose in reports with ambiguous or borderline terminology, especially when describing lesions with evolving or inconclusive characteristics. While the healthy class generally showed higher precision and recall, particularly in the English datasets, there were some fluctuations across models (notably in NB and TF-IDF embeddings) where healthy cases were occasionally misclassified due to scarce or inconsistent textual cues.

Table 3. 3-class classification results obtained on mammography reports using TF-IDF.

	Class	Exclude stopwords	DT		RF		NB		LR	
			Romanian	English	Romanian	English	Romanian	English	Romanian	English
Accuracy	Average	TRUE	0.59	0.73	**0.8**	*0.82*	0.69	0.71	0.76	0.67
		FALSE	0.61	0.67	0.78	0.76	0.69	0.76	**0.8**	0.71
Precision	Benign	TRUE	0.5	0.67	**0.81**	*0.82*	0.59	0.61	0.72	0.57
		FALSE	0.5	0.59	0.8	0.75	0.59	0.68	0.78	0.62
	Healthy	TRUE	0.71	0.5	0.8	*1*	0	0	**1**	*1*
		FALSE	0.67	0.67	0.8	0.75	0	*1*	**1**	*1*
	Malignant	TRUE	0.64	*0.84*	**0.8**	0.81	0.78	0.79	0.77	0.74
		FALSE	0.68	0.74	0.77	0.77	0.78	0.82	**0.8**	0.76
	Average	TRUE	0.59	0.73	0.8	*0.84*	0.61	0.62	0.78	0.7
		FALSE	0.61	0.67	0.79	0.76	0.61	0.79	**0.81**	0.73
Recall	Benign	TRUE	0.55	0.6	0.65	0.7	**0.7**	0.7	0.65	0.65
		FALSE	0.5	0.65	0.6	0.6	**0.7**	*0.75*	**0.7**	0.65
	Healthy	TRUE	**0.83**	*0.67*	0.67	0.5	0	0	0.5	0.17
		FALSE	0.67	*0.67*	0.67	0.5	0	0.17	0.5	0.17
	Malignant	TRUE	0.56	0.84	**0.96**	*1*	0.84	0.88	0.92	0.8
		FALSE	0.68	0.68	**0.96**	0.96	0.84	0.92	0.96	0.88
	Average	TRUE	0.59	0.73	**0.8**	*0.82*	0.69	0.71	0.76	0.67
		FALSE	0.61	0.67	0.78	0.76	0.69	0.76	**0.8**	0.71
F1-score	Benign	TRUE	0.52	0.63	0.72	*0.76*	0.64	0.65	0.68	0.6
		FALSE	0.5	0.62	0.69	0.67	0.64	0.71	**0.74**	0.63
	Healthy	TRUE	**0.77**	0.57	0.73	*0.67*	0	0	*0.67*	0.29
		FALSE	0.67	*0.67*	0.73	0.6	0	0.29	*0.67*	0.29
	Malignant	TRUE	0.6	0.84	**0.87**	*0.89*	0.81	0.83	0.84	0.77
		FALSE	0.68	0.71	0.86	0.86	0.81	0.87	**0.87**	0.81
	Average	TRUE	0.59	0.73	**0.8**	*0.81*	0.65	0.66	0.76	0.65
		FALSE	0.61	0.67	0.77	0.75	0.65	0.74	**0.8**	0.68

For improved readability, the highest performance scores for each embedding and classifier combination are summarized visually in Figs. 1 and 2, for the Romanian and English versions of the reports for the diagnosis task, respectively. This allows a clearer overview of the observed trends across languages.

4.4   BI-RADS Scoring

We now turn our attention to the BI-RADS scoring task, which involves classifying each case into one of five BI-RADS categories, based on the most serious abnormality present in the mammography report.

Tables 7, 8, 9 and 10 contain the results of the experiments performed when considering the BI-RADS score assigned to a mammography report as its label. The first obvious observation is that, because the *6* class is severely underrepresented, classification is not performed correctly, leading to all performance

Table 4. 3-class classification results obtained on mammography reports using LSI.

	Class	Exclude stopwords	DT		RF		NB		LR	
			Romanian	English	Romanian	English	Romanian	English	Romanian	English
Accuracy	Average	TRUE	0.63	0.69	0.71	0.71	0.59	0.65	0.73	0.75
		FALSE	0.65	0.59	**0.76**	0.67	0.63	0.76	0.75	*0.76*
Precision	Benign	TRUE	0.57	0.58	0.69	0.61	0.47	0.56	0.65	0.67
		FALSE	0.57	0.47	**0.8**	0.57	0.55	*0.83*	0.67	0.7
	Healthy	TRUE	0.33	0.67	0.5	0.67	0.5	0.6	0.5	*1*
		FALSE	0.57	0.33	0.57	0.67	0.5	0.56	**1**	*1*
	Malignant	TRUE	**0.8**	*0.81*	0.79	0.8	0.67	0.71	0.79	0.79
		FALSE	0.74	0.78	0.79	0.77	0.7	0.8	0.79	0.8
	Average	TRUE	0.66	0.7	0.72	0.71	0.57	0.64	0.7	0.77
		FALSE	0.65	0.61	**0.77**	0.68	0.62	*0.78*	**0.77**	*0.78*
Recall	Benign	TRUE	0.4	*0.7*	0.45	*0.7*	0.4	0.5	0.65	*0.7*
		FALSE	0.6	0.45	0.6	0.65	0.3	0.5	**0.7**	*0.7*
	Healthy	TRUE	0.67	0.67	**0.83**	*0.83*	0.33	0.5	0.17	0.17
		FALSE	0.67	0.5	0.67	0.67	**0.83**	*0.83*	0.17	0.17
	Malignant	TRUE	0.8	0.68	0.88	0.8	0.8	0.8	**0.92**	0.92
		FALSE	0.68	0.72	**0.92**	0.68	0.84	*0.96*	**0.92**	*0.96*
	Average	TRUE	0.63	0.69	0.71	0.71	0.59	0.65	0.73	0.75
		FALSE	0.65	0.59	**0.76**	0.67	0.63	*0.76*	0.75	*0.76*
F1-score	Benign	TRUE	0.47	0.64	0.55	0.65	0.43	0.53	0.65	0.68
		FALSE	0.59	0.46	**0.69**	0.6	0.39	0.62	0.68	*0.7*
	Healthy	TRUE	0.44	*0.67*	**0.62**	0.44	0.4	0.55	0.25	0.29
		FALSE	**0.62**	0.4	0.62	*0.67*	0.62	*0.67*	0.29	0.29
	Malignant	TRUE	0.8	0.74	0.83	0.8	0.73	0.75	**0.85**	0.85
		FALSE	0.71	0.75	0.85	0.72	0.76	0.84	**0.85**	*0.87*
	Average	TRUE	0.63	0.69	0.69	0.7	0.57	0.64	0.7	0.72
		FALSE	0.65	0.6	**0.76**	0.67	0.6	*0.75*	0.72	0.74

Table 5. 3-class classification results obtained on mammography reports and age using TF-IDF.

	Class	Exclude stopwords	DT		RF		NB		LR	
			Romanian	English	Romanian	English	Romanian	English	Romanian	English
Accuracy	Average	TRUE	0.53	0.67	**0.78**	*0.78*	0.69	0.71	0.69	0.65
		FALSE	0.53	0.71	**0.78**	0.75	0.69	0.73	0.75	0.69
Precision	Benign	TRUE	0.43	0.59	**0.74**	*0.74*	0.58	0.61	0.6	0.54
		FALSE	0.43	0.64	0.71	0.68	0.58	0.64	0.67	0.59
	Healthy	TRUE	0.5	0.29	**1**	*1*	0	0	0	0
		FALSE	0.5	0.67	**1**	0.67	0	0	0	0
	Malignant	TRUE	0.64	*0.86*	0.8	0.8	0.78	0.79	0.74	0.74
		FALSE	0.62	0.78	**0.81**	0.79	0.78	0.79	0.8	0.76
	Average	TRUE	0.54	0.69	**0.8**	*0.8*	0.61	0.62	0.6	0.58
		FALSE	0.54	0.71	**0.8**	0.74	0.61	0.64	0.65	0.6
Recall	Benign	TRUE	0.5	0.65	0.7	*0.7*	0.7	*0.7*	0.6	0.65
		FALSE	0.5	*0.7*	**0.75**	0.65	0.7	*0.7*	0.7	0.65
	Healthy	TRUE	**0.5**	0.33	0.33	0.33	0	0	0	0
		FALSE	0.33	*0.67*	**0.5**	0.33	0	0	0	0
	Malignant	TRUE	0.56	0.76	**0.96**	*0.96*	0.84	0.88	0.92	0.8
		FALSE	0.6	0.72	0.88	0.92	0.84	0.92	**0.96**	0.88
	Average	TRUE	0.53	0.67	**0.78**	*0.78*	0.69	0.71	0.69	0.65
		FALSE	0.53	0.71	**0.78**	0.75	0.69	0.73	0.75	0.69
F1-score	Benign	TRUE	0.47	0.62	0.72	*0.72*	0.64	0.65	0.6	0.59
		FALSE	0.47	0.67	**0.73**	0.67	0.64	0.67	0.68	0.62
	Healthy	TRUE	0.5	0.31	0.5	0.5	0	0	0	0
		FALSE	0.4	*0.67*	**0.67**	0.44	0	0	0	0
	Malignant	TRUE	0.6	0.81	**0.87**	*0.87*	0.81	0.83	0.82	0.77
		FALSE	0.61	0.75	0.85	0.85	**0.87**	0.85	**0.87**	0.81
	Average	TRUE	0.53	0.68	0.77	*0.77*	0.65	0.66	0.64	0.61
		FALSE	0.53	0.71	**0.78**	0.73	0.65	0.68	0.7	0.64

Table 6. 3-class classification results obtained on mammography reports and age using LSI.

	Class	Exclude stopwords	DT		RF		NB		LR	
			Romanian	English	Romanian	English	Romanian	English	Romanian	English
Accuracy	Average	TRUE	0.65	0.65	**0.75**	0.65	0.65	*0.76*	0.73	0.73
		FALSE	0.71	0.63	0.73	0.75	0.63	0.75	**0.75**	*0.76*
Precision	Benign	TRUE	0.55	0.54	**0.79**	0.54	0.56	*0.75*	0.64	0.64
		FALSE	0.65	0.52	0.71	0.65	0.54	0.71	0.67	0.7
	Healthy	TRUE	0.57	0.38	0.5	0.75	0.8	*1*	**1**	0
		FALSE	0.5	0.4	0.5	*1*	0.43	0.75	**1**	*1*
	Malignant	TRUE	0.75	*0.89*	0.79	0.74	0.67	0.74	0.79	0.79
		FALSE	**0.81**	0.76	0.79	0.8	0.71	0.77	0.79	0.8
	Average	TRUE	0.65	0.7	0.76	0.66	0.64	*0.78*	0.75	0.64
		FALSE	0.71	0.63	0.73	0.77	0.61	0.74	**0.77**	*0.78*
Recall	Benign	TRUE	0.55	0.65	0.55	0.65	0.45	0.6	**0.7**	0.7
		FALSE	0.55	0.55	0.5	*0.75*	0.35	0.6	0.7	0.7
	Healthy	TRUE	**0.67**	0.5	**0.67**	0.5	**0.67**	*0.67*	0.17	0
		FALSE	**0.67**	0.33	**0.67**	0.5	0.5	0.5	0.17	0.17
	Malignant	TRUE	0.72	0.68	**0.92**	0.68	0.8	**0.92**	0.88	0.92
		FALSE	0.84	0.76	**0.92**	0.8	0.88	0.92	**0.92**	*0.96*
	Average	TRUE	0.65	0.65	**0.75**	0.65	0.65	*0.76*	0.73	0.73
		FALSE	0.71	0.63	0.73	0.75	0.63	0.75	**0.75**	*0.76*
F1-score	Benign	TRUE	0.55	0.59	0.65	0.59	0.5	0.67	0.67	0.67
		FALSE	0.59	0.54	0.59	0.7	0.42	0.65	**0.68**	*0.7*
	Healthy	TRUE	0.62	0.43	0.57	0.6	**0.73**	*0.8*	0.29	0
		FALSE	0.57	0.36	0.57	0.67	0.46	0.6	0.29	0.29
	Malignant	TRUE	0.73	0.77	**0.85**	0.71	0.73	0.82	0.83	0.85
		FALSE	0.82	0.76	**0.85**	0.8	0.79	0.84	**0.85**	*0.87*
	Average	TRUE	0.65	0.66	**0.74**	0.65	0.64	*0.76*	0.7	0.68
		FALSE	0.7	0.63	0.72	0.74	0.61	0.73	0.72	0.74

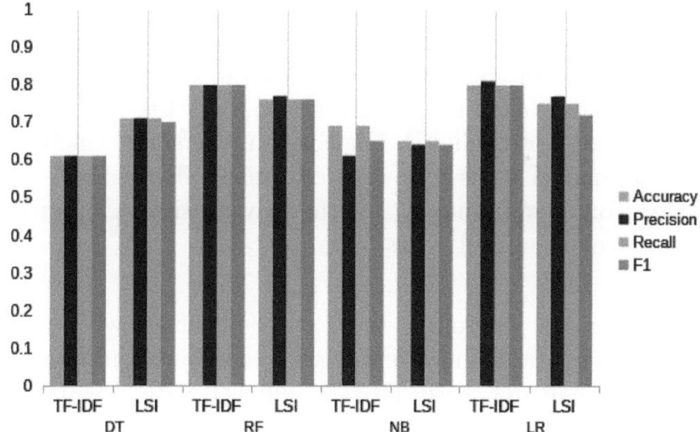

Fig. 1. Best performance metrics for the diagnosis task across embeddings and classifiers, for Romanian reports.

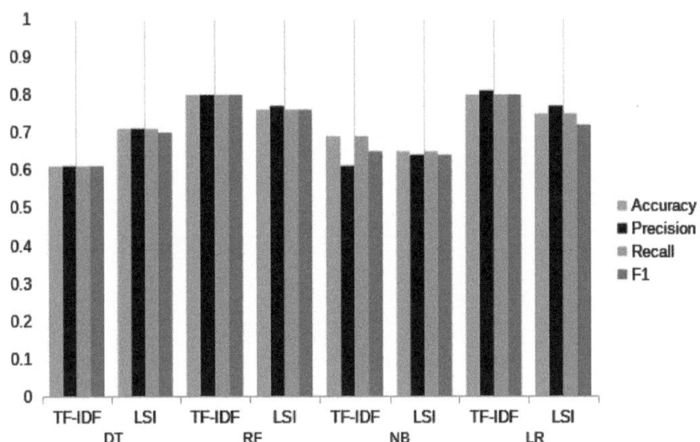

Fig. 2. Best performance metrics for the diagnosis task across embeddings and classifiers, for English translations.

indicators being 0. Following, similar scenarios to those observed in the diagnosis classification task are encountered. There is no definitive conclusion regarding stopwords, language or age influence.

Stopwords. In terms of stopwords exclusion, their removal does not have any effect on NB and LR for TF-IDF embeddings for Romanian, yet it benefits NB when applied to the reports in English concatenated with the patient's age. For LSI, stopwords appear to contribute to the overall classification in Romanian, but once the age is also taken into consideration, only LR obtains better performance when they are not excluded. For English, the same conclusion holds for LR – stopwords removal leads to lower performance indicators – thus, for this type of classification as well, making LR the only consistent algorithm across all representations, languages and types of data.

Language. As for the language, the results obtained with Romanian and English data are comparable, however, only two combinations obtain better performance on Romanian: RF with TF-IDF and DT with LSI, both applied on reports alone.

Age. The concatenation of age does help towards a better classification for DT and LR applied on both TF-IDF and LSI representations for Romanian, while for English, these algorithms yield a higher accuracy when age is not considered.

Representation. From the same tables, we can conclude that TF-IDF algorithm is better-suited for RF, NB and LR classifiers for Romanian. For English, on the other hand, only NB work better with TF-IDF, the rest of the algorithms yielding better performance on the LSI representations.

Classification. Nonetheless, the best performance metrics for Romanian are achieved with the same combination as for the diagnosis classification: RF applied on TF-IDF representations of reports (without the age), excluding stopwords: 73% accuracy, 77% precision, 73% recall and 67% F1-score. We re-iterate the fact that the class corresponding to BI-RADS score 6 is under-represented, thus causing a decrease in performance compared to the diagnosis task. For English, however, the best combination seems to be between NB and LSI embeddings of reports and ages, without excluding the stopwords (75% accuracy and recall, 74% precision, 73% F1-score).

For the BI-RADS scoring task, misclassifications were primarily observed between adjacent BI-RADS categories, especially between BI-RADS 4 and 5. This reflects both the clinical difficulty of differentiating borderline cases and the inherent ambiguity often found in report language when describing evolving

Table 7. 5-class classification results obtained on mammography reports using TF-IDF.

	Class	Exclude stopwords	DT		RF		NB		LR	
			Romanian	English	Romanian	English	Romanian	English	Romanian	English
Accuracy	Average	TRUE	0.45	0.61	**0.73**	0.69	0.59	0.67	0.67	0.69
		FALSE	0.47	0.59	0.69	*0.71*	0.59	0.67	0.67	0.63
Precision	1	TRUE	0.5	0.5	0.67	*1*	0	0	1	*1*
		FALSE	0.6	0.56	0.57	0.75	0	0	1	*1*
	2	TRUE	0.69	*0.8*	**0.86**	*0.8*	0.62	0.65	0.79	0.73
		FALSE	0.69	0.78	0.83	0.79	0.62	0.65	0.79	0.67
	4	TRUE	0.14	0.44	1	0.5	0.17	0.5	0	0.67
		FALSE	0.12	0.44	1	*0.75*	0.17	0.5	0	0.6
	5	TRUE	0.5	0.62	**0.67**	0.63	**0.67**	*0.76*	0.59	0.64
		FALSE	0.62	0.58	0.65	0.66	**0.67**	*0.76*	0.59	0.6
	6	TRUE	0	0	0	0	0	0	0	0
		FALSE	0	0	0	0	0	0	0	0
	Average	TRUE	0.48	0.62	**0.77**	0.69	0.47	0.58	0.58	0.7
		FALSE	0.54	0.6	0.74	*0.71*	0.47	0.58	0.58	0.65
Recall	1	TRUE	**0.67**	0.67	**0.67**	0.5	0	0	0.5	0.33
		FALSE	0.5	*0.83*	**0.67**	0.5	0	0	0.5	0.17
	2	TRUE	0.6	0.53	0.8	0.8	**0.87**	*0.87*	0.73	0.73
		FALSE	0.6	0.47	0.67	0.73	**0.87**	*0.87*	0.73	0.67
	4	TRUE	**0.22**	0.44	0.11	0.33	0.11	*0.56*	0	0.44
		FALSE	0.22	0.44	0.11	0.33	0.11	*0.56*	0	0.33
	5	TRUE	0.4	0.75	1	0.85	0.8	0.8	1	0.9
		FALSE	0.5	0.7	1	*0.95*	0.8	0.8	1	0.9
	6	TRUE	0	0	0	0	0	0	0	0
		FALSE	0	0	0	0	0	0	0	0
	Average	TRUE	0.45	0.61	**0.73**	0.69	0.59	0.67	0.67	0.69
		FALSE	0.47	0.59	0.69	*0.71*	0.59	0.67	0.67	0.63
F1-score	1	TRUE	0.57	0.57	**0.67**	*0.67*	0	0	**0.67**	0.5
		FALSE	0.55	*0.67*	0.62	0.6	0	0	**0.67**	0.29
	2	TRUE	0.64	0.64	**0.83**	*0.8*	0.72	0.74	0.76	0.73
		FALSE	0.64	0.58	0.74	0.76	0.72	0.74	0.76	0.67
	4	TRUE	0.17	0.44	**0.2**	0.4	0.13	*0.53*	0	*0.53*
		FALSE	0.16	0.44	*0.2*	0.46	0.13	*0.53*	0	0.43
	5	TRUE	0.44	0.68	**0.8**	0.72	0.73	*0.78*	0.74	0.75
		FALSE	0.56	0.64	0.78	*0.78*	0.73	*0.78*	0.74	0.72
	6	TRUE	0	0	0	0	0	0	0	0
		FALSE	0	0	0	0	0	0	0	0
	Average	TRUE	0.46	0.6	**0.67**	0.67	0.52	0.62	0.59	0.66
		FALSE	0.5	0.58	0.63	*0.68*	0.52	0.62	0.59	0.59

Table 8. 5-class classification results obtained on mammography reports using LSI.

	Class	Exclude stopwords	DT		RF		NB		LR	
			Romanian	English	Romanian	English	Romanian	English	Romanian	English
Accuracy	Average	TRUE	0.53	0.53	0.63	0.69	0.43	0.51	0.61	0.63
		FALSE	0.61	0.55	**0.67**	0.63	0.57	0.69	0.61	0.67
Precision	1	TRUE	0.38	0.4	0.33	0.67	0.5	0.5	0.5	0.5
		FALSE	0.67	0.4	0.67	0.56	0.55	0.56	**1**	1
	2	TRUE	0.55	0.69	0.64	0.71	0.55	0.7	0.69	0.69
		FALSE	0.69	0.78	0.83	0.91	**1**	0.89	0.69	0.71
	4	TRUE	0.42	0.25	**0.57**	0.56	0.1	0.22	0	0
		FALSE	0.25	0.31	0.4	0.36	0.14	0.56	0	1
	5	TRUE	0.65	0.61	**0.71**	0.73	0.5	0.54	0.59	0.61
		FALSE	0.67	0.71	0.64	0.65	0.59	0.71	0.58	0.62
	6	TRUE	0	0	0	0	0	0	0	0
		FALSE	0	0	0	0	0	0	0	0
	Average	TRUE	0.53	0.53	0.61	0.67	0.43	0.52	0.49	0.5
		FALSE	0.59	0.61	**0.65**	0.65	0.62	0.7	0.55	0.75
Recall	1	TRUE	0.5	0.33	0.33	0.33	0.17	0.5	0.17	0.17
		FALSE	0.67	0.67	0.67	0.83	**1**	0.83	0.17	0.17
	2	TRUE	0.4	0.73	0.6	0.8	0.4	0.47	**0.73**	0.73
		FALSE	0.6	0.47	0.67	0.67	0.4	0.53	**0.73**	0.8
	4	TRUE	**0.56**	0.33	0.44	0.56	0.11	0.22	0	0
		FALSE	0.22	0.56	0.22	0.44	0.11	0.56	0	0.11
	5	TRUE	0.65	0.55	0.85	0.8	0.7	0.7	**0.95**	1
		FALSE	0.8	0.6	0.9	0.65	0.8	0.85	**0.95**	1
	6	TRUE	0	0	0	0	0	0	0	0
		FALSE	0	0	0	0	0	0	0	0
	Average	TRUE	0.53	0.53	0.63	0.69	0.43	0.51	0.61	0.63
		FALSE	0.61	0.5	**0.67**	0.63	0.57	0.69	0.61	0.67
F1-score	1	TRUE	0.43	0.36	0.33	0.44	0.25	0.5	0.25	0.25
		FALSE	0.67	0.53	0.67	0.67	**0.71**	0.67	0.29	0.29
	2	TRUE	0.46	0.71	0.62	0.75	0.46	0.56	0.71	0.71
		FALSE	0.64	0.58	**0.74**	0.77	0.57	0.67	0.71	0.75
	4	TRUE	0.48	0.29	**0.5**	0.56	0.11	0.22	0	0
		FALSE	0.24	0.4	0.29	0.4	0.12	0.56	0	0.2
	5	TRUE	0.65	0.58	**0.88**	0.76	0.58	0.61	0.73	0.75
		FALSE	0.73	0.65	0.75	0.65	0.68	0.77	0.72	0.77
	6	TRUE	0	0	0	0	0	0	0	0
		FALSE	0	0	0	0	0	0	0	0
	Average	TRUE	0.53	0.53	0.61	0.67	0.41	0.5	0.52	0.53
		FALSE	0.59	0.56	**0.64**	0.63	0.54	0.68	0.52	0.67

or suspicious findings. The BI-RADS 1 and 2 classes were generally classified with reasonable precision, particularly in the English dataset, while performance on BI-RADS 4 remained the most inconsistent across models and embeddings. Across all experiments, the BI-RADS 6 category was never correctly predicted, due to its absence or severe under-representation in the dataset, confirming the challenge posed by highly imbalanced classes.

Figures 3 and 4 provide a visual summary of the highest performance scores for the BI-RADS scoring task for Romanian and English reports, respectively. This visualization emphasizes the relative performance differences between the Romanian and English datasets.

Table 9. 5-class classification results obtained on mammography reports and age using TF-IDF.

	Class	Exclude stopwords	DT		RF		NB		LR	
			Romanian	*English*	*Romanian*	*English*	*Romanian*	*English*	*Romanian*	*English*
Accuracy	*Average*	TRUE	0.45	0.55	**0.65**	0.65	0.59	0.69	0.61	0.65
		FALSE	0.55	0.49	0.59	0.63	0.59	*0.69*	0.61	0.61
Precision	*1*	TRUE	0.5	0.5	**0.75**	*1*	0	0	0	0
		FALSE	0.6	0.17	0.4	0.33	0	0	0	*1*
	2	TRUE	0.62	*0.73*	**0.79**	*0.73*	0.62	0.65	0.65	0.65
		FALSE	0.62	0.58	0.69	0.67	0.59	0.65	0.65	0.67
	4	TRUE	0.3	0.36	**0.33**	0.5	0.17	0.56	0	*0.67*
		FALSE	0.25	0.4	0	0.5	0.17	0.56	0	0.5
	5	TRUE	0.46	0.57	0.6	0.6	0.67	*0.77*	0.59	0.64
		FALSE	0.65	0.59	0.59	0.66	**0.7**	*0.77*	0.59	0.58
	6	TRUE	0	0	0	0	0	0	0	0
		FALSE	0	0	0	0	0	0	0	0
	Average	TRUE	0.48	0.56	**0.61**	*0.66*	0.47	0.59	0.42	0.56
		FALSE	0.55	0.49	0.48	0.59	0.48	0.59	0.42	0.63
Recall	*1*	TRUE	**0.67**	*0.5*	0.5	0.33	0	0	0	0
		FALSE	0.5	0.17	0.33	0.17	0	0	0	0.17
	2	TRUE	0.33	0.53	0.73	0.73	**0.87**	*0.87*	0.73	0.73
		FALSE	0.53	0.47	0.6	0.67	**0.87**	*0.87*	0.73	0.67
	4	TRUE	**0.33**	0.44	0.11	0.22	0.11	*0.56*	0	0.44
		FALSE	0.22	0.47	0	0.22	0.11	*0.56*	0	0.22
	5	TRUE	0.55	0.65	0.9	0.9	0.8	0.85	**1**	0.9
		FALSE	0.75	0.65	0.95	*0.95*	0.8	0.85	**1**	0.9
	6	TRUE	0	0	0	0	0	0	0	0
		FALSE	0	0	0	0	0	0	0	0
	Average	TRUE	0.45	0.55	**0.65**	0.65	0.59	*0.69*	0.61	0.65
		FALSE	0.55	0.49	0.59	0.63	0.59	*0.69*	0.61	0.61
F1-score	*1*	TRUE	0.57	*0.5*	**0.6**	*0.5*	0	0	0	0
		FALSE	0.55	0.17	0.36	0.22	0	0	0	0.29
	2	TRUE	0.43	0.62	**0.76**	0.73	0.72	*0.74*	0.69	0.69
		FALSE	0.57	0.52	0.64	0.67	0.7	*0.74*	0.69	0.67
	4	TRUE	**0.32**	0.4	0.17	0.31	0.13	*0.56*	0	0.53
		FALSE	0.24	0.42	0	0.31	0.13	*0.56*	0	0.31
	5	TRUE	0.5	0.6	0.72	0.72	0.73	*0.81*	**0.74**	0.75
		FALSE	0.7	0.62	0.73	0.78	**0.74**	0.72	**0.74**	0.71
	6	TRUE	0	0	0	0	0	0	0	0
		FALSE	0	0	0	0	0	0	0	0
	Average	TRUE	0.45	0.55	**0.61**	0.61	0.52	*0.63*	0.49	0.59
		FALSE	0.55	0.49	0.52	0.58	0.52	*0.63*	0.49	0.56

Table 10. 5-class classification results obtained on mammography reports and age using LSI.

	Class	Exclude stopwords	DT		RF		NB		LR	
			Romanian	English	Romanian	English	Romanian	English	Romanian	English
Accuracy	Average	TRUE	0.47	0.55	**0.69**	0.71	0.51	0.65	0.57	0.63
		FALSE	0.39	0.49	0.59	0.71	0.47	*0.75*	0.59	0.67
Precision	1	TRUE	0.5	0.38	**0.67**	0.6	0.57	*1*	0	0
		FALSE	0.33	0.33	0.57	0.8	0.5	0.8	0	*1*
	2	TRUE	**0.86**	0.67	0.77	0.8	0.78	0.8	0.65	0.67
		FALSE	0.5	0.65	0.73	0.8	0.73	*0.85*	0.65	0.71
	4	TRUE	0.18	0.38	**0.6**	0.56	0	0.25	0	0
		FALSE	0.2	0.29	0.25	0.5	0.21	0.71	0	*1*
	5	TRUE	0.48	*0.73*	**0.67**	*0.73*	0.52	0.59	0.58	0.61
		FALSE	0.56	0.6	0.64	0.7	0.5	0.69	0.58	0.62
	6	TRUE	0	0	0	0	0	0	0	0
		FALSE	0	0	0	0	0	0	0	0
	Average	TRUE	0.53	0.59	**0.67**	0.69	0.5	0.63	0.42	0.43
		FALSE	0.44	0.52	0.58	0.69	0.51	0.74	0.42	*0.75*
Recall	1	TRUE	**1**	0.5	0.67	0.5	0.67	0.5	0	0
		FALSE	0.5	0.17	0.67	*0.67*	0.5	*0.67*	0	0.17
	2	TRUE	0.4	0.53	0.67	*0.8*	0.47	*0.8*	**0.73**	*0.8*
		FALSE	0.33	0.73	0.53	*0.8*	0.53	0.73	**0.73**	*0.8*
	4	TRUE	0.22	*0.67*	**0.33**	0.56	0	0.11	0	0
		FALSE	**0.33**	0.44	0.22	0.44	**0.33**	0.56	0	0.11
	5	TRUE	0.5	0.555	0.9	0.8	0.75	0.85	0.9	*1*
		FALSE	0.45	0.45	0.8	0.8	0.5	0.9	**0.95**	*1*
	6	TRUE	0	0	0	0	0	0	0	0
		FALSE	0	0	0	0	0	0	0	0
	Average	TRUE	0.47	0.55	**0.69**	0.71	0.51	0.65	0.57	0.63
		FALSE	0.39	0.49	0.59	0.71	0.47	*0.75*	0.59	0.67
F1-score	1	TRUE	**0.67**	0.43	**0.67**	0.55	0.62	0.67	0	0
		FALSE	0.4	0.22	0.62	*0.73*	0.5	*0.73*	0	0.29
	2	TRUE	0.555	0.59	**0.71**	*0.8*	0.58	*0.8*	0.69	0.73
		FALSE	0.4	0.69	0.62	*0.8*	0.62	0.79	0.69	0.75
	4	TRUE	0.2	0.48	**0.43**	0.56	0	0.15	0	0
		FALSE	0.25	0.35	0.24	0.47	0.26	*0.62*	0	0.2
	5	TRUE	0.49	0.63	0.71	0.76	0.61	0.69	0.71	0.75
		FALSE	0.5	0.51	0.71	0.74	0.5	*0.78*	**0.72**	0.77
	6	TRUE	0	0	0	0	0	0	0	0
		FALSE	0	0	0	0	0	0	0	0
	Average	TRUE	0.47	0.56	**0.66**	0.7	0.48	0.61	0.48	0.51
		FALSE	0.4	0.49	0.57	0.7	0.48	*0.73*	0.48	0.59

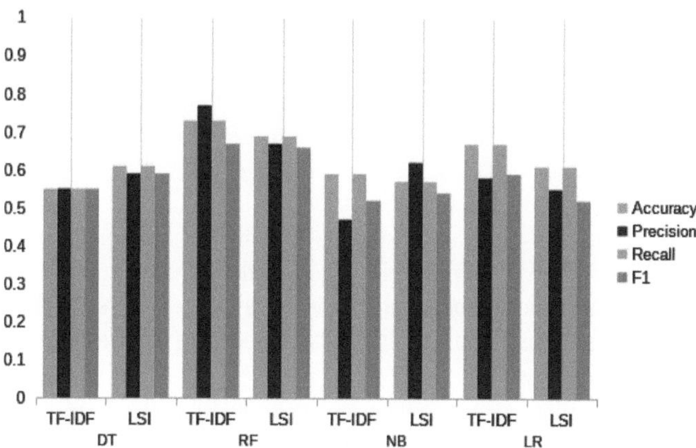

Fig. 3. Best performance metrics for the BI-RADS scoring task across embeddings and classifiers, for Romanian reports.

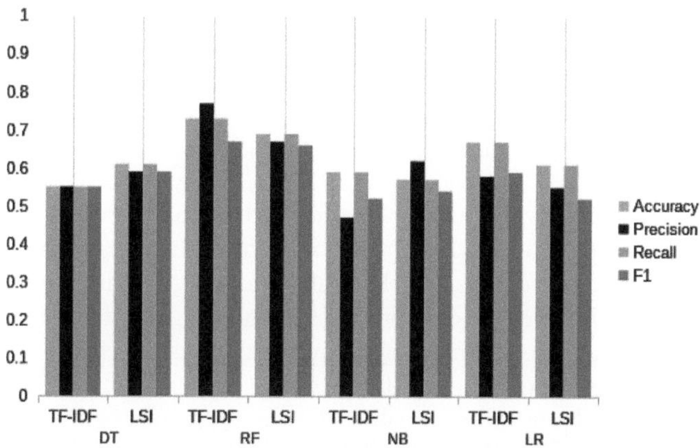

Fig. 4. Best performance metrics for the BI-RADS scoring task across embeddings and classifiers, for English translations.

5 Conclusions and Future Work

In this study, we investigated the performance of four machine learning classifiers (Logistic Regression, Decision Tree, Random Forest, and Naive Bayes) in conjunction with two embedding strategies (TF-IDF and LSI) for classifying mammography reports from the RDBMC dataset. The experiments were conducted on 24 distinct data configurations: Romanian and English report texts, with or without age concatenation, and with or without stopword removal. We addressed two classification tasks: 1. estimating the patient's overall condition (i.e., healthy, benign, or malignant), and 2. predicting the BI-RADS score, resulting in a 3-class and a 5-class classification problem, respectively.

The experimental results indicate that the classification of patient diagnosis yields better overall performance than BI-RADS scoring, which is likely due to the severe class imbalance in the latter task, particularly the underrepresentation of BI-RADS score 6. Despite these challenges, several consistent trends have emerged.

The most robust configuration across both classification tasks for Romanian texts involved using TF-IDF embeddings (with stopwords removed) and the Random Forest classifier. In the diagnosis task, this setup achieved 80% accuracy. For BI-RADS classification, although the scores decreased slightly due to class imbalance, the same combination still produced the highest performance for Romanian (73% accuracy).

In the case of English reports, the optimal setup varied between tasks. For diagnosis, the best results were again obtained using Random Forest with TF-IDF embeddings (excluding stopwords), yielding improvements over the Romanian results: 82% accuracy. For BI-RADS classification, however, the combination of Naive Bayes with LSI embeddings (including age and stopwords) provided the best performance: 75% accuracy.

Although the English variants yield slightly higher scores overall, the system shows greater robustness for Romanian, with consistent performance across tasks and configurations. Moreover, the differences between Romanian and English results remain small, which highlights the generalizability of the approach across languages.

Given their reliability and strong performance, the selected models can be effectively integrated into a multi-modal computer-aided diagnosis system, complementing image-based analysis with natural language understanding of mammography reports. Such a system could assist radiologists in clinical workflows, especially in low-resource settings, by providing robust decision support based on textual and structured patient information.

Our primary goal was to establish a clear and reproducible baseline for the automated classification of Romanian-language mammography reports, addressing a gap in existing research on medical NLP for low-resource languages. We believe that these baseline results can serve as a valuable reference point for the development of more advanced approaches in this field.

As future work, we aim to explore more advanced language models (e.g., transformer-based architectures) and multi-modal fusion strategies that combine textual reports with imaging features. Furthermore, we plan to develop explainability modules for clinical interpretability, enabling the system to justify its predictions in human-understandable terms.

Acknowledgments. This research is supported by the project "Romanian Hub for Artificial Intelligence - HRIA", Smart Growth, Digitization and Financial Instruments Program, 2021–2017, MySMIS no. 334906.

References

1. Radiology assistant. Bi-RADS for mammography and ultrasound 2013 (2014). https://radiologyassistant.nl/breast/bi-rads/bi-rads-for-mammography-and-ultrasound-2013. Accessed 11 June 2025
2. Centers for disease control and prevention. about mammograms (2024). https://www.cdc.gov/breast-cancer/about/mammograms.html. Accessed 15 Apr 2025
3. World health organization. breast cancer (2024). https://www.who.int/news-room/fact-sheets/detail/breast-cancer. Accessed 13 May 2025
4. American cancer society. survival rates for breast cancer (2025). https://www.cancer.org/cancer/types/breast-cancer/understanding-a-breast-cancer-diagnosis/breast-cancer-survival-rates.html. Accessed 20 Apr 2025
5. Boroumandzadeh, M., Parvinnia, E.: Automated classification of Bi-RADS in textual mammography reports. Turk. J. Electr. Eng. Comput. Sci. **29**(2), 632–647 (2021)
6. Breiman, L.: Random forests. Mach. Learn. **45**, 5–32 (2001)
7. Briciu, A., Călin, A.D., Miholca, D.L., Moroz-Dubenco, C., Petracu, V., Dascălu, G.: Machine-learning-based approaches for multi-level sentiment analysis of Romanian reviews. Mathematics **12**(3), 456 (2024)
8. Burlăcioiu, C., Boboc, C., Mitre, B., Dragne, I.: Text mining in business. A study of Romanian client's perception with respect to using telecommunication and energy apps. Econ. Comput. Econ. Cybern. Stud. Res. **57**(1), 221–234 (2023)
9. Cañete, J., Chaperon, G., Fuentes, R., Ho, J.H., Kang, H., Pérez, J.: Spanish pretrained BERT model and evaluation data. arXiv preprint arXiv:2308.02976 (2023)
10. Castro, S.M., et al.: Automated annotation and classification of Bi-RADS assessment from radiology reports. J. Biomed. Inform. **69**, 177–187 (2017)
11. Coita, I.F., Cioban, S., Mare, C.: Is trust a valid indicator of tax compliance behaviour? A study on taxpayers' public perception using sentiment analysis tools. In: Dima, A.M., Kelemen, M. (eds.) Digitalization and Big Data for Resilience and Economic Intelligence. Springer Proceedings in Business and Economics, pp. 99–108. Springer, Cham (2022). https://doi.org/10.1007/978-3-030-93286-2_7
12. Conneau, A., et al.: Unsupervised cross-lingual representation learning at scale. arXiv preprint arXiv:1911.02116 (2019)
13. Cramer, J.S.: The origins of logistic regression. Technical report, Tinbergen Institute (2002)
14. Ervik, M., Lam, F., L.M.C.M.F.J.M.F.A.B.F.: Global cancer observatory: cancer over time. International Agency for Research on Cancer, Lyon, France (2024). https://gco.iarc.who.int/overtime. Accessed 09 June 2025
15. Frank, E., Witten, I.H.: Generating accurate rule sets without global optimization (1998)
16. Kamel, H., Abdulah, D., Al-Tuwaijari, J.M.: Cancer classification using gaussian naive bayes algorithm. In: 2019 International Engineering Conference (IEC), pp. 165–170. IEEE (2019)
17. Lafferty, J., McCallum, A., Pereira, F., et al.: Conditional random fields: probabilistic models for segmenting and labeling sequence data. In: ICML, Williamstown, MA, vol. 1, p. 3 (2001)
18. Lehman, C.D., et al.: National performance benchmarks for modern screening digital mammography: update from the breast cancer surveillance consortium. Radiology **283**(1), 49–58 (2017)

19. López-Úbeda, P., Martín-Noguerol, T., Luna, A.: Automatic classification and prioritisation of actionable Bi-RADS categories using natural language processing models. Clin. Radiol. **79**(1), e1–e7 (2024)
20. Neagu, D.C., Rus, A.B., Grec, M., Boroianu, M.A., Bogdan, N., Gal, A.: Towards sentiment analysis for Romanian twitter content. Algorithms **15**(10), 357 (2022). https://doi.org/10.3390/a15100357, https://www.mdpi.com/1999-4893/15/10/357
21. Quinlan, J.R.: Induction of decision trees. Mach. Learn. **1**, 81–106 (1986)
22. Russu, R.M., Dinsoreanu, M., Vlad, O.L., Potolea, R.: An opinion mining approach for Romanian language. In: 2014 IEEE 10th International Conference on Intelligent Computer Communication and Processing (ICCP), pp. 43–46 (2014). https://doi.org/10.1109/ICCP.2014.6936978
23. Sippo, D.A., et al.: Automated extraction of Bi-RADS final assessment categories from radiology reports with natural language processing. J. Digit. Imaging **26**, 989–994 (2013)

Towards Self-learning AI-Based Quality Rules for Statistical Process Control

Marcel Dechert[1]([✉]) [iD], Fulya Horozal[1] [iD], Sebastian Scholze[1] [iD], Marcel Wabo[2], and Ratan Kotipalli[2]

[1] Institut für angewandte Systemtechnik Bremen GmbH, Bremen, Germany
{dechert,horozal,scholze}@atb-bremen.de
[2] Continental Automotive Technologies GmbH, Babenhausen, Germany
{marcel.wabo,raja.venkata.ratan.kotipalli}@continental-corporation.com

Abstract. Amid the immense complexity of modern Industry 4.0 manufacturing, traditional Statistical Process Control (SPC) approaches, which are reliant on manually defined thresholds and expertise-driven control charts, are proving increasingly insufficient. The vast amount of data, continual evolution of products and processes and the necessity for timely, meaningful alerts have exposed key limitations in existing methods, including scalability, adaptability and transparency. This paper introduces a novel automated, AI-driven methodology called *Self-learning AI-based Q-Rules* (SLAQ) to overcome these challenges. The SLAQ methodology combines optimization techniques with reinforcement learning to dynamically generate and refine quality rules (Q-Rules) for anomaly detection in production data. Initial Q-Rules are derived from historical data via optimization algorithms. These rules are subsequently refined through reinforcement learning using human feedback to enhance detection performance and adaptability to address changing process conditions.

Keywords: statistical process control · data science · AI · reinforcement learning

1 Introduction

In the era of Industry 4.0 and Big Data, traditional Statistical Process Control (SPC) has become increasingly unfeasible, particularly for large-scale manufacturers, due to its reliance on manually defined thresholds and static control charts. Managing thousands of products and variants across multiple production lines and stations has become unmanageable for humans alone, especially given the massive volume of process measurements generated. As products and production facilities continuously evolve, the statistical properties of process measurements also shift over time, which requires additional effort to maintain the accuracy of existing monitoring. Moreover, in real-world industrial applications,

C. Chira et al. (Eds.): InnoComp 2025, CCIS 2793, pp. 243–254, 2026.
https://doi.org/10.1007/978-3-032-12478-4_15

timely and meaningful alerts are essential for any SPC system to gain acceptance.

Since its introduction in the 1920s, SPC has undergone significant evolution leading to advancements such as multivariate, real-time and adaptive SPC techniques. However, there is still a need for more flexible, scalable and interpretable approach that can accommodate different processes and parameters. Manufacturing companies frequently use off-the-shelf solutions [11] to implement SPC typically using data analysis tools that allow quality engineers to monitor processes by creating control charts to identify process variations, data trends and detect out-of-control conditions. These tools require a lot of expertise from both a statistical and a process knowledge perspective in order to define the control chart and interpret them accordingly.

Machine learning (ML) models have been shown to be more proficient at detecting anomalies due to their capacity to process complex, non-linear patterns [10]. However, creating, maintaining and scaling production-ready machine learning models remains a time-consuming endeavor, which often involves manual data processing and model training tasks. Consequently, a trade-off must be made between better detection rates and efforts spent on setting up and maintaining the models. Even ML models that have proven to produce valuable outcomes frequently fail to be implemented on the shop floor. This is typically due to cost-efficiency considerations taking precedence, with expenses relating to maintenance, cloud providers and license fees often outweighing the anticipated benefits [14]. Auto-ML tools hold the promise to reduce the necessity of human expertise, but have so far struggled to fulfill expectations. Expert input is still often required for critical steps, and the lack of human-in-the-loop features and high cost of computational resources have prevented widespread adoption [1].

To address these challenges of scalability, cost and interpretability, we propose a novel AI-driven methodology called *Self-learning AI-based Q-Rules* (SLAQ), which is currently being developed as part of the German-funded DIAZI project focused on the digitalization of industrial processes [3]. The SLAQ methodology consists of two main stages:

1. **Algorithmic Q-Rule Generation:** Generating baseline quality rules (Q-Rules) from historical production data using optimization algorithms calibrated to specific deviation rates.
2. **AI-enhanced Q-Rules Generation:** Incorporating human feedback to iteratively refine the algorithmically generated baseline Q-Rules through reinforcement learning.

The first stage of the SLAQ methodology has been implemented. In this phase, data from the production database of a highly automated automotive production line is processed in near real-time. Q-Rules are automatically generated, checked, and visualized through a web-based GUI application (AI Q-Rules Dashboard), which provides anomaly notifications and a traffic-light status indicator to give users an overview of the current production state. The second stage is currently in early development and has not been fully implemented. These two stages are illustrated in Fig. 1 along with additional concepts that can be added to SLAQ, such as timeseries prediction and forecasting of potential future deviations using the generated Q-Rules.

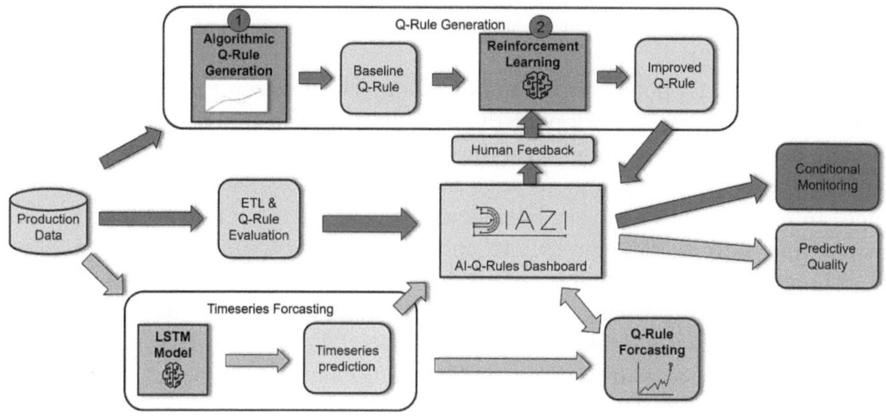

Fig. 1. Self-learning AI-based Q-Rules Overview

The paper is organized as follows: Sect. 2 discusses state-of-art and related work. Section 3.1 introduces the notion of quality rules (Q-Rules). The algorithmic Q-Rules generation is presented in Sect. 3.2. The concept behind the AI-enhanced Q-Rules generation using reinforcement learning is described in Sect. 3.3. Section 4 provides concluding remarks and outlines directions for future work.

2 Related Work

Statistical Process Control (SPC) is a well-known concept for monitoring the condition of a process over time with the objective of detecting anomalous process behavior that affects performance [12]. The primary objective of SPC charts is to detect out-of-control signals indicating the existence of special causes that disrupt process stability, while distinguishing them from common cause variations. Its formal beginning is often attributed to Walter A. Shewhart of the Bell Telephone Laboratories, who developed the statistical control chart concept in the 1920s. The original SPC rules devised by Shewhart were particularly effective for identifying large shifts in the process. In the 1950s, the Western Electric Company extended these foundational rules by introducing the Western Electric rules, which improved chart sensitivity and enabled the detection of smaller process shifts [18]. Building on this, Lloyd Nelson later refined and broadened the set of rules to recognize a wider range of abnormal patterns [13]. Historically, this often involved basic control charts focusing on univariate measurements. Over the decades, SPC methodologies have evolved significantly. More advanced techniques, such as Cumulative Sum (CUSUM) and Exponentially Weighted Moving

Average (EWMA) control charts were developed offering improved sensitivity for detecting smaller shifts and handling data characteristics like autocorrelation. There has also been a notable shift from univariate post-process SPC to Multivariate Statistical Process Control (MSPC), with increasing interest in the identification of concurrent patterns, where multiple abnormal patterns may appear simultaneously [2].

Goecks et al. [7] point out the major differences between *traditional SPC* and *smart SPC* systems: Traditional SPC uses manual or semi-automated data collection, has limited system integration, requires manual adjustments and focuses on monitoring current processes with historical data. Smart SPC features real-time, automated data analysis, integrates with IoT systems, uses AI to self-adjust parameters, and includes predictive and forecasting capabilities. Our proposed SLAQ methodology is in line with this definition of smart SPC, as the framework addresses the most pressing shortcomings of these established methods, namely the need to manually define the control limits ("Estimation Error Challenge") and the challenge of adapting control charts to changing processes [19] with the generative and self-learning features.

Many ML-based approaches have been applied to improve process control, including (semi-) supervised methods such as Support Vector Machines (SVMs), Random Forests (RFs) etc., as well as unsupervised techniques such as Clustering algorithms like K-Means, Gradient Boosting Machines (GBM), and many more. These methods often provide superior results over traditional SPC control charts and optimization methods in terms of accuracy, speed, and adaptability for manufacturing processes. ML models can have the ability to handle complex, unstructured, and big data, capture intricate patterns for higher accuracy, and offer faster computation for complex optimization tasks compared to traditional methods that rely on explicit mathematical models and assumptions [9, 10].

Reinforcement Learning (RL) differs from the above-mentioned machine learning techniques in particular because an RL agent can learn by a trial-and-error principle in direct interaction with its environment without requiring any pre-collected data or prior (human) knowledge. Through this, it has the ability to adapt flexibly to uncertain conditions [15]. This makes it a well-suited technology to be applied under Industry 4.0 conditions. RL is starting to significantly impact various areas within the manufacturing sector by addressing complex optimization challenges of Production Scheduling, Inventory Management, Maintenance Planning, and SPC [5]. Viharos et al. point out that RL is not yet widely used in the field of SPC and show an application of RL for SPC in manufacturing, where a learning agent interacts with a validated production simulation environment, taking production control actions to influence the process, and receiving costs as negative rewards to learn a policy that minimizes production cost while maintaining a high ratio of good products [17]. It has been shown that Reinforcement learning can be improved by employing human feedback by addressing the challenges of limited data and poorly defined reward functions in real-world applications. The author examines three primary ways humans can assist agents: providing demonstrations of desired behavior, offering online evaluative feedback, and defining a curriculum of tasks to guide the learning process [16].

3 The SLAQ Concept: Self-learning AI-Based Q-Rules

3.1 From Control Charts to Q-Rules

Various statistical trends and patterns in production data can reveal process issues that may lead to reduced quality or scrap. The classical process control rules mentioned in Sect. 2 (e.g. Western Electric and Nelson rules) were designed to detect such out-of-control conditions, but are less effective at detecting smaller shifts or adaptation over time. To address these limitations, we identified several generalized statistical rules used to ensure quality, and refer to them as *quality rules* (*Q-Rules*). Unlike traditional control charts that use fixed control limits, typically set at 3 standard deviations from the mean (i.e., 3 sigma), Q-Rules offer greater flexibility by adapting to specific process measurements, and enabling the detection of a wider range of abnormal patterns. Based on our initial statistical rules, we created Q-Rule templates to capture common deviation patterns using a single or a combination of multiple Q-Rules, see Table 1. This approach is extensible, so further patterns can always be added.

Table 1. Deviation Patterns & Matching Q-Rules

Pattern	Description	Possible Causes [8,12]	Q-Rule Templates
Trend	Sequential increase or decrease	Tool wear, drifting calibration	1, 2, 3
Shift	Sudden upward or downward change in the average level of the process	Changes in material, machine, or operator	1, 2
Freak	Single greatly deviating measurement	Noisy process	1
Cycle	Repeating up/down pattern, Periodic variations	Shift schedules, temperature cycles, operator fatigue, raw material deliveries	4
Systematic	Predictable point-to-point fluctuations e.g. "Zig-zag", lacking the unsystematic variation of a normal pattern	Same as cyclic patterns	6, 7
Stratification	Clustered near center line with low variation	Can hint at data collection issue	5
Mixture	Points tend to fall near the high and low edges of the pattern, with an absence of points near the middle	Overcontrol, where operators make process adjustments too often	4, 5, 7

In the literature, there is a distinction between *single control chart patterns* or *basic CCPs*, as opposed to *concurrent control chart patterns*, where multiple single patterns are happening all at once [6]. In this paper, we focus on the detection of single patterns. Usually, a combination of Q-Rules is required to fully capture each pattern. The list of Q-Rules templates for the first iteration of the SLAQ concept is as follows:

1. X points more than Y standard deviations from center line.
2. X points in a row on the same side of center line (above or below).
3. X points in a row all increasing or decreasing.
4. X out of Y points more than Z standard deviations from center line.
5. X out of Y points less than Z standard deviations from center line.
6. X points in a row more than Y standard deviations from center line (above or below).
7. X points in a row alternating up and down from point to point.

3.2 Algorithmic Q-Rule Generation

The goal of the initial (baseline) Q-Rule generation is to automatically identify a well-fitting Q-Rule for the given process parameters to start and scale the monitoring without requiring expert process or statistical knowledge. The first step in the Q-Rule generation pre-processing performed prior to determining the optimal control limits. If a process parameter is unsuitable for statistical analysis, such as having little or no variation in the measurements, no Q-Rule will be generated for it. Since traditional SPC rules are designed under the assumption that process data follow a Gaussian (normal) distribution, the distribution of process measurements are evaluated, and if necessary, e.g., in case of skewness, appropriate transformations such as Box-Cox [4] are applied to approximate a Gaussian distribution. This way, the validity of sigma-based control limits is ensured.

To generate a meaningful baseline, we start from the end-user perspective. A frequent problem with SPC systems is their tendency to produce undesired notifications, in which process measurements show a deviating pattern outside of the specified limits, yet no actual problem can be identified on the shop floor (false positives). On the other hand, other deviation patterns might be missed, where a notification would have been justified (false negatives). As a result, such notifications (e.g., an email arriving every 10 min) are often ignored. To address this, and in consultation with domain experts, we established that an appropriate baseline Q-Rule should strike a balance deviating neither too frequently nor too infrequently. We define the *deviation rate* as the ratio of total deviations to the number of time intervals (windows), formally expressed as:

$$\text{Deviation Rate} = \frac{\text{Total Deviations}}{\text{Number of Windows}} \tag{1}$$

A deviation rate of 1 indicates exactly one deviation per interval. The deviation rate serves as the objective function for the optimization. We employ a suite of optimization algorithms, namely Grid Search (Brute force), gradient-based, and Bayesian Optimization, to determine the optimal values for the number of data points and the sigma parameter needed to achieve the target deviation rate of 1, which means a perfect fit of the Q-Rule to the maximum number of notifications. Among these, gradient-based optimization proves to be the most computationally efficient in most scenarios. [1] The procedure illustrated in Fig. 2 is based on the first Q-Rule template, which was selected for our proof-of-concept implementation. Other Q-Rule templates will follow a similar flow. The resulting Q-Rule serves as a baseline for next stage of AI-enhanced Q-Rule refinement using reinforcement learning.

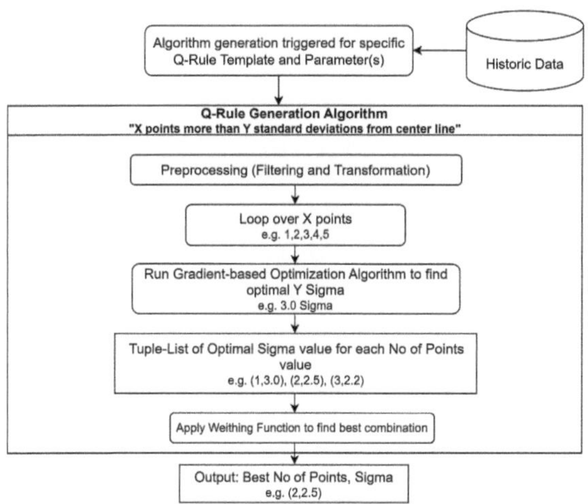

Fig. 2. Algorithm to generate Baseline Q-Rule.

3.3 AI-Enhanced Q-Rules: Reinforcement Learning with Human in-the-Loop

Framed as a reinforcement learning (RL) problem, our objective is to train a model-free, single-agent system that learns how to produce optimal Q-Rules to keep processes under control. We define the possible Q-Rule configurations as a discrete action space; for instance, in the case of Rule 1 from Sect. 3.1, the Q-Rule parameters X (number of points) and Y (sigma threshold) are both bounded. For our purposes, steps of 1 point and 0.1 sigma are sufficient. This enables us to construct the action set as all reasonable (X, Y) pairs and to employ simpler, more robust RL algorithms, such as Q-Learning, compared to those used for a continuous action space, e.g., Deep Deterministic Policy Gradient (DDPG).

---

[1] Benchmarking the algorithms on 60 days of training data for a specific process parameter, the algorithm based on the grid search from SciPy library (https://scipy.org) took 3.73 s, the gradient-based algorithm from the same package took 0,63 s and Bayesian Optimization (https://github.com/bayesian-optimization/BayesianOptimization) needed 1.83 s to produce the results.

Within the SLAQ methodology, the key elements of the reinforcement learning process are mapped as follows: The *agent* functions as a Q-Rule Generator, learning to define optimal Q-Rule configurations. Each *action* corresponds to the selection of a specific Q-Rule parameterization for specific process measurement data. The *environment* comprises the production data and the evaluation system, which applies the chosen Q-Rule. The *state* represents the contextual information available prior to rule selection, such as the historical data and related baseline Q-Rules. The *reward* is formulated as a composite metric that combines expert-driven insights with outcome-based signals to effectively capture the quality and utility of each applied Q-Rule. Figure 3 illustrates the interaction between these components.

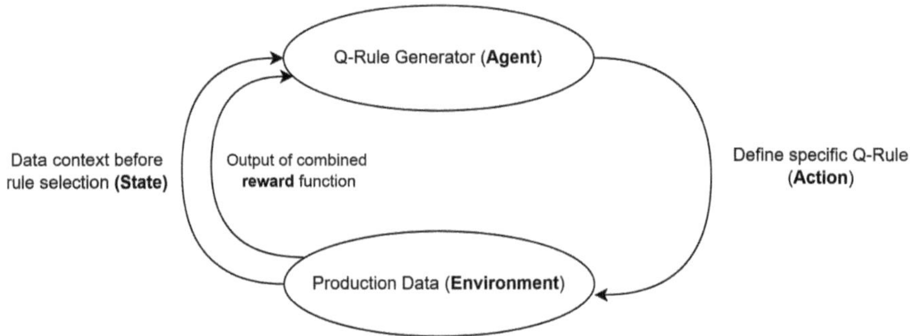

Fig. 3. Reinforcement Learning Components mapped to Q-Rules Generator.

An initial version of the agent can be trained using our baseline data. Agent training can start "offline" before human input is collected and will continue to be trained "online" based on collected feedback from human experts from a real production plant, enabling iterative refinement and improved Q-Rule generation over time.

To gather meaningful user feedback, users must be able to assess a Q-Rule based on a clear understanding of the process measurement data and its interaction with the Q-Rules during the review period. To facilitate this, the AI Q-Rules Dashboard offers a plot of the data and a slider that allows manipulation of the Q-Rule parameters (in this case, the sigma value) and the percentage of outliers (i.e., the amount of data points above or below the threshold lines), see Fig. 4). Afterwards, the expert users can fill in the feedback form (Fig. 5).

Human feedback is incorporated into the learning process via a unified reward function that combines multiple signals:

- **Explicit expert ratings:** Numerical evaluations of a Q-Rule's perceived effectiveness (1–5 scale).
- **Strong disagreement:** Experts can explicitly reject unsuitable Q-Rules, which are heavily penalized in learning.

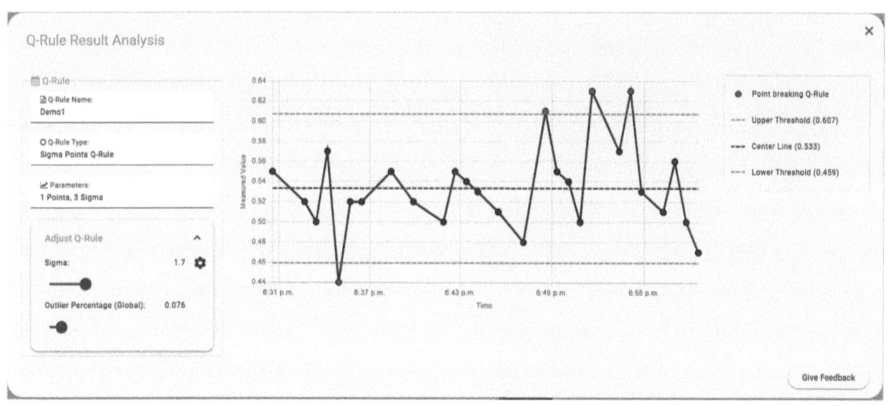

Fig. 4. Dashboard with Q-Rule Slider.

- **Outcome-based signals:** Reflect the real-world consequences of the Q-Rule, such as successful detections versus false or missed alarms (true/false positive, true/false negative).
- **Improvement suggestions:** Direct qualitative feedback from domain experts.

These diverse feedback types are integrated into a single reward that steers the agent's learning. The agent iteratively explores and refines its Q-Rules to maximize long-term performance while aligning with expert expectations. This human-in-the-loop reinforcement learning setup allows the agent to converge on optimal or near-optimal Q-Rules adapted to the process dynamics. The reward function is formally defined as:

$$R = \alpha \cdot \left(\frac{r - r_{\min}}{r_{\max} - r_{\min}} \right) + (1-\alpha) \cdot \left(\frac{w_{\mathrm{TP}} \cdot \mathrm{TP} + w_{\mathrm{TN}} \cdot \mathrm{TN} - w_{\mathrm{FP}} \cdot \mathrm{FP} - w_{\mathrm{FN}} \cdot \mathrm{FN}}{\mathrm{TP} + \mathrm{FP} + \mathrm{TN} + \mathrm{FN}} \right) \tag{2}$$

The reward function is designed as a weighted sum, with the first term capturing human insights and the second term reflecting the observed performance of the Q-Rule. The variable r represents the expert's numerical rating for the Q-Rule and is normalized between the minimum and maximum rating values r_{min} and r_{max}, to ensure comparability with the outcome-based term. The second term weights the outcome-based performance, where TP (true positives), TN (true negatives), FP (false positives), and FN (false negatives) represent counts for correct and incorrect anomaly detections made by the Q-Rule. Corresponding weights w_{TP}, w_{TN}, w_{FP}, w_{FN} allow prioritizing certain types of outcomes (e.g., penalizing false alarms more heavily). We use a single weighting coefficient $\alpha(0 \leq \alpha \leq 1)$, representing the expert judgment. It can be tuned to reflect the desired balance between more subjective expert feedback and objective outcome-based empirical results. We will start with a reasonable default that emphasizes expert input (e.g., $\alpha > 0.5$) and plan to adjust these weights dynamically as more outcome-based data becomes available.

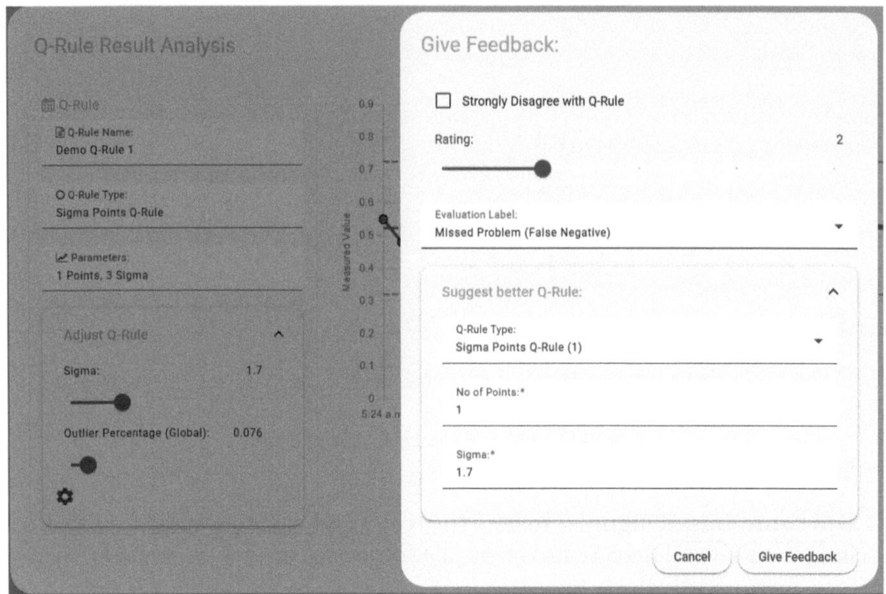

Fig. 5. Dashboard with Q-Rule Expert Feedback Form.

4 Conclusion and Future Work

In this paper, we presented a new approach to combine classic SPC with modern AI techniques to assemble a reliable, scalable, yet interpretable system for detecting deviations in diverse process data. We introduced a set of generic Q-Rules templates for detecting various deviation patterns and an algorithmic approach to generate instances of the Q-Rule templates for given process data. The full implementation of all Q-Rule templates in code is still in progress. We developed a dashboard for managing Q-Rules, viewing and analyzing results, and collecting expert feedback to customize the process control to the user needs, and enhance anomaly detection. We outlined how to integrate expert input with outcome-based evaluation to enhance the Q-Rule generation through reinforcement learning, which is still to be implemented.

However, the reinforcement learning training requires a large amount of data, which has yet to be collected. A substantial number of interactions between the agent and the environment, i.e. the Q-Rules generated for specific process measurements and the corresponding human feedback, is required to adequately explore the space of Q-Rule configurations in order to ensure the agent to learn robust policies and avoid overfitting on limited data.

Future work will focus on demonstrating the effectiveness of our approach in large-scale, real-world automotive manufacturing processes. As part of the validation process, we plan to compare the performance of the SLAQ approach against traditional SPC methods and other AI-driven approaches to evaluate its real-world applicability. A structured evaluation will be conducted to assess the benefits and limitations of SLAQ across multiple dimensions including initial setup effort, data requirements, computational efficiency, robustness, maintainability (adaptation effort), and detection accuracy. Additionally, we plan to extend SLAQ with Long Short-Term Memory (LSTM) time series prediction, using Q-Rules as guiding constraints to predict potential future process deviations and enable earlier intervention. Moreover, scalability and flexibility must be addressed not only within the AI framework but also through a coherent model training, deployment, and system architecture using open-source software to avoid license fees and vendor lock-in, combined with Kubernetes-based orchestration and MLOps best practices to enhance traceability and reliability.

Finally, SPC approaches focus solely on process measurements without directly linking to actual process outcomes. We plan to investigate the relationship between the SLAQ-based process control results and the scrap or rework occurring in the process, advancing from conditional monitoring toward predicting scrap.

Acknowledgments. The research project DIAZI leading to these results received funding from the German Federal Ministry for Economic Affairs and Energy (BMWE) with grant number 13IK012A.

Disclosure of Interests. The authors have no competing interests to declare that are relevant to the content of this article.

References

1. Azevedo, K., Quaranta, L., Calefato, F., Kalinowski, M.: A multivocal literature review on the benefits and limitations of automated machine learning tools (2024). https://doi.org/10.48550/arXiv.2401.11366, arXiv:2401.11366
2. Biegel, T., Helm, P., Jourdan, N., Metternich, J.: SSMSPC: self-supervised multivariate statistical in-process control in discrete manufacturing processes. J. Intell. Manuf. (2023). https://doi.org/10.1007/s10845-023-02156-7
3. BMWE, F.M.f.E.A.a.E.: Zukunftsinvestitionen Fahrzeughersteller und Zulieferindustrie (2024). https://www.bundeswirtschaftsministerium.de/Redaktion/DE/Publikationen/Industrie/zukunftsinvestitionen-fahrzeughersteller-und-zulieferindustrie.html
4. Box, G.E.P., Cox, D.R.: An analysis of transformations. J. Roy. Stat. Soc.: Ser. B (Methodol.) **26**(2), 211–243 (1964). https://doi.org/10.1111/j.2517-6161.1964.tb00553.x
5. Farooq, A., Iqbal, K.: A survey of reinforcement learning for optimization in automation. In: 2024 IEEE 20th International Conference on Automation Science and Engineering (CASE), pp. 2487–2494 (2024). https://doi.org/10.1109/CASE59546.2024.10711718, arXiv:2502.09417
6. García, E., Peñabaena-Niebles, R., Jubiz-Diaz, M., Perez-Tafur, A.: Concurrent control chart pattern recognition: a systematic review. Mathematics **10**(6), 934 (2022). https://doi.org/10.3390/math10060934

7. Goecks, L., Habekost, A., Coruzzolo, A., Sellitto, M.: Industry 4.0 and smart systems in manufacturing: guidelines for the implementation of a smart statistical process control. Appl. Syst. Innov. **7**, 24 (2024). https://doi.org/10.3390/asi7020024
8. Hachicha, W., Ghorbel, A.: A survey of control-chart pattern-recognition literature (1991–2010) based on a new conceptual classification scheme. Comput. Industr. Eng. **63**(1), 204–222 (2012). https://doi.org/10.1016/j.cie.2012.03.002
9. Kausik, A.K., Rashid, A.B., Baki, R.F., Jannat Maktum, M.M.: Machine learning algorithms for manufacturing quality assurance: a systematic review of performance metrics and applications. Array **26**, 100393 (2025). https://doi.org/10.1016/j.array.2025.100393
10. Khoza, S.C., Grobler, J.: Comparing machine learning and statistical process control for predicting manufacturing performance. In: Moura Oliveira, P., Novais, P., Reis, L.P. (eds.) EPIA 2019. LNCS (LNAI), vol. 11805, pp. 108–119. Springer, Cham (2019). https://doi.org/10.1007/978-3-030-30244-3_10
11. Minitab, LLC: Minitab. https://www.minitab.com
12. Montgomery, D.C.: Statistical Quality Control: A Modern Introduction. Wiley, 7th edn. (2014). https://www.perlego.com/book/3866383/statistical-quality-control-a-modern-introduction-pdf
13. Nelson, L.S.: The shewhart control chart-tests for special causes. J. Qual. Technol. **16**(4), 237–239 (1984). https://doi.org/10.1080/00224065.1984.11978921
14. Paleyes, A., Urma, R.G., Lawrence, N.D.: Challenges in deploying machine learning: a survey of case studies. ACM Comput. Surv. **55**(6), 114:1–114:29 (2022). https://doi.org/10.1145/3533378, https://dl.acm.org/doi/10.1145/3533378
15. Sutton, R.S., Barto, A.G.: Reinforcement Learning: An Introduction, 2nd edn. The MIT Press (2018). http://incompleteideas.net/book/the-book-2nd.html
16. Taylor, M.E.: Improving reinforcement learning with human input. In: Proceedings of the Twenty-Seventh International Joint Conference on Artificial Intelligence, IJCAI-18, pp. 5724–5728 (2018). https://doi.org/10.24963/ijcai.2018/817
17. Viharos, Z.J., Jakab, R.: Reinforcement learning for statistical process control in manufacturing. Measurement **182**, 109616 (2021). https://doi.org/10.1016/j.measurement.2021.109616
18. Western Electric Company: Statistical Quality Control Handbook. American Telephone and Telegraph Company, Chicago (1956)
19. Woodall, W.H., Montgomery, D.C.: Some current directions in the theory and application of statistical process monitoring. J. Qual. Technol. **46**(1), 78–94 (2014). https://doi.org/10.1080/00224065.2014.11917955, https://www.tandfonline.com/doi/full/10.1080/00224065.2014.11917955 publisher: Informa UK Limited publisher: Informa UK Limited

Laura Andreica$^{(\boxtimes)}$ [iD], Mara Macelaru [iD], Ioan Alin Danci, Mario Viman [iD], and Mihai Crisan [iD]

Technical University of Cluj-Napoca, Cluj-Napoca, Romania
`laura.andreica@campus.utcluj.ro`

Modeling Regional Smart Specialization via Value Chains: A Stochastic Approach Using Markov Chains for Smart Development Planning

Abstract. This paper presents an innovative, data-driven framework for determining the Economic Smart Specialization (ESS) of local communities through the use of discrete-time Markov chains. By synthesizing three critical layers of information—intrinsic industrial specialization derived from NACE-based revenue data, inter-industry supply chain flows–the proposed model captures the complex interplay between economic structure and geographic proximity. Specialization is modeled as a stochastic process, enabling simulation of how sectoral strengths evolve and diffuse across regions.

The framework yields a normalized ESS matrix that provides a probabilistic index of specialization at the locality level, informed by both upstream and downstream industrial relationships. Through Principal Component Analysis (PCA) and relative standard deviation metrics, the model quantifies the diversity and complexity of economic specialization under multiple transition scenarios. Applied to more than 3000 localities in Romania, the methodology demonstrates high scalability and adaptability, offering policy makers a powerful tool to design targeted, resilient, and context-sensitive smart specialization strategies.

This research bridges regional economic modeling and machine learning, contributing a reproducible, interpretable, and policy-relevant approach to enhancing territorial innovation potential in line with the principles of the Smart Specialization Strategy (S3).

Keywords: Markov Chains · Stochastic Algorithms · Economic Graph Modeling · Matrix Factorization · Specialization Inference · Data-Driven Policy Modeling

1 Introduction

The economic transformation of regions increasingly relies on evidence-based strategies that can capture complex patterns of specialization, innovation potential, and systemic interdependencies. In the European Union, the Smart Specialization Strategy (S3) has emerged as a cornerstone of regional development

C. Chira et al. (Eds.): InnoComp 2025, CCIS 2793, pp. 255–273, 2026.
https://doi.org/10.1007/978-3-032-12478-4_16

policy, promoting the identification and reinforcement of the unique assets, capabilities, and competitive advantages of each region. However, translating these policy aspirations into operational tools that are driven by data remains a persistent challenge, particularly in environments where economic dynamics are influenced by both structural industrial factors and spatial proximity effects.

Recent advances in computational modeling, particularly in the domain of stochastic processes and machine learning, offer new opportunities to address this gap. Markov chains, widely used in probabilistic modeling and machine learning applications, provide a robust mathematical framework for capturing the evolution of systems in which the next state depends only on the current configuration. When applied to economic ecosystems, Markov models allow the simulation of industrial transitions and the propagation of specialization patterns through interconnected value chains and geographic networks.

In this context, we propose a novel methodological framework that leverages discrete-time Markov chains to estimate the Economic Smart Specialization (ESS) of local communities. Our approach is based on three core layers of data: (1) intrinsic industrial specialization derived from local economic indicators and NACE revenue distributions, and (2) inter-industry dependencies modeled via national and regional value chain matrices.

The innovation of our framework lies in its ability to algorithmically integrate these layers into a unified stochastic transition model. The resulting ESS matrix offers a probabilistic index of sectoral importance across regions, reflecting both the internal economic structure and the external systemic influences. By performing principal component analysis (PCA) in multiple transition scenarios, intra-, upstream, downstream, and combined, we assess how specialization evolves in various value chain configurations. We further quantify the diversity of economic specialization using relative standard deviation metrics, highlighting how interconnectedness fosters more resilient and diversified regional profiles.

This research contributes to the fields of machine learning and regional economics by introducing a scalable, reproducible, and interpretable model that aligns economic modeling techniques with policy-relevant outcomes. From a machine learning perspective, our work exemplifies the use of Markov-based learning to uncover latent structure in high-dimensional economic data. From a smart specialization point of view, it provides a decision support tool that operationalize the principles of S3 with computational precision and spatial awareness.

The next parts of the paper are organized as follows: Sect. 1.2 examines previous studies on smart specialization, value chains, and the application of Markov chains in economic modeling. Section 3 presents the theoretical foundation of the model, emphasizing the integration of value chains and Markov processes. Section 2 defines the multilayered computational framework, encompassing the development of the intrinsic specialization, VCV, VCC, and contagion matrices. Section 4 highlights the experimental design and assessment methodology, integrating PCA analysis and standard deviation measurements. In the end, Sect. 5 summarises the findings, revisits the research questions, and establishes paths for future research.

1.1   Research Objective and Questions

The primary objective of this research is to develop and validate a computational framework for determining the Economic Smart Specialization (ESS) of local communities by integrating intrinsic industrial activity, inter-sectoral value chain dynamics, and spatial-demographic proximity into a unified stochastic model based on discrete-time Markov chains. The goal is to provide a scalable, data-driven tool that can support regional policymakers in identifying and improving local competitive advantages, facilitating targeted investment, and informing smart specialization strategies grounded in systemic economic interdependencies.

This research aims to explore how local economic specialization can be modeled as a stochastic process influenced by industrial interdependencies and spatial proximity. The study is guided by the following research questions:

- **RQ1:** How can Markov chains be used to integrate intrinsic specialization (IS), value chain interdependencies, and spatial proximity into a unified framework to assess economic smart specialization at the locality level?
- **RQ2:** To what extent do upstream and downstream flows in the value chain modify the relative importance of industries within a locality's specialization profile?
- **RQ3:** What are the measurable differences in specialization diversity and structural complexity when comparing intrinsic specialization with supply chain adjusted models?
- **RQ4:** How scalable and adaptable is the proposed ESS model for applications in regional development policies, especially in heterogeneous national contexts such as Romania?

1.2   Related Work

Markov processes and chains are essential in the smart specialization of communities, providing a framework for decision-making and resource allocation based on data-driven insights. Smart specialization strategies, as discussed in reputable studies [1,2,9], aim to stimulate regional growth and economic development by capitalizing on the unique strengths and potentials of each region. These strategies involve a systematic process to identify key areas of specialization and innovation.

In this context, Markov processes are employed to model the dynamics of regional development and analyze the transition probabilities between different states of specialization. Using Markov chains, communities can evaluate the likelihood of transitioning between specialization areas and make well-informed decisions about investment and policy formulation [10,13]. This approach enables a more strategic and adaptive method for smart specialization, aligning resources with the evolving needs and opportunities within a region.

Moreover, the decentralized and bottom-up nature of smart specialization, as emphasized in studies [1,3], can benefit from the flexibility and scalability offered by Markov processes. These models can adapt to different levels of resolution in defining communities and specialization areas, allowing stakeholders to customize strategies to various contexts and scales.

Overall, integrating Markov processes and chains into smart specialization initiatives empowers communities to navigate complex decision landscapes, prioritize activities based on regional potentials, and promote sustainable growth and innovation [11,15]. By embracing data-driven methodologies supported by Markov modeling, communities can enhance their competitiveness, resilience, and long-term development prospects.

The concept of smart specialization as a data-driven regional innovation strategy is further elaborated by McCann and Ortega-Argilés [9], who underscore the importance of aligning regional capabilities with global value chains and knowledge flows. Wu et al. [16] extend this by reviewing smart supply chains, showing how digitized and interconnected systems can improve specialization results through adaptive logistics and information exchange.

Matei et al. [8] propose an algorithm for determining the intrinsic economic specialization of a locality. Analyzing the value chain for defining smart specialization strategies for communities involves a multidimensional approach. Romão [12] highlights the importance of the diversity of services in the destination value chain, particularly in the tourism sector, which can play a central role in smart specialization strategies. Tommaso and Pinsky [5] further emphasize the importance of redefining products and markets, redefining productivity in the value chain, and developing clusters with local communities as strategies to implement shared value (SV) strategies. This aligns with the idea of using the value chain to create shared value within communities.

In addition, Wu et al. [16] discuss the essence of smart supply chain management driven by advanced data analytics throughout the value chain. This data-driven approach is crucial to effectively define and implement smart specialization strategies. Djekic et al. [6] introduce a methodology that synergies value chain promotion with quality development approaches, indicating the importance of integrating quality aspects into the value chain analysis process.

Sayadi et al. [14] emphasize the translation of consumer requirements into optimal practices to enhance product quality within the value chain, underscoring the importance of aligning value chain activities with customer needs.

This body of literature highlights a gap that the present work addresses: the lack of a unified computational model that combines intrinsic industrial data and value chain flows within a Markov framework to estimate the economic smart specialization potential.

2   Methodology

This study proposes a multilayered methodology that integrates economic activity data and inter-industry value chain into a coherent framework grounded in Markov chain theory. The central objective is to estimate the Economic Smart Specialization (ESS) of localities by simulating how industrial strengths evolve and diffuse within and across regions.

The methodology unfolds in several stages:

- **Data integration:** The model begins by compiling localized economic data from publicly available sources. Specifically, gross value added (GVA) figures and industrial revenues are aggregated at the level of individual localities and grouped by two-digit NACE codes. This serves as the foundation for assessing each locality's baseline specialization profile.
- **Modeling industrial dynamics:** Next, the industrial interactions within the economy are modeled using value chain data that describe upstream and downstream linkages between sectors. These interdependencies are integrated into the model to account for indirect specialization effects, how the performance of a sector in one locality is influenced by its suppliers or clients elsewhere.
- **Specialization index computation:** The final output is an ESS matrix that reflects the adjusted specialization profile of each locality. This matrix incorporates both local economic structure and systemic interactions, providing a probabilistic representation of sectoral strengths across space.
- **Computation of the importance of economic activities in each scenario:** As a final step, we compute the principal component analysis (PCA) on each matrix to determine the importance of each activity in the givne scenario. The more important an activity, the higher the PCA value associated with it.

Table 1 summarizes the principal notation used throughout the modeling framework.

Table 1. Notation for matrices and parameters used in the ESS framework

Symbol	Description
m	Number of localities (e.g., communes, towns, or municipalities)
n	Number of economic activities (two-digit NACE codes)
$IS \in \mathbb{R}^{m \times n}$	Intrinsic specialization matrix: normalized industrial weights per locality
s_{ij}	Raw sectoral revenue or GVA for locality i and sector j (pre-normalization)
S_{ij}	Normalized specialization weight for locality i and sector j
σ_j	Normalization coefficient for sector j: $\sigma_j = \frac{1}{\sum_{i=1}^{m} s_{ij}}$
$SC \in \mathbb{R}^{n \times n}$	Supply Chain matrix: proportion of inputs sourced from sector i to j
$DC \in \mathbb{R}^{n \times n}$	Demand Chain matrix: proportion of outputs sold from sector i to j
$VCV = IS \cdot SC$	Value chain propagation matrix (upstream impact)
$VCC = IS \cdot DC$	Value chain contagion matrix (downstream impact)
$ESS = IS \cdot SC \cdot DC$	Final Economic Smart Specialization matrix
PCA	Principal Component Analysis applied to IS, VCV, VCC, and ESS matrices to reduce dimensionality
$PC1$	First principal component representing dominant economic specialization dimension
SDR	Relative standard deviation of PCA scores: $\frac{\text{stdev}}{\text{mean}}$ (proxy for diversity)

3   Value Chains and Markov Processes

The integration of economic value chains with stochastic models such as Markov processes provides a powerful analytical framework for regional specialization analysis. Value chains capture inter-industry dependencies and are central to understanding the structural and functional flow of economic activity. Romão [12] emphasizes the diversity of services and interconnections within destination value chains as key enablers of competitive regional specialization, particularly in tourism-driven economies.

From a computational perspective, Markov chains offer a principal method for modeling transitions between states, such as industries or specialization levels, based on current conditions without dependence on historical paths. Ching and Ng [4] demonstrate the broad applicability of Markov models in real-world systems, providing foundational techniques for their use in decision-making and modeling of dynamic environments.

3.1   State Space and Stationary Probabilities in the Markov Model

In the context of this study, the state space of the Markov process is defined as the set of all possible industrial activities (sectors) identified by two-digit NACE codes, denoted as $\mathcal{S} = \{1, 2, ..., n\}$, where n is the total number of economic sectors considered. Each state corresponds to an economic activity, and transitions between states represent inter-industry flows of economic value, either through upstream (supply-side) or downstream (demand-side) linkages. For each locality, the normalized intrinsic specialization vector can be interpreted as an initial probability distribution over this state space, where the probability mass reflects the relative weight of each industry in the local economic structure.

The transition dynamics are encoded in the supply chain matrix (SC) and demand chain matrix (DC), both of which are column- and row-stochastic matrices, respectively. These matrices define how the specialization profile evolves when influenced by inter-sectoral dependencies. Under classical Markov theory, a transition matrix T is said to admit a stationary distribution π if there exists a vector such that $\pi T = \pi$ and $\sum_{i=1}^{n} \pi_i = 1$. Economically, a stationary distribution in this context represents a long-run equilibrium structure of specialization in which the relative importance of each industry remains stable under repeated propagation through the value chain.

While the model presented here does not explicitly simulate repeated Markov transitions over time, the composite matrix $SC \cdot DC$ can be interpreted as a one-step approximation of a Markov process that captures both upstream and downstream structural effects. The existence of a unique stationary distribution is guaranteed if the resulting transition matrix is irreducible and aperiodic, which generally holds under mild assumptions when all sectors are interconnected either directly or indirectly through the national value chain.

However, in this study, the goal is not to compute or analyze asymptotic convergence properties, but rather to use the Markov transition structure as a computational mechanism for capturing and amplifying the systemic economic influence of inter-sectoral relationships. Thus, the ESS matrix resulting from $IS \cdot SC \cdot DC$ can be seen as a transformed specialization distribution that reflects both initial economic capacity and structural potential, without requiring explicit convergence to a stationary state.

3.2   Intrinsic Specialization Weight Matrix

The economic specialization of an area is reflected in the industry and the gross value added to the whole ecosystem. Based on the Nomenclature of Economic Activities (NACE). The first matrix in Eq. 1 is an organized system of all localities and their specific industry, given the revenue and the population number. s_{NM} is the intrinsic specialization of the locality N and the specific industry M. To normalize the value of each revenue, a diagonal matrix with σ_i is applied, so that after multiplying the two matrices, the sum of the values on each line of the new matrix is equal to 1. The entire transition is similar to the stochastic matrix used in the Markov chain [4], except that in this model each line of the new matrix is a stochastic vector. After having the normalized matrix, with $\sum weights = 1$, we can determine the impact of each specialization in the area studied.

$$
IS = \begin{bmatrix} s_{11} & s_{12} & s_{13} & ... & s_{1M} \\ s_{21} & s_{22} & s_{23} & ... & s_{2M} \\ s_{31} & s_{32} & s_{33} & ... & s_{3M} \\ ... & & & & \\ s_{N1} & s_{N2} & s_{N3} & ... & s_{NM} \end{bmatrix} \cdot \begin{bmatrix} \sigma_1 & 0 & 0 & ... & 0 \\ 0 & \sigma_2 & 0 & ... & 0 \\ 0 & 0 & & ... & 0 \\ ... & & & & \\ 0 & 0 & 0 & ... & \sigma_M \end{bmatrix} = \begin{bmatrix} S_{11} & S_{12} & S_{13} & ... & S_{1M} \\ S_{21} & S_{22} & S_{23} & ... & S_{2M} \\ S_{31} & S_{32} & S_{33} & ... & S_{3M} \\ ... & & & & \\ S_{N1} & S_{N2} & S_{N3} & ... & S_{NM} \end{bmatrix}
$$

$$(1)$$

where $\sigma_i = \dfrac{1}{\sum\limits_{j=1}^{M}(s_{ij})}$ is the normalization factor that represents the sum of intrin-

sic specializations $s_{i1}, s_{i2}, ..., s_{i,M}$ of the locality i.

3.3 Value Chain Matrix

The economic value chain is modeled through two distinct matrices: the *supply chain matrix (SC)* and the *demand chain matrix (DC)*. Both matrices capture inter-industry flows between NACE-coded sectors and are essential to understanding how industries are interconnected through inputs and outputs.

The supply chain matrix SC is a square matrix where each element SC_{ij} represents the share of the total purchases made by industry j that come from industry i. The columns in SC are normalized, reflecting the fact that each column j represents the full structure of the input suppliers of the industry j'.

In contrast, the demand chain matrix DC captures the proportion of the output of industry i that is sold to industry j. Each row in DC is normalized, so that the sum of all demand destinations for a given producing industry is equal to 100%.

In both matrices, diagonal elements are allowed and represent intra-industry flows (i.e., when a sector both produces and consumes within itself). These matrices serve as the core building blocks for simulating specialization dynamics in the proposed Markov framework, Further a representation of a supply value chain matrix is mathematically expressed in Eq. 2. Similar is also the demand value chain matrix.

$$SC = \begin{bmatrix} SC_{11} & SC_{12} & SC_{13} & ... & SC_{1M} \\ SC_{21} & SC_{22} & SC_{23} & ... & SC_{2M} \\ SC_{31} & SC_{32} & SC_{33} & ... & SC_{3M} \\ ... & & & & \\ SC_{M1} & C_{M2} & SC_{M3} & ... & SC_{MM} \end{bmatrix} \tag{2}$$

3.4 Economic Smart Specialization Weight Matrix

The Economic Smart Specialization (ESS) matrix is derived through a multistep process that integrates intrinsic industrial structure with the topology of the national value chain. This approach captures both the current specialization and the structural potential of each locality.

The analysis begins with the **Intrinsic Specialization Matrix (IS)**, which captures the existing economic profile of each locality based on the sectoral revenue distribution. A preliminary PCA is applied to this matrix, and the resulting scores—corresponding to each CAEN code—are presented in the second column of the evaluation Table 2, column named `PCA_Instrinsic`.

This matrix is first multiplied with the **Supply Chain Matrix (SC)**—representing the upstream flow of goods—resulting in a new matrix denoted as **VCV**. This matrix is then analyzed using Principal Component Analysis (PCA) to identify the most structurally influential CAEN codes from a supplier's perspective. The results obtained are detailed in Table 2, column named `PCA_Vcv`.

In parallel, the same intrinsic specialization matrix IS is multiplied by the **Demand Chain Matrix (DC)**—capturing downstream industrial links—to obtain the **VCC** matrix. Again, PCA is applied to VCC to extract the dominant demand-oriented industrial specializations, as depicted in Table 2, column named `PCA_Vcc`.

Finally, to compute the overall specialization potential, all three matrices are combined into a single propagation sequence.

$$
\underbrace{\begin{bmatrix} ESS_{11} & ESS_{12} & \cdots & ESS_{1n} \\ ESS_{21} & ESS_{22} & \cdots & ESS_{2n} \\ \vdots & \vdots & \ddots & \vdots \\ ESS_{m1} & ESS_{m2} & \cdots & ESS_{mn} \end{bmatrix}}_{ESS\ (m \times n)} = \underbrace{\begin{bmatrix} IS_{11} & IS_{12} & \cdots & IS_{1n} \\ IS_{21} & IS_{22} & \cdots & IS_{2n} \\ \vdots & \vdots & \ddots & \vdots \\ IS_{m1} & IS_{m2} & \cdots & IS_{mn} \end{bmatrix}}_{IS\ (m \times n)} \cdot \underbrace{\begin{bmatrix} SC_{11} & SC_{12} & \cdots & SC_{1n} \\ SC_{21} & SC_{22} & \cdots & SC_{2n} \\ \vdots & \vdots & \ddots & \vdots \\ SC_{n1} & SC_{n2} & \cdots & SC_{nn} \end{bmatrix}}_{SC\ (n \times n)} \cdot \underbrace{\begin{bmatrix} DC_{11} & DC_{12} & \cdots & DC_{1n} \\ DC_{21} & DC_{22} & \cdots & DC_{2n} \\ \vdots & \vdots & \ddots & \vdots \\ DC_{n1} & DC_{n2} & \cdots & DC_{nn} \end{bmatrix}}_{DC\ (n \times n)}
\tag{3}
$$

This multiplication aggregates both upstream and downstream effects, producing a refined matrix that estimates the economic specialization potential across all CAEN codes for each region.

Then a final PCA is performed on the ESS matrix to generate standardized impact scores and rank CAEN codes according to their forward-looking contribution to regional smart specialization. This ranking provides a data-driven foundation for strategic policy targeting and industrial prioritization.

Once the ESS matrix is constructed, it serves as the input for a Principal Component Analysis (PCA), applied across CAEN sectors for each region. The goal is to reduce the dimensionality of the specialization profile and extract a single, interpretable score that captures the regional potential for smart specialization.

Each row in the ESS matrix represents a region's distribution of specialization weights across all sectors. By applying PCA to these rows, the model identifies the principal components that explain most of the variance in the specialization space. The first principal component (PC1) is retained as a one-dimensional indicator of economic potential, aggregating upstream, downstream, and intrinsic effects.

The resulting scores are standardized and used to:

- Rank counties by their overall smart specialization potential;
- Identify regions with the most coherent and concentrated specialization structure;
- Highlight underperforming areas with fragmented or low-impact industrial bases.

In parallel, PCA is also applied column-wise on the ESS matrix to assess which CAEN codes exhibit the highest overall contribution to regional specialization. This provides a dual perspective: both region-focused (rows) and sector-focused (columns), enabling comparative analysis across territories and industries simultaneously.

Table 2. PCA-based specialization scores for CAEN sectors. Complete CAEN sector PCA scores and standardized values with conditional coloring and formatted values.

CAEN	PCA_Int	STD_t	PCA_Vcv	STD_Vcv	PCA_Vcc	STD_Vcc	PCA_Pot	STD_Pot
CAEN_01	0.0075	-0.1507	0.4341	0.1941	0.2750	0.1061	0.7050	0.5282
CAEN_02	0.0012	-0.1569	0.2331	-0.0069	0.3740	0.2050	0.5946	0.4177
CAEN_03	0.0001	-0.1581	0.4642	0.2241	0.3733	0.2044	0.5831	0.4062
CAEN_05	0.0000	-0.1581	0.5716	0.3315	0.3795	0.2105	0.5019	0.3250
CAEN_06	0.2469	0.0888	0.3430	0.1030	0.2930	0.1241	0.4101	0.2333
CAEN_07	0.0001	-0.1580	0.0404	-0.1996	0.3196	0.1506	0.4615	0.2846
CAEN_08	0.0005	-0.1576	0.1323	-0.1077	0.2803	0.1113	0.4558	0.2789
CAEN_09	0.0004	-0.1577	0.1382	-0.1018	0.3687	0.1998	0.4873	0.3105
CAEN_10	0.0008	-0.1573	0.7502	0.5102	0.3389	0.1700	0.7935	0.6166
CAEN_11	0.0061	-0.1521	0.7491	0.5091	0.3715	0.2025	0.7420	0.5651
CAEN_12	0.0017	-0.1564	0.9473	0.7073	0.2409	0.0720	0.7316	0.5547
CAEN_13	0.0014	-0.1567	0.3882	0.1481	0.2611	0.0922	0.6053	0.4285
CAEN_14	0.0007	-0.1575	0.8152	0.5752	0.2203	0.0513	0.6752	0.4983
CAEN_15	0.0004	-0.1578	1.0000	0.7600	0.1533	-0.0157	0.5654	0.3885
CAEN_16	0.0029	-0.1552	0.5114	0.2714	0.2195	0.0505	0.4707	0.2938
CAEN_17	0.0003	-0.1579	0.4873	0.2472	0.1941	0.0251	0.4430	0.2661
CAEN_18	0.0002	-0.1579	0.7494	0.5094	0.2270	0.0581	0.6787	0.5018
CAEN_19	0.0019	-0.1563	0.5327	0.2927	0.5364	0.3674	0.5380	0.3611
CAEN_20	0.0005	-0.1576	0.2502	0.0102	0.3494	0.1805	0.5759	0.3990
CAEN_21	0.0010	-0.1571	0.5508	0.3108	0.2271	0.0582	0.5494	0.3726
CAEN_22	0.0086	-0.1496	0.2518	0.0118	0.2579	0.0889	0.5458	0.3690
CAEN_23	0.0008	-0.1574	0.3928	0.1527	0.4183	0.2494	0.4684	0.2916
CAEN_24	0.0038	-0.1543	0.0794	-0.1607	0.1604	-0.0085	0.3095	0.1326
CAEN_25	0.0016	-0.1566	0.1723	-0.0677	0.1575	-0.0114	0.2933	0.1164
CAEN_26	0.0003	-0.1578	0.2199	-0.0201	0.0679	-0.1010	0.3829	0.2060
CAEN_27	0.0038	-0.1544	0.4546	0.2146	0.0696	-0.0994	0.2360	0.0592
CAEN_28	0.0004	-0.1578	0.2944	0.0544	0.0658	-0.1031	0.2493	0.0724
CAEN_29	0.0037	-0.1544	0.1766	-0.0634	0.0680	-0.1009	0.2888	0.1120
CAEN_30	0.0009	-0.1573	0.0001	-0.2399	0.0904	-0.0786	0.2381	0.0613
CAEN_31	0.0015	-0.1566	0.5470	0.3070	0.0629	-0.1060	0.5507	0.3738
CAEN_32	0.0003	-0.1578	0.5291	0.2891	0.0952	-0.0738	0.3463	0.1694
CAEN_33	0.0000	-0.1581	0.0218	-0.2182	0.2478	0.0789	0.5192	0.3423
CAEN_35	0.5314	0.3733	0.5213	0.2812	0.2871	0.1182	0.4859	0.3090
CAEN_36	0.0004	-0.1578	0.3701	0.1300	0.1961	0.0272	0.3401	0.1633
CAEN_37	0.0000	-0.1581	0.3345	0.0944	0.1657	-0.0033	0.3218	0.1449
CAEN_38	0.0003	-0.1579	0.4057	0.1657	0.1672	-0.0017	0.3434	0.1665
CAEN_39	0.0000	-0.1581	0.1457	-0.0943	0.1244	-0.0446	0.3402	0.1633
CAEN_41	0.1879	0.0298	0.0649	-0.1751	0.1541	-0.0148	0.4890	0.3122
CAEN_42	0.0007	-0.1574	0.1667	-0.0734	0.1561	-0.0129	0.4398	0.2629
CAEN_43	0.0006	-0.1575	0.1742	-0.0658	0.1846	0.0156	0.4915	0.3147
CAEN_45	0.1851	0.0270	0.6524	0.4123	0.0019	-0.1670	0.5471	0.3702

(continued)

Table 2. (*continued*)

CAEN	PCA_Int	STD_t	PCA_Vcv	STD_Vcv	PCA_Vcc	STD_Vcc	PCA_Pot	STD_Pot
CAEN_46	1.0000	0.8419	0.6524	0.4123	0.0019	-0.1670	0.5502	0.3733
CAEN_47	0.9431	0.7850	0.0219	-0.2181	0.5771	0.4081	0.7739	0.5971
CAEN_49	0.0046	-0.1535	0.6358	0.3958	0.1591	-0.0098	0.5009	0.3240
CAEN_50	0.0001	-0.1580	0.4941	0.2541	0.1956	0.0266	0.4715	0.2946
CAEN_51	0.0000	-0.1581	0.5423	0.3023	0.2351	0.0661	0.5078	0.3309
CAEN_52	0.0051	-0.1531	0.5025	0.2625	0.2949	0.1259	0.6853	0.5085
CAEN_53	0.0010	-0.1571	0.5973	0.3573	0.2944	0.1254	0.7596	0.5827
CAEN_55	0.0003	-0.1578	0.1089	-0.1311	0.4603	0.2913	0.9494	0.7726
CAEN_56	0.0007	-0.1575	0.0935	-0.1465	0.2451	0.0761	1.0000	0.8231
CAEN_58	0.0000	-0.1581	0.6642	0.4242	0.2059	0.0369	0.7320	0.5552
CAEN_59	0.0000	-0.1581	0.0737	-0.1663	0.2062	0.0373	0.3574	0.1806
CAEN_60	0.0000	-0.1581	0.0940	-0.1460	0.2372	0.0683	0.3744	0.1975
CAEN_61	0.0996	-0.0585	0.2412	0.0012	0.2703	0.1013	0.4561	0.2792
CAEN_62	0.1376	-0.0205	0.2094	-0.0306	0.2867	0.1177	0.5805	0.4037
CAEN_63	0.0001	-0.1581	0.1009	-0.1391	0.2521	0.0832	0.3983	0.2214
CAEN_64	0.0004	-0.1578	0.1713	-0.0688	0.2777	0.1087	0.2839	0.1070
CAEN_65	0.0000	-0.1581	0.1274	-0.1126	0.1621	-0.0068	0.2241	0.0473
CAEN_66	0.0000	-0.1581	0.1174	-0.1226	0.1056	-0.0634	0.1621	-0.0148
CAEN_68	0.0018	-0.1563	0.4804	0.2404	0.1497	-0.0193	0.0752	-0.1017
CAEN_69	0.0001	-0.1581	0.2197	-0.0203	0.1842	0.0153	0.2504	0.0735
CAEN_70	0.0885	-0.0697	0.1918	-0.0482	0.1602	-0.0088	0.2782	0.1014
CAEN_71	0.0005	-0.1576	0.1772	-0.0628	0.1895	0.0205	0.2312	0.0544
CAEN_72	0.0008	-0.1574	0.0752	-0.1648	0.1854	0.0164	0.2572	0.0804
CAEN_73	0.0002	-0.1579	0.3975	0.1575	0.3424	0.1735	0.3699	0.1930
CAEN_74	0.0000	-0.1581	0.2027	-0.0373	0.1530	-0.0160	0.4099	0.2331
CAEN_75	0.0003	-0.1579	0.2589	0.0189	0.3486	0.1796	0.6134	0.4365
CAEN_77	0.0007	-0.1574	0.0635	-0.1765	0.3873	0.2183	0.4459	0.2691
CAEN_78	0.0000	-0.1581	0.3633	0.1233	0.1586	-0.0104	0.2778	0.1010
CAEN_79	0.0000	-0.1581	0.0583	-0.1817	0.0000	-0.1690	0.3046	0.1278
CAEN_80	0.0006	-0.1576	0.4635	0.2235	0.4275	0.2586	0.3847	0.2079
CAEN_81	0.0000	-0.1581	0.4000	0.1600	0.3903	0.2213	0.4557	0.2788
CAEN_82	0.0021	-0.1561	0.2374	-0.0027	0.4854	0.3165	0.3744	0.1975
CAEN_84	0.0001	-0.1581	0.0288	-0.2112	0.2131	0.0441	0.2834	0.1066
CAEN_85	0.0000	-0.1581	0.0363	-0.2038	0.4746	0.3057	0.5330	0.3561
CAEN_86	0.0001	-0.1581	0.1940	-0.0460	0.4486	0.2796	0.5440	0.3671
CAEN_87	0.0000	-0.1581	0.1544	-0.0857	0.3760	0.2070	0.7003	0.5235
CAEN_88	0.0000	-0.1581	0.0529	-0.1872	0.5251	0.3561	0.4328	0.2559
CAEN_90	0.0000	-0.1581	0.0002	-0.2398	0.4608	0.2918	0.3812	0.2043
CAEN_91	0.0000	-0.1581	0.0001	-0.2399	0.4535	0.2845	0.5561	0.3793
CAEN_92	0.0001	-0.1581	0.3709	0.1309	0.4918	0.3229	0.3125	0.1356
CAEN_93	0.0000	-0.1581	0.3594	0.1194	0.5300	0.3610	0.3519	0.1751
CAEN_94	0.0000	-0.1581	0.0116	-0.2284	0.4207	0.2517	0.3632	0.1864
CAEN_95	0.0001	-0.1581	0.6550	0.4150	0.4543	0.2853	0.3923	0.2154
CAEN_96	0.0000	-0.1581	0.4491	0.2091	0.7387	0.5697	0.4283	0.2514
CAEN_97	0.0000	-0.1581	0.0000	-0.2400	1.0000	0.8310	0.5006	0.3237
CAEN_98	0.0000	-0.1581	0.0886	-0.1515	0.6533	0.4843	0.6405	0.4636
CAEN_99	0.0000	-0.1581	0.0165	-0.2235	0.0355	-0.1335	0.0000	-0.1769
STD	0.1581		0.2400		0.1690		0.1769	

3.5 Dimensionality Reduction via Principal Component Analysis

To evaluate and compare the economic specialization profiles of localities across multiple scenarios, Principal Component Analysis (PCA) is employed as a dimensionality reduction technique. The rationale for using PCA is twofold: (i) to transform the high-dimensional sectoral data into a compact and interpretable form that captures the principal axes of variation, and (ii) to quantify the relative importance of each economic activity in driving regional specialization patterns under different structural configurations.

In this framework, PCA is applied to a series of matrices—namely, the intrinsic specialization matrix (IS), the upstream-adjusted matrix $(VCV = IS \cdot SC)$, the downstream-adjusted matrix $(VCC = IS \cdot DC)$, and the fully integrated specialization matrix $(ESS = IS \cdot SC \cdot DC)$. Each of these matrices is of size $m \times n$, where m represents the number of localities and n denotes the number of economic activities classified by two-digit NACE codes.

Conceptually, each row of these matrices corresponds to a locality's normalized specialization vector across n sectors, effectively capturing the probabilistic distribution of industrial weight within that locality. PCA is then applied row-wise to decompose the covariance structure among economic activities across all localities. The principal components derived from this decomposition represent orthogonal directions of maximal variance in the data, where the first component (PC1) accounts for the highest proportion of variance and serves as a unidimensional indicator of specialization potential.

The loadings associated with each component indicate the contribution of individual economic sectors to that latent dimension, while the component scores quantify the alignment of each locality's specialization profile with the identified patterns. This approach enables both cross-sectoral and cross-territorial comparisons, facilitating the identification of dominant industries and economically coherent regional configurations. By isolating the most informative dimensions of variability, PCA supports a policy-relevant analysis of smart specialization grounded in both structural dependencies and spatial heterogeneity.

4 Experimental Setup

To evaluate the effectiveness and scalability of the proposed model, we applied it to the case of Romania, a country composed of 41 counties and over 3,000 localities. The analysis was conducted at the locality level, using economic activity data categorized by two-digit NACE Rev. 2 codes. This level of granularity was selected to ensure relevance and interpretability while avoiding the excessive detail and data sparsity that often arise when using four-digit classifications.

The number of industrial activities included in the model corresponds to the complete set of available NACE two-digit codes represented in national and local statistical records. These activities form the state space of the Markov process and are used as the primary dimensions in the specialization analysis.

The core of the experimental analysis involves tracking how the relevance of each economic activity changes under different transition scenarios:

- **Baseline specialization:** Figure 1 illustrates the distribution of intrinsic economic potential based solely on the initial industrial profiles of each locality.
- **Downstream impact analysis:** Figure 2 presents the influence of downstream value chain interactions (i.e., sales from one industry to another). This reveals how activities gain or lose significance when their role as providers in the supply chain is considered.
- **Upstream impact analysis:** Figure 3 shows the effect of upstream value chain connections (i.e., acquisitions from other industries), highlighting the extent to which sectors rely on external inputs and the diffusion of economic value through supplier relationships.
- **Combined impact analysis:** Figure 4 depicts the adjusted specialization landscape after accounting for both upstream and downstream flows. This represents the comprehensive economic smart specialization profile for each locality.

To quantify the distributional diversity of the specialization in each scenario, we calculated the *relative standard deviation* of the values of the principal components extracted through PCA. This metric provides a normalized measure of concentration versus diversification in the economic structure.

$$SDR_i = \frac{\text{stdev}(\text{PCA}_i)}{\text{avg}(\text{PCA}_i)}$$

where SDR_i denotes the relative standard deviation for the i-th scenario, and PCA_i represents the set of scores of principal components corresponding to economic activities.

As noted in the previous literature [7], a lower SDR indicates greater economic diversity. Table 3 presents the values calculated for each experimental setting. The results show that the highest level of diversity occurs when both upstream and downstream effects are jointly considered, suggesting that comprehensive value chain integration promotes more balanced and diversified specialization patterns across localities.

Table 3. Relative standard deviation of PCA scores across specialization scenarios

Intrinsic	Downstream	Upstream	Combined
3.98	0.50	0.56	**0.34**

These findings empirically validate the modeling approach and demonstrate its capacity to capture both structural dependencies and spatial-economic diffusion in a meaningful and analytically robust way.

4.1 Case Studies and Sectoral Interpretations Based on PCA Indicators

By correlating the analysis of the local economy with the sectoral links in the value chains, a robust model emerges to identify domains with smart specialization potential. The indicators used - PCA_Vcv (sales), PCA_Vcc (purchases)

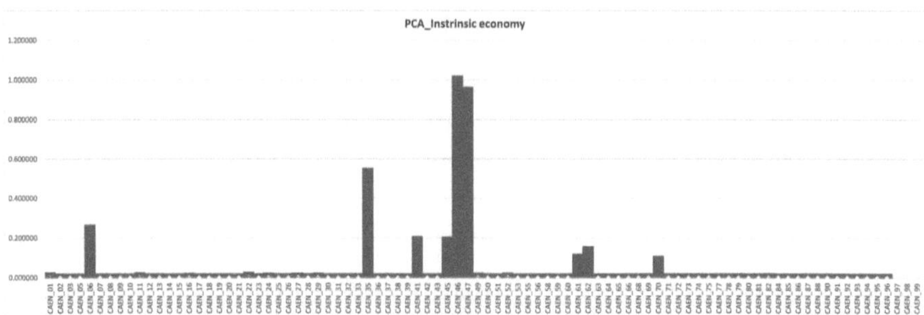

Fig. 1. Intrinsic NACE codes economic potential

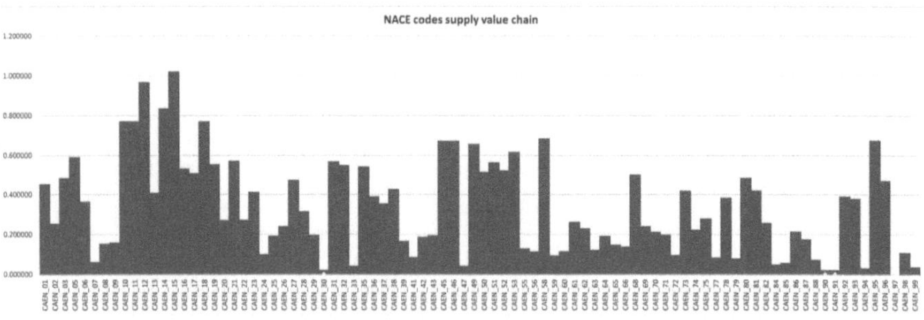

Fig. 2. NACE codes relevance after value chain downstream impact

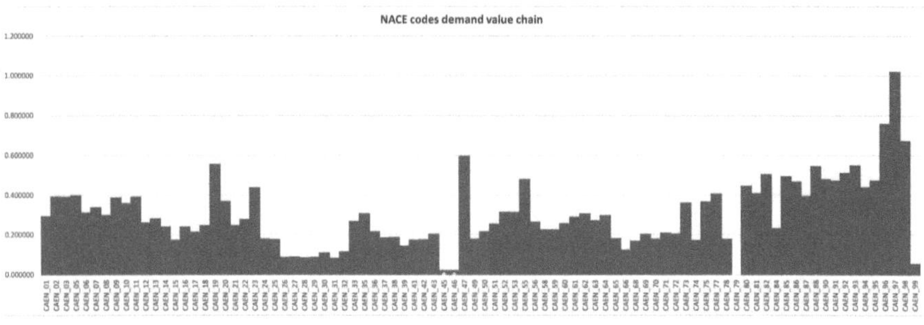

Fig. 3. NACE codes relevance after value chain upstream impact

and PCA_potential - help us to understand which sectors already act as economic engines and which have real chances of becoming such through local integration and strategic development.

A first relevant example is the food industry (NACE 10). Although it has a modest economic presence in many counties (low PCA_Intrinsic scores), data show that it is extremely well connected on both the demand and supply sides (high PCA_Vcv and PCA_Vcc), with a total potential exceeding 0.79. In counties with a strong agricultural profile – such as Vaslui, Teleorman, or Olt – there

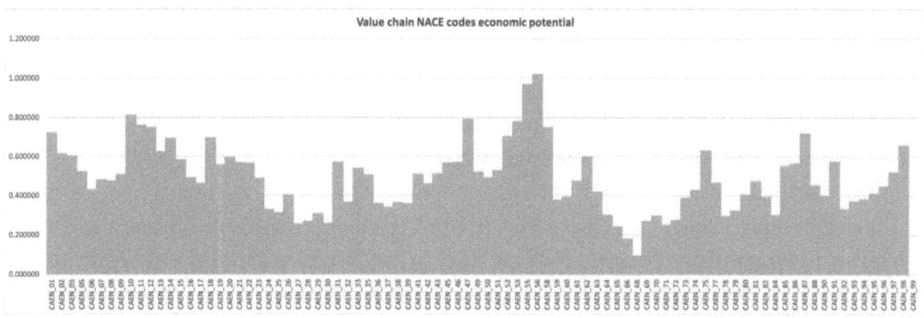

Fig. 4. NACE codes economic potential based on the complete value chain

is a clear need for local processing units that could valorize primary production and generate added value locally.

Public food services (NACE 56) offer another compelling example. With a maximum PCA_potential score (1.0), this sector demonstrates the ability to connect agricultural production, local processing, and tourism. In counties like Sibiu or Maramureș, this potential has already been tapped by building territorial brands based on local gastronomy, hospitality and agritourism - a model of functional and sustainable micro-economy.

In contrast, wholesale trade (NACE 46) – although dominant in terms of economic mass (PCA_Intrinsic = 1.0) – is poorly connected to local production (PCA_Vcc 0.002). This imbalance shows that, in the absence of strategic interventions such as regional logistics development or support for local suppliers, the sector remains isolated with limited impact on smart development. The potential of 0.55 confirms a middle pat, promising only if properly supported.

Another emerging example is the beverage industry (NACE 11), which combines all the characteristics of a viable smart specialization sector: high scores on all indicators and a potential exceeding 0.78. In areas with traditions in viticulture or orchards, such as Vrancea or Arad, this can generate products with local identity, high added value, and export potential.

Last but not least, the energy industries (NACE 35) and construction (NACE 41–43) show moderate but balanced scores. These are stable sectors that can support the development of other industrial branches. In counties undergoing an energy transition, like Gorj, they can become pillars of economic transformation through investments in green technologies and modern infrastructure.

Smart specialization does not involve simply identifying the largest or most visible economic sectors. What matters is seeing beyond the figures and understanding which activities can be integrated into local production and distribution networks and which can stimulate real and sustainable added value.

5   Conclusions

This study proposes a novel methodological framework for assessing economic smart specialization (ESS) at the locality level by integrating intrinsic industrial data, inter-sectoral value chain relationships, and spatial-demographic proximity into a unified model based on discrete-time Markov chains. The framework is designed to simulate the dynamic evolution of regional economic profiles by treating industrial specialization as a stochastic process influenced by both internal capacities and systemic external factors.

The computational model introduced here enables the derivation of a normalized ESS matrix that reflects the probabilistic significance of each industrial activity in geographic regions. By incorporating both upstream and downstream components of the supply chain, the model accounts for the reciprocal influence of inter-industry dependencies.

The experimental application to the Romanian context demonstrates the empirical robustness, scalability, and relevance of the model for regional policy making. Quantitative results derived from principal component analysis (PCA) and relative standard deviation (SDR) metrics confirm that the combined incorporation of value chain data significantly enhances the diversity and structural complexity of local specialization profiles. This highlights the critical role of systemic interdependencies in fostering resilient and diversified economic ecosystems.

In general, the proposed ESS model provides a reproducible, data-driven decision support tool capable of guiding targeted investment strategies and informing regional development policies in line with the principles of smart specialization. Its flexibility and adaptability to different national and subnational contexts underscore its potential utility for evidence-based economic planning in heterogeneous territorial settings.

5.1 Answering the Research Questions

RQ1: How can Markov chains be used to integrate intrinsic specialization, interdependencies of the value chain, and spatial proximity into a unified framework to assess economic smart specialization at the locality level?

The study introduces a methodological framework in which discrete-time Markov chains are used to model the economic smart specialization (ESS) of local communities. This is operationalized through the construction of transition matrices that reflect three core dimensions: (i) intrinsic specialization, derived from localized economic output data classified by NACE codes and (ii) interindustry relationships, represented through upstream and downstream flows in the value chain. The probabilistic nature of Markov chains facilitates the dynamic simulation of transitions in industrial importance, while matrix operations enable the integration of heterogeneous data layers into a cohesive model. The final ESS matrix, normalized for comparability, provides a scalable and data-driven representation of sectoral importance across regions.

RQ2: To what extent do upstream and downstream flows in the value chain modify the relative importance of industries within a locality's specialization profile?

The model explicitly accounts for the structural interdependencies between industries by incorporating value chain matrices into the economic specialization computation. The downstream matrix (SC) and its transpose (SC^T), representing the upstream flows, are used to adjust the intrinsic specialization matrix of the baseline (IS). Principal Component Analysis (PCA) applied to these transformed matrices reveals notable shifts in sectoral relevance. Specifically, the simulation results demonstrate that industries that serve as key suppliers or beneficiaries within the value chain experience significant revaluation in their strategic importance, thereby uncovering latent economic potential not observable from intrinsic data alone.

RQ3: What are the measurable differences in specialization diversity and structural complexity when comparing intrinsic specialization with supply chain adjusted models?

To evaluate changes in specialization diversity, the study uses the relative standard deviation (SDR) of the PCA-derived component scores. The intrinsic model exhibits a high SDR value (3.98), indicating a concentrated economic structure. In contrast, the incorporation of downstream and upstream flows results in substantially lower SDR values (0.50 and 0.56, respectively), suggesting an increase in sectoral diversification. The most balanced and complex specialization profile emerges in the combined scenario (SDR = 0.34), which integrates both directions of the value chain. These findings underscore the importance of systemic interdependencies in shaping resilient and diversified regional economic structures.

RQ4: How scalable and adaptable is the proposed ESS model for applications in regional development policies, especially in heterogeneous national contexts such as Romania?

The ESS framework was empirically validated using comprehensive economic and demographic data for more than 3000 localities in Romania. Its design accommodates different geographic and administrative levels, making it suitable for diverse regional planning contexts. Reliance on publicly accessible economic data, combined with modular matrix computations, ensures both scalability and adaptability. The model's capacity to simulate localized economic dynamics through Markov-based transitions positions it as a robust decision support tool for policymakers seeking to implement targeted smart specialization strategies. Its flexibility also allows for extension to other national contexts with varying degrees of economic heterogeneity.

5.2   Future Work

Several directions for future research emerge from the current study. First, while the model effectively integrates economic and spatial-demographic data, it can be further enriched by incorporating additional dimensions from the quadruple helix framework, including governance quality, academic activity, and environmental indicators. Such an integration would provide a more holistic representation of the determinants of regional innovation and specialization.

Second, future studies may explore clustering techniques to group localities based not only on geographic proximity but also on economic profile similarity. This could facilitate comparative analysis across regions and support the design of macro-regional strategies aligned with European territorial typologies. Comparative analyzes with existing administrative divisions (e.g. NUTS regions) could provide information on the efficacy and alignment of current policy structures.

Third, methodological extensions could consider the temporal evolution of specialization profiles by adopting higher-order Markov models or incorporating temporal dynamics explicitly. This would allow for the forecasting of specialization trajectories and a more proactive policy formulation.

Lastly, the model generalization should be tested in other national contexts with varying levels of economic development and data granularity. Cross-country validation would enhance its credibility as a transferable policy tool and may reveal structural variations in how smart specialization is manifested across different institutional and economic landscapes.

Acknowledgements. This work was supported by the following projects:

– "New Tool Based on Intelligent Algorithms and Machine Learning for Detecting Plant Diseases from Images", funded through the National Research Grant Competition – GNaC ARUT 2023.

– NetZeRoCities competence center, funded by the European Union, NextGenerationEU, and Romanian Government under the National Recovery and Resilience Plan for Romania, contract no. 760007/30.12.2022, with the Romanian Ministry of Research, Innovation, and Digitalization, through specific research project P1: Smart Governance.

– "Collaborative Framework for Smart Agriculture"—COSA that received funding from Romania's National Recovery and Resilience Law (PNRR-III-C9) 2022-I8, under grant agreement 760070.

References

1. Akberdina, V.: Smart specialization and sustainable regional growth. In: E3S Web Conference, vol. 250, p. 04009 (2021)
2. Amosha, O., Lyakh, O., Soldak, M., Cherevatskyi, D.: Institutional determinants of implementation of the smart specialisation concept: case for old industrial coal-mining regions in Ukraine, vol 17, no. 3, pp. 305–322 (2018)
3. Balland, P.A., Boschma, R., Crespo, J., Rigby, D.L.: Smart specialization policy in the European union: relatedness, knowledge complexity and regional diversification. Reg. Stud. **53**(9), 1252–1268 (2019)
4. Ching, W.K., Ng, M.K.: Markov chains. Models Algorithms Appl. **650** (2006)
5. De Tommaso, S.F.N., Pinsky, V.: Creating shared value: the case of innovability at Suzano in Brazil. Innov. Manag. Rev. (2021)
6. Djekic, I., Skunca, D., Nastasijevic, I., Tomovic, V., Tomasevic, I.: Transformation of quality aspects throughout the chicken meat supply chain. Br. Food J. **120**(5), 1132–1150 (2018)

7. Matei, O., Andreica, L., Danci, I.A., Avram, A., Vaduva, B.: Using Markov chains for determining the proximity contagion of smart specialization of localities. In: Soft Computing Models in Industrial and Environmental Applications, pp. 117–126 (2024)
8. Matei, O., Andreica, L., Danci, I.A., Avram, A., Tudor, F.: Using machine learning for identifying the intrinsic economic specializations of localities. In: Quintián, H., et al. (eds.) SOCO 2024. LNNS, vol. 889, pp. 137–146. Springer, Cham (2024). https://doi.org/10.1007/978-3-031-75010-6_14
9. McCann, P., Ortega-Argilés, R.: Smart specialization, regional growth and applications to European union cohesion policy. Reg. Stud. **49**(8), 1291–1302 (2015)
10. Patelli, A., Gabrielli, A., Cimini, G.: Generalized Markov stability of network communities (2019)
11. Piccardi, C.: Finding and testing network communities by lumped Markov chains. PLoS ONE **6**(11), e27028 (2011)
12. Romão, J.: Variety, smart specialization and tourism competitiveness. Sustainability **12**(14), 5765 (2020)
13. Salehifar, M., Akyol, E., Viswanatha, K., Rose, K.: On optimal coding of hidden Markov sources. In: 2014 Data Compression Conference. IEEE (2014)
14. Sayadi, S., Erraach, Y., Parra-López, C.: Translating consumer's olive-oil quality-attribute requirements into optimal olive-growing practices. Br. Food J. **119**(1), 190–214 (2017)
15. Visvizi, A., Lytras, M.D.: Sustainable smart cities and smart villages research: rethinking security, safety, well-being, and happiness. Sustainability **12**(1), 215 (2019)
16. Wu, L., Yue, X., Jin, A., Yen, D.C.: Smart supply chain management: a review and implications for future research. Int. J. Logist. Manag. **27**(2), 395–417 (2016)

Analysis of Economic Proximity Contagion and Regional Level Using Eigenvectors

Laura Andreica$^{(\boxtimes)}$ ⬩, Mara Macelaru ⬩, Ioan Alin Danci, Gabriel Andreica ⬩, and Patrick Axinte ⬩

Technical University of Cluj-Napoca, Cluj-Napoca, Romania
laura.andreica@campus.utcluj.ro

Abstract. This paper introduces a novel spectral framework for analyzing the spatial dynamics of Smart Specialization (S3) diffusion across Romanian regions. While S3 has become a central pillar of EU cohesion policy, the mechanisms underlying inter-regional innovation contagion remain poorly quantified. We address this gap by integrating Principal Component Analysis (PCA) with eigenvalue–eigenvector decomposition to model economic proximity contagion among 42 counties and over 3,000 localities. A contagion matrix is constructed using population, Gross Value Added (GVA), and Geographic Distance to estimate influence potential and structural centrality within the national innovation ecosystem. Our results uncover non-obvious clusters of innovation diffusion, highlight Bucharest and Cluj-Napoca as central propagation hubs, and demonstrate how structural positioning and connectivity can outweigh raw economic mass in determining territorial impact. The methodology offers a scalable and reproducible approach for identifying regional leaders and followers in smart specialization, with clear implications for multilevel governance and targeted policy design. This work contributes to the operationalization of S3 contagion analysis through unsupervised learning and network-based modeling in transitional economies.

Keywords: PCA · Eigenvectors · Machine Learning · Smart Specialization · Economic Contagion · Regional Innovation Systems

1 Introduction

Smart specialization (S3) has emerged as the cornerstone of the European Union's cohesion policy, promoting place-based innovation strategies tailored to regional strengths, capabilities and potential for excellence. Although considerable efforts have been dedicated to formulating and implementing S3 strategies at national and regional levels, the dynamics of cross-regional influence, here conceptualized as contagion, remains insufficiently explored. This is particularly relevant in the context of Romania, where regional disparities in innovation capacity, institutional maturity, and absorptive capability create a complex landscape for the inter-territorial diffusion of smart specialization practices [6,11].

C. Chira et al. (Eds.): InnoComp 2025, CCIS 2793, pp. 274–289, 2026.
https://doi.org/10.1007/978-3-032-12478-4_17

From a systems perspective, the development of S3 trajectories in one administrative region may propagate or influence neighboring or structurally similar regions. This phenomenon of policy contagion is driven by a combination of institutional mimicry, competitive alignment, and knowledge spillover. Previous studies suggest that the success of innovation ecosystems depends not only on endogenous factors but also on the quality and density of interregional links [4]. Quantifying such interregional dependencies requires a methodological shift from classical econometric models toward more expressive, high-dimensional approaches that can uncover latent structures of influence.

In this paper, we introduce a data-driven framework to analyze contagion patterns in smart specialization across Romanian counties and cities. Our approach leverages Principal Component Analysis (PCA) and eigenvalue-eigenvector decomposition to identify underlying dimensions of regional innovation performance and to detect the most structurally influential regions in the S3 landscape. PCA allows us to reduce the dimensionality of a diverse set of innovation and development indicators, while spectral analysis highlights the directions in which variance and influence are most concentrated [3].

By mapping Romanian regions into a latent innovation space, we provide empirical evidence of influence hierarchies and contagion clusters. These insights not only enhance the analytical depth of regional innovation studies but also offer practical value for policymakers aiming to coordinate and align S3 interventions across administrative boundaries.

This contribution is novel in two respects. First, it introduces spectral methods and unsupervised learning to the modeling of S3 contagion, a domain traditionally dominated by qualitative assessments and static benchmarking. Second, it provides a reproducible and quantitative foundation for identifying regional leaders and followers in the smart specialization ecosystem, with potential scalability to cross-national or longitudinal studies.

1.1 Research Objective and Questions

The main objective of this study is to create and evaluate a spectrum analytical framework to assess and model economic contagion and structural effect on Smart Specialization (S3) at both regional and local levels in Romania. By integrating principal component analysis (PCA) and eigenvalue-eigenvector decomposition, the research aims to identify latent structures of innovation diffusion and assess the hierarchical relevance of Romanian counties and localities in the wider economic and spatial network. This approach advances the quantitative modeling of cross-regional influence mechanisms, enabling more precise planning and alignment of Smart Specialization Strategies across administrative boundaries.

In pursuit of this objective, the paper addresses the following research questions:

1. **RQ1:** How can spectral methods, specifically PCA and eigenvector centrality, be used to quantify structural influence and contagion in the context of Smart Specialization across regions?

2. **RQ2:** Which Romanian counties and localities exhibit the highest influence potential in terms of economic contagion, and what factors contribute to their centrality within the national innovation network?
3. **RQ3:** How does the incorporation of proximity contagion–based on population, gross value added, and geographic distance–modify the hierarchical landscape of innovation influence as derived from raw economic indicators?

1.2 Structure of the Article

This article is structured into six sections.

Section 2 examines the current literature on Smart Specialization, Contagion Modeling in Economic Geography, and Spectral Approaches, including Principal Component research (PCA) and eigenvalue decomposition, in regional research.

Section 3 presents the research design, including the methodological framework adopted to model regional influence. Details the construction of the proximity-based contagion matrix and the spectral approach used to extract structural centralities. Section 4 formalizes the contagion model and introduces the mathematical formulation that integrates population, gross value added (GVA), and geographic distance as core variables influencing regional interactions. Section 5 outlines the experimental implementation of the model, including data preprocessing, normalization techniques, and the application of PCA. It also includes comparative results before and after contagion modeling, as well as visualizations of structural influence across counties and localities. Section 5.3 interprets the empirical findings, emphasizing the emergence of influence hubs and the structural anomalies revealed by the contagion-adjusted hierarchy. It also discusses the policy implications for spatial planning and regional innovation alignment.

Finally, Sect. 6 summarizes the key findings of the study, answers the research questions posed, and outlines the potential directions for future research in this field.

2 Related Work

Smart specialization strategies (S3) aim to enhance competitiveness by fostering innovation through the identification of unique regional assets. The European Union's approach reflects a shift toward experimentalist governance, as outlined by Radosevic et al. [13], who underscore the tension between procedural accountability and stakeholder-driven discovery processes.

Carayannis et al. [5] advance a "3P" framework–posture, propensity, and performance–as a multidimensional evaluative model consistent with the architecture of S3. Their integration of Multiple Criteria Decision Analysis (MCDA) provides a complementary base for PCA to identify the latent structures of regional innovation capacity.

Matei et al. [10] use machine learning to determine the most relevant instrinisc economic activities for localities. This can be combines with the contagion matrix, as defined in [9] for a more complex overview of the impact of local economies on the macro-economy.

While traditionally employed in financial systems, contagion matrices can model inter-urban dependencies and diffusion effects. Almeida et al. [1] introduced a contagion matrix based on CoVaR to detect systemic risk, a method adaptable to urban analysis for capturing spillovers among cities. Extending this logic, Matei et al. [9] utilized Markov Chains to assess the proximity contagion of smart specialization, demonstrating dynamic interactions among localities and offering a novel contribution to urban network theory.

PCA serves as a fundamental tool for dimensionality reduction and feature extraction in multivariate analyses. Jolliffe et al. [8] provide a foundational overview, while applications such as those in [5] and [9] illustrate PCA's utility in capturing latent structures across economic and innovation indicators.

In contexts such as Ukraine, smart specialization is posited as a key mechanism for regional reconstruction and economic renewal. Pidorycheva et al. [12] and Antoniuk et al. [2] argue for recalibrating development strategies through innovation-focused policies. These cases reinforce the need for analytical tools like PCA to guide resource allocation and policy focus in dynamic and uncertain environments.

3   Research Methodology and Methods

3.1   Methodological Approach

This study adopts a quantitative unsupervised learning approach rooted in spectral analysis to explore the structures of contagion and influence embedded in the Smart Specialization (S3) profiles of Romanian counties and cities. The methodology is designed to extract latent patterns of interregional similarity and structural centrality from high-dimensional regional innovation datasets. Specifically, we integrate Principal Component Analysis (PCA) for dimensionality reduction and eigenvalue-eigenvector decomposition for influence quantification and structural mapping.

The research design is cross-sectional and comparative, drawing on regionally disaggregated indicators representative of innovation capacity, economic specialization, institutional engagement, human capital development, and S3 implementation maturity. These indicators are organized into a normalized multivariate dataset suitable for spectral analysis.

3.2   Principal Component Analysis (PCA)

PCA is used to transform the original data matrix $\mathbf{X} \in \mathbb{R}^{n \times p}$, where n denotes the number of regions (counties and cities) and p the number of S3-related variables, into a lower dimensional space that maximizes variance preservation [7]. The covariance matrix is defined as:

$$\Sigma = \frac{1}{n-1}\mathbf{X}^\top\mathbf{X}$$

We compute the eigenvalues $\lambda_1 \geq \lambda_2 \geq \ldots \geq \lambda_p$ and corresponding eigenvectors $\mathbf{v}_1, \mathbf{v}_2, \ldots, \mathbf{v}_p$ of Σ. The first principal components k are selected such that the cumulative explained variance exceeds a predefined threshold (e.g. 85%), allowing for the identification of dominant latent dimensions that drive regional differentiation.

3.3 Spectral Decomposition and Eigenvector Centrality

To evaluate regional influence, we perform a spectral decomposition of the adjacency or similarity matrix $\mathbf{A}$, constructed from pairwise correlations or proximity in the PCA-reduced space. The spectral representation is given by:

$$\mathbf{A} = \mathbf{Q}\mathbf{\Lambda}\mathbf{Q}^{-1}$$

where $\mathbf{\Lambda}$ is the diagonal matrix of eigenvalues and $\mathbf{Q}$ contains the corresponding eigenvectors. The eigenvector associated with the largest non-trivial eigenvalue (the principal eigenvector) is interpreted as an indicator of structural centrality in the smart specialization influence network.

The eigenvector centrality assigns to each region a score proportional to the sum of the centralities of its neighbors, thereby capturing recursive influence dynamics. High centrality scores signal regions that not only exhibit strong individual performance, but also exert significant influence on structurally similar territories.

4 Model of the Economic Contagion

As defined by Matei et al. [9], the contagion matrix, mathematically expressed in Eq. 1, represents the influence of population number and distance between i and y localities, calculated by the Eq. 2.

The main diagonal is 1, as its influence does not alter the status of the existing smart specialization. The matrix is organized in columns, with each column denoting the condition of a particular locality and the effects from all other localities. The complete matrix represents a combination of all places concerning GVA, population figures, and the distances among them. When applied to actual data, we will obtain values that vary, either sub-unit or super-unit quantities.

$$C = \begin{bmatrix} 1 & C_{12} & C_{13} & \ldots & C_{1N} \\ C_{21} & 1 & C_{23} & \ldots & C_{2N} \\ C_{31} & C_{32} & 1 & \ldots & C_{3N} \\ \ldots & & & & \\ C_{N1} & C_{N2} & C_{N3} & \ldots & 1 \end{bmatrix} \tag{1}$$

The principal elements in defining proximity contagion are **population size**, **distance**, and **gross value added**. The collective behavior of a crowd can influence and dictate the subsequent actions of individuals. Gross value added (GVA) is an economic indicator that quantifies the value of goods and services generated within a certain area, industry, or sector of an economy. The contagion refers to behaviour, living environment, cognition, emotions, and particularly to economic activities. The transmission of smart specialization from locality j to locality i is defined as:

$$Cij = \frac{P_j \cdot GVA_j \cdot P_i \cdot GVA_i}{d_{ij}^2} \qquad (2)$$

Here, P_k represents the population of the locality k, GVA_k means the gross value added of the locality and d_{ij} indicates the distance between localities i and j. Following numerous experiments, we modified the formula outlined in [9] by multiplying the GVA and populations of the two localities instead of dividing them.

In Eq. 2, the population P is expressed in thousands of individuals, the GVA is expressed in billions of money units, and the distance d is measured in kilometers. Equation 2 intuitively suggests that contagion escalates with the influence of businesses and people propagating economic results, in relation to the robustness of local enterprises and the resident population. The contagionÂǎdecreases with increasing distance between the two localities.

5   Experimental Setup

The empirical analysis was conducted at two distinct levels of territorial aggregation, reflecting both administrative relevance and the granularity of the dynamics of smart specialization in Romania.

First, at the regional level, the study includes all 41 counties in Romania, along with the municipality of Bucharest, resulting in a total of 42 NUTS-3 units. These administrative entities represent the core units of regional governance and are directly responsible for the design and implementation of Smart Specialization Strategies (RIS3), as mandated by the European Cohesion Policy framework.

Second, to capture fine-grained spatial patterns and bottom-up innovation signals, the analysis was extended to the level of all localities (*lower-tier administrative-territorial units*), covering a total of over 3,000 municipalities, towns and communes. This granular scale allows for the detection of microclusters, peripheral contagion dynamics, and localized leadership effects that may not be visible at aggregated county level.

By examining both the meso (county) and micro (locality) scales, the study provides a comprehensive view of the structural influence and contagion potential across the Romanian territorial system. This dual-scale approach enhances the robustness of the spectral modeling and supports multilevel policy interpretation.

The result of the processing is a square contagion matrix $\mathbf{C}$, where each entry C_{ij} quantifies the economic influence of the locality j on the locality i.

5.1 Development of PCA Contagion Model

The dataset comprises innovation and development indicators for Romanian counties and local administrative-territorial units (municipalities, towns, and communes). Indicators include demographic data (population size), economic performance (Gross Value Added), and distances between the counties and the localities. Data was sourced from national statistical institutions, and geospatial repositories providing coordinates for calculating inter-locality distances.

Following steps were carried out:

1. computation of population and GVA values for all regions
2. calculation of pairwise distances between all entities
3. generation of the full contagion matrix C, where each element C_{ij} represents the directional influence of region j on region i

The matrix was structured such that each column encodes the external influences acting on a particular region, while the diagonal values were set to 1 to reflect self-influence. The contagion scores derived from this matrix were subsequently used to adjust the PCA-derived rankings, enabling a more accurate estimation of each region's structural influence in a spatially connected innovation ecosystem.

The dataset used to determined the principal component analysis was normalized using min-max scaling to ensure compatibility across different measurement scales. Additionally, missing values were handled via regional imputation or through removal if data sparsity exceeded a predefined threshold. The resulting data matrix includes over 3,000 localities and 42 counties, ready for input into PCA and contagion modeling.

5.2 Comparative Results Pre vs. Post Contagion

In the first phase of analysis, raw data was processed without applying any contagion model. For each locality and county, an economic strength indicator was calculated by multiplying the population by its respective GVA. This product serves as a proxy for the endogenous economic potential of each unit.

To ensure consistency between indicators and improve the comparability of values, all variables were normalized within the range [0, 1]. This normalization step was essential prior to applying Principal Component Analysis (PCA), which was used to assign weights and extract the dominant latent structures in the data.

As depicted in Fig. 1 and Table 1, we obtained the initial economic scores for all 42 Romanian counties. All counties in the table are sorted alphabetically, Δ Position means Before Rank minus After Rank, and the direction symbols indicate movement if the county is growing or shrinking. The representation of the results in Fig. 1 and Fig. 2 means that the darkest areas are also the most influential and powerful counties in the country, those that can change and inspire other counties or TUAs.

Table 1. Rank Shifts Of Counties Before And After Economic Contagion

County	Rank Before Contagion	Rank After Contagion	Δ Position	textbfDirection
Alba	19	21	-2	↓
Arad	17	18	-1	↓
Arges	7	5	2	↑
Bacau	10	8	2	↑
Bihor	11	23	-12	↓
Bistrita-Nasaud	26	33	-7	↓
Botosani	27	34	-7	↓
Braila	33	4	29	↑
Brasov	8	25	-17	↓
Bucuresti	1	1	0	→
Buzau	21	9	12	↑
Calarasi	34	41	-7	↓
Caras-Severin	40	27	13	↑
Cluj	3	7	-4	↓
Constanta	6	11	-5	↓
Covasna	41	31	10	↑
Dambovita	20	6	14	↑
Dolj	12	15	-3	↓
Galati	16	17	-1	↓
Giurgiu	38	10	28	↑
Gorj	31	37	-6	↓
Harghita	29	28	1	↑
Hunedoara	25	36	-11	↓
Ialomita	36	30	6	↑
Iasi	9	14	-5	↓
Ilfov	2	2	0	→
Maramures	18	29	-11	↓
Mehedinti	42	42	0	→
Mures	13	13	0	→
Neamt	22	24	-2	↓
Olt	23	20	3	↑
Prahova	5	3	2	↑
Salaj	37	35	2	↑
Satu Mare	22	39	-17	↓
Sibiu	14	16	-2	↓
Suceava	15	19	-4	↓
Teleorman	35	22	13	↑
Timis	4	12	-8	↓
Tulcea	39	40	-1	↓
Valcea	24	32	-8	↓
Vaslui	30	23	7	↑
Vrancea	32	30	2	↑

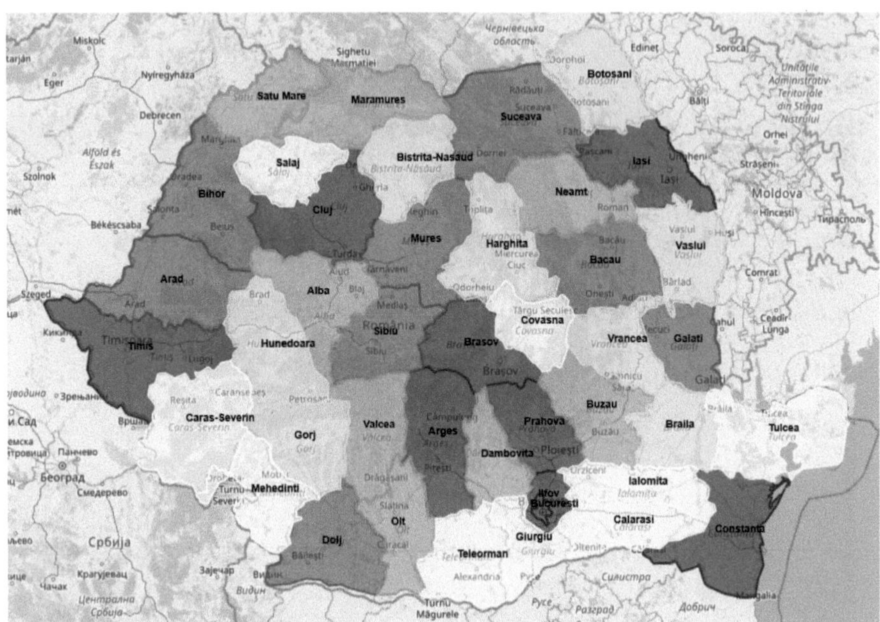

Fig. 1. Raw influence scores of the most relevant counties. Darker colors indicate higher economic influence

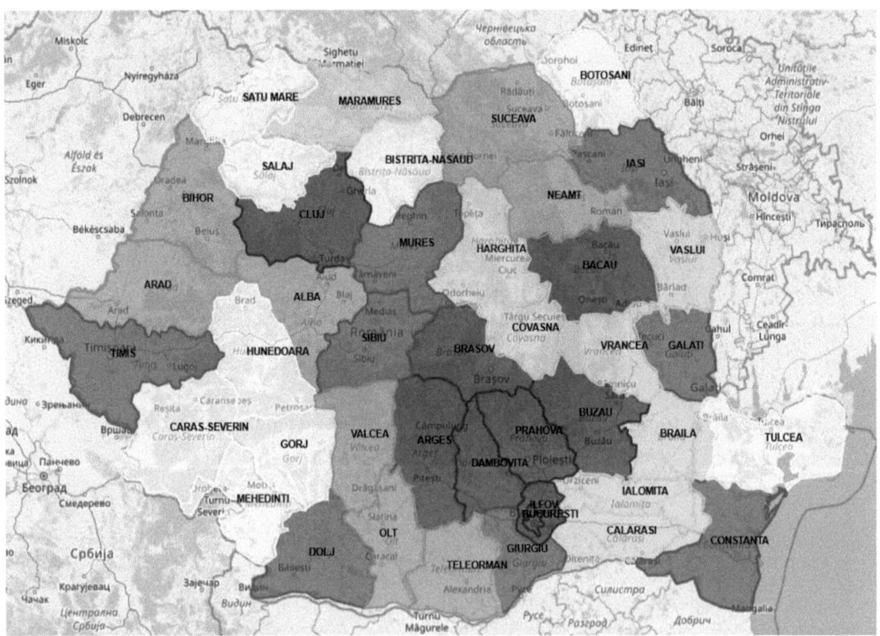

Fig. 2. Contagion influence scores of the most relevant counties. Darker colors indicate higher economic influence

The contagion model was applied at a micro-territorial scale, encompassing all 3,118 localities across Romania. The objective was to assess how each locality's economic influence is reshaped through proximity-driven diffusion mechanisms, incorporating factors such as innovation potential, population density, and economic output–evaluated through a combination of PCA and eigenvector centrality.

Figure 3 illustrates the top 50 localities based on their normalized economic scores before the contagion process, offering a concise view of the pre-contagion economic hierarchy at the local level. In this visualization, each locality is represented by a blue circle, with the circle's size indicating the magnitude of its intrinsic economic potential.

In contrast, the post-contagion landscape is depicted in Fig. 4, where red circles denote the diffused contagion influence of each locality. As with the pre-contagion figure, the size of each red circle reflects the relative strength of economic contagion potential that a locality exerts over others. A striking concentration of large red circles appears near the capital, underlining Bucharest's dominant role in the national economic network following the diffusion process.

This locality-level analysis confirms the network nature of territorial influence: localities that are economically weaker but embedded within influential ecosystems (e.g., Ilfov communes near Bucharest) exhibit higher post-contagion centrality than better-developed but isolated areas.

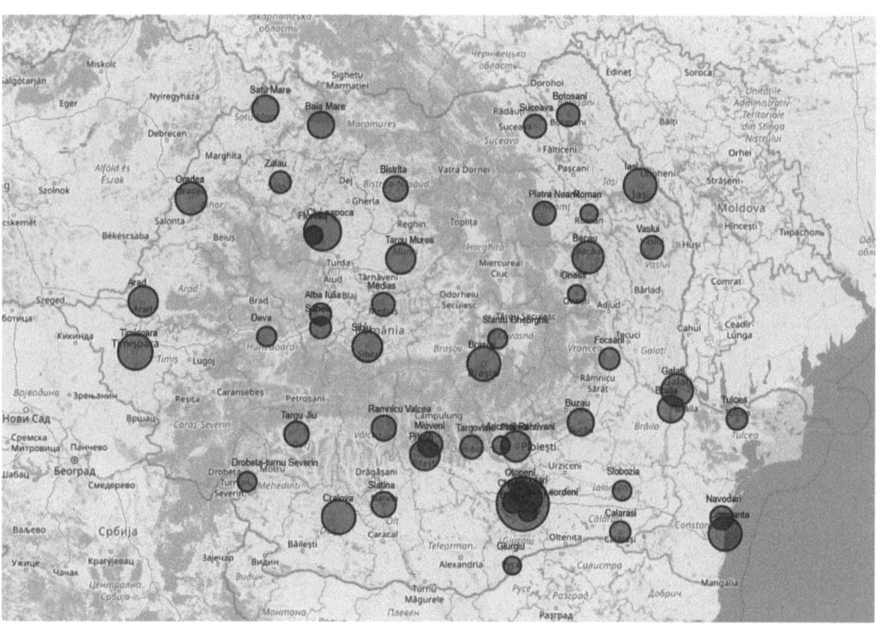

Fig. 3. Raw influence scores of the most relevant localities

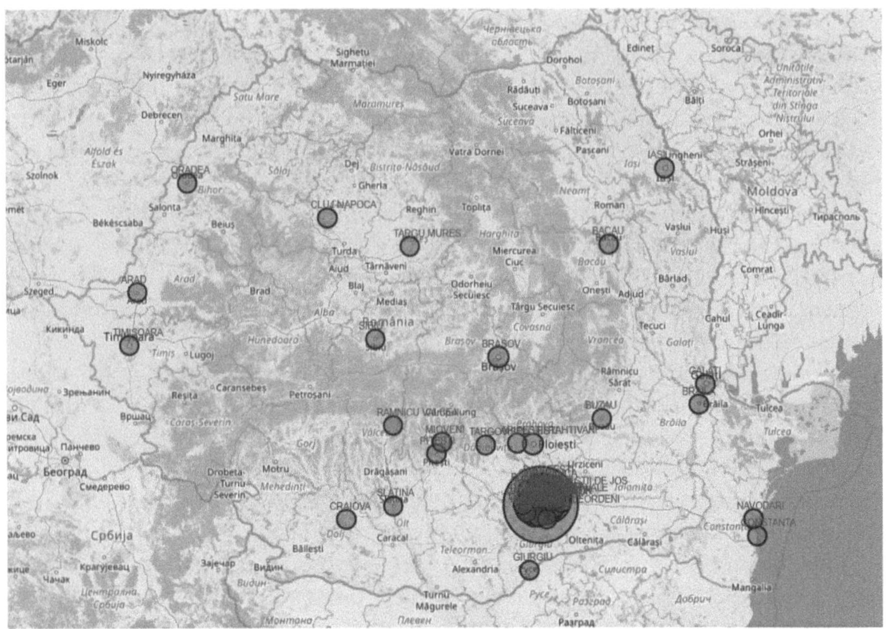

Fig. 4. Contagion influence scores of the most relevant localities

5.3   Discussion of the Results

To assess external validity, we correlated the contagion-based centrality scores with independent innovation indicators such as county-level R&D expenditure and patent applications (sourced from Eurostat and the Romanian National Institute of Statistics). The Pearson correlation between post-contagion eigenvector centrality and patent activity was 0.62, suggesting a moderate-to-strong alignment between the structural influence estimated by our model and empirical innovation outcomes.

The application of the territorial contagion model to Romania's Administrative-Territorial Units (ATUs) offers a more nuanced view of how economic influence is distributed across the national space. The capital and its metropolitan area form the indisputable core of Romania's territorial system. The municipality of Bucharest holds a maximum PCA score of 1, reflecting the highest concentration of demographic, administrative, and economic resources. The ATUs neighboring Bucharest in Ilfov County, such as Voluntari (0.546), Popești-Leordeni (0.091), Otopeni (0.052), and Chiajna (0.049), show exceptional contagion values, being not only influenced by but also dependent on the capital. This ensemble forms a metropolitan network with a high capacity to diffuse economic development to the rest of the country. This concentration of influence is explained by the developed infrastructure, large population, central administrative importance, and attraction of the majority of investments in the area. Furthermore, it can be observed that many localities in Ilfov heavily rely on the capital's fiscal resources, raising legitimate questions about their actual capacity to make independent decisions.

Next in the hierarchy of influence are municipalities like Ploiești (0.049), Brașov (0.017), Pitești (0.0067), Cluj-Napoca (0.0062), Constanța (0.0048), and Craiova and Iași (each with 0.0033). These act as regional growth poles, contributing to the diffusion of development in their immediate regions. Brașov benefits from a diversified industry, air connectivity, and a central location. Cluj has consolidated itself as a university center and innovation hub, while Pitești and Ploiești capitalize on their existing industrial infrastructure. However, these development poles face concrete challenges that limit their expansion potential. Rapid growth has generated problems such as urban traffic congestion, tensions in the labor market—where demand often exceeds the supply of qualified labor—and deficiencies in transport infrastructure that would effectively connect central and peripheral areas.

A notable case is Bacău County, which scores higher in contagion than Iași County, despite the latter's established status as a top-tier university, administrative, and technological hub in northeastern Romania. This apparent contradiction can be explained by a set of structural and functional factors that favor Bacău within the contagion model's logic. irstly, Bacău County benefits from a polycentric urban network, with dynamic secondary centers such as Onești, Moinești, and Comănești. These localities, through their economic activity and territorial connections, contribute to the extension of the county's economic influence beyond its administrative center. In contrast to Iași County's more monocentric structure dominated by its county seat, Bacău diffuses influence through a more balanced territorial distribution of activities. Secondly, its central geographic location within Moldova and its road and logistical connections—especially the E85 route and the future A7 motorway—enhance the county's capacity to generate and attract commercial flows. Thus, Bacău functions as an essential intermediate node in regional and national mobility networks, amplifying economic diffusion effects.

Lastly, the county's economic profile, with a solid industrial tradition (chemical industry, wood, aeronautical equipment) and a developing logistics component, ensures greater diversification of commercial exchanges. In this logic, Bacău not only concentrates economic activity but also effectively disperses it within the territorial network, as faithfully reflected by its contagion score. Therefore, Bacău's advantage over Iași does not contradict the absolute economic hierarchy but highlights that territorial influence, as defined by the contagion model, depends on connectivity, network positioning, and economic intermediation capacity—not solely on the size of an urban center. In contrast, although more economically developed, Iași is more geographically peripheral and concentrates flows within its own metropolitan area, with fewer lateral connections relevant to inter-county diffusion.

Other significant examples support the observation that territorial influence is not solely determined by economic size but more by strategic positioning within a functional network and the degree of territorial connectivity.

Thus, Argeș County scores higher in contagion than Constanța County, despite Constanța hosting an international logistics hub. The explanation lies in Pitești's proximity to Bucharest, its integration into the Pan-European Corridor A1, and its highly connected automotive industry within global supply chains. Although Constanța has a maritime strategic advantage, it remains relatively isolated from the rest of the country, connected mainly via the Bucharest–seaside axis, limiting the diffusion of its influence within the national network.

Similarly, Arad County records greater influence than Dolj County, even though Craiova is a larger urban center with a higher population and GDP. Arad, however, enjoys a privileged position as Romania's western gateway, located at the intersection of Pan-European road and rail corridors, allowing it to disseminate influence over long distances, including cross-border. Dolj, on the other hand, suffers from relatively low connectivity—the highway connecting it to Bucharest is still under development—and its southwestern geographical position limits its economic network impact.

Another revealing case is Sibiu County, which surpasses more populated and economically stronger counties such as Dolj or even Iași in terms of contagion. Although smaller demographically, Sibiu is centrally located, connected to modern transport networks (including the A1 motorway), benefits from an international airport, and exerts a multifaceted regional influence—economic, cultural, and touristic. Additionally, Sibiu County is well integrated into a dense regional network, functionally collaborating with neighboring Transylvanian counties and attracting foreign investment through easy access to the EU's western market.

6   Conclusions

These examples eloquently confirm that, in the logic of the applied contagion model, geographic accessibility, position in the national network, and the ability to radiate influence through infrastructure and functional connections can generate more influence than simple economic mass. These examples reinforce the idea that position in the economic network and geographic accessibility can outweigh economic size in impact. The correlation model consistently identifies the country's main economic centers (Bucharest, Cluj, Timișoara, Iași, Brașov), suggesting an empirical validation of Romania's urban hierarchy. Moreover, its ability to highlight significant anomalies (e.g., Bacău) confirms the method's utility in identifying emerging nodes and territorial inefficiencies. The results indicate that investments in transport infrastructure, support for polycentric urban networks, and stimulation of inter-county connections can contribute to increased positive territorial contagion. Additionally, the analysis can underpin cohesion and spatial planning policies that go beyond strict administrative logic, aiming at real functional networks.

The methodology proposed in this paper is applicable beyond Romania and can be adapted to other transitional or EU regions. Provided that comparable data on population, economic output, and geospatial positioning is available, the spectral contagion model can support regional innovation diagnostics and policy alignment in contexts such as Poland, Bulgaria, or Ukraine. This opens pathways for comparative S3 contagion studies at the cross-national level.

6.1 Answers to the Research Questions

RQ1: How can spectral methods, specifically PCA and eigenvector centrality, be used to quantify structural influence and contagion in the context of Smart Specialization across regions?

The study demonstrates that Principal Component Analysis (PCA) effectively captures the latent variance among Romanian regions by compressing multiple economic and innovation indicators into dominant principal components. This dimensionality reduction enables the isolation of key patterns of differentiation in regional development. Subsequently, the spectral decomposition of a contagion-based similarity matrix, built from geographic and economic proximities, allows for the computation of eigenvector centrality scores. These scores quantify the recursive influence of each region in the national innovation network, providing a robust, data-driven estimation of structural centrality. Together, PCA and eigenvector centrality facilitate a scalable and interpretable framework for modeling cross-regional influence in S3 contexts.

RQ2: Which Romanian counties and localities exhibit the highest influence potential in terms of economic contagion, and what factors contribute to their centrality within the national innovation network?

The results show that Bucharest and Cluj-Napoca are the most structurally influential nodes within the Romanian smart specialization network. Their centrality is not only a function of economic mass—measured by population and gross value added (GVA)—but also of functional positioning within transport and innovation networks. Other influential regions include Brașov, Ploiești, Pitești, and Sibiu, each benefiting from polycentric urban structures, strategic logistics corridors, or integration into broader regional innovation ecosystems. Notably, Bacău County outperforms Iași in contagion ranking, due to its balanced urban distribution, superior logistic connectivity, and inter-regional linkage capacity, despite Iași's greater economic scale.

RQ3: How does the incorporation of proximity contagion—based on population, gross value added, and geographic distance—modify the hierarchical landscape of innovation influence as derived from raw economic indicators?

The incorporation of the proximity contagion model substantially alters the raw economic hierarchy. While initial PCA scores reflect endogenous potential (population × GVA), the adjusted contagion matrix reveals a more network-sensitive influence structure. Regions that may be smaller or less economically dominant in isolation–such as Arad, Bacău, or Sibiu–emerge as disproportionately influential due to their strategic geographic placement, transport inter-connectivity, and economic spillover effects. Conversely, some high-GVA regions like Constanța or Iași exhibit limited contagion due to geographic peripherality or monocentric urban systems. This reconfiguration confirms that regional influence in S3 ecosystems is relational and mediated by spatial-functional proximity, rather than being solely determined by economic mass.

6.2 Licitations

A key limitation of this study is the reliance on deterministic contagion scores derived from static economic indicators. While min-max normalization improves comparability, it may disproportionately amplify the influence of high-GVA urban cores. Moreover, PCA assumes linear relationships, which may not capture complex nonlinear innovation dynamics. Future work could incorporate nonlinear embeddings or probabilistic contagion modeling to address these concerns.

6.3 Future Work

While this study provides a robust static analysis of regional influence and economic contagion in the context of Smart Specialization (S3), future work should focus on extending the proposed framework along several methodological and empirical dimensions.

First, a natural progression would involve the temporal generalization of the contagion model by integrating panel data that tracks the evolution of key indicators (e.g., GVA, population, investment flows, innovation outputs) over time. This would allow the computation of time-varying contagion matrices and dynamic PCA embeddings, offering insight into how regional influence and structural centrality evolve under the impact of economic shocks, policy interventions, or infrastructure development.

Second, future research should explore the formulation of dynamic graph models, where nodes (regions or localities) and weighted edges (contagion intensities) vary over discrete or continuous time intervals. This would enable the application of temporal eigenvector centrality, graph neural networks (GNNs), or Markov switching models to detect structural transitions, path dependencies, or contagion acceleration effects.

Third, the model can be tested and validated in other national or transnational contexts (e.g., cross-border regions within the EU) to assess its transferability and generalizability. Such comparative studies would enrich the evidence base for S3 policy diffusion across heterogeneous institutional environments.

Finally, incorporating real-time spatial data sources (e.g., satellite-derived economic activity proxies, mobility patterns, or regional patenting rates) could enhance the granularity and reactivity of the model, making it suitable for adaptive policymaking and real-time investment prioritization.

These future extensions will not only improve the explanatory power of the proposed framework but will also contribute to the theoretical consolidation of contagion-based models in regional innovation systems.

Acknowledgements. This research has been supported by the following projects:

– NetZeRoCities competence center, funded by the European Union, NextGenerationEU, and Romanian Government under the National Recovery and Resilience

Plan for Romania, contract no. 760007/30.12.2022, with the Romanian Ministry of Research, Innovation, and Digitalization, through specific research project P1: Smart Governance.

– "Collaborative Framework for Smart Agriculture"—COSA that received funding from Romania's National Recovery and Resilience Law (PNRR-III-C9) 2022-I8, under grant agreement 760070.

– Romanian National Authority for Scientific Research and Innovation, CCCDI - UEFISCDI, project number ERANET-M-3-ERANET-REPLACER 17/2024, within PNCDI IV.

References

1. Almeida, A., Frascaroli, B., da Cunha, D.R.: Risk measures and contagion matrix: an application of CoVaR for the Brazilian financial market. Braz. Rev. Financ. **10**, 551–584 (2012)

2. Antoniuk, V.: The need to rebuild the economy of the regions of Ukraine and the role of smart specialization in its provision. Herald Econ. Sci. Ukr. (2024)

3. Balland, P.A., Rigby, D.L., Boschma, R.: Complex economic activities concentrate in large cities. Nat. Hum. Behav. **6**(1), 88–95 (2022)

4. Camagni, R., Capello, R.: Territorial capital and regional development. Reg. Stud. **53**(1), 1–10 (2019)

5. Carayannis, E., Grigoroudis, E.: Toward a new impact assessment approach of smart specialization strategies using multiple criteria decision analysis. IEEE Trans. Eng. Manage. **71**, 8795–8808 (2024)

6. Foray, D.: Smart specialization strategies and the transformation of regional innovation systems. Camb. J. Reg. Econ. Soc. **11**(1), 1–16 (2018)

7. Greenacre, M., Groenen, P.J., Hastie, T., d'Enza, A.I., Markos, A., Tuzhilina, E.: Principal component analysis. Nat. Rev. Methods Primers **2**(1), 100 (2022)

8. Jolliffe, I., Cadima, J.: Principal component analysis: a review and recent developments. Philos.Trans. Roy. Soc. A **374** (2016)

9. Matei, O., Andreica, L., Danci, I.A., Avram, A., Vaduva, B.: Using Markov chains for determining the proximity contagion of smart specialization of localities. In: Quintián, H., et al. (eds.) SOCO 2024. LNNS, vol. 889, pp. 117–126. Springer, Cham (2024)

10. Matei, O., Andreica, L., Danci, I.A., Avram, A., Tudor, F.: Using machine learning for identifying the intrinsic economic specializations of localities. In: Quintián, H., et al. (eds.) SOCO 2024. LNNS, vol. 889, pp. 137–146. Springer, Cham (2024). https://doi.org/10.1007/978-3-031-75010-6_14

11. McCann, P., Soete, L.: The EU smart specialisation policy: challenges and opportunities for regional innovation governance. Reg. Stud. **54**(10), 1327–1339 (2020)

12. Pidorycheva, I., Bash, A.: Smart specialization of industrial regions of Ukraine: organizational and economic support. Econ. Ind. (2024)

13. Radosevic, S., Zoretic, T.: Eu smart specialization policy between experimentation and accountability: dynamic policy cycle perspective. Eur. Plan. Stud. **32**, 1693–1712 (2024)

Comparison of Multi Decision Criteria Analysis Methods in Molecular Dynamics of Tritium Interaction with Beryllium at Tokamak Fusion Reactors

Alper Pahsa[✉][iD]

Havelsan Inc., Mustafa Kemal Mah. Şehit Öğretmen Şenay Aybüke Yalçın Cad. No:39, 06510 Ankara, Turkey
apahsa@havelsan.com.tr

Abstract. The understanding plasma-structural material interaction in Tokamak fusion reactors is tough. This topic has been studied extensively, including molecular dynamic simulations. Plasma material's interaction with high temperature fluctuations changes the fusion reactor wall's surface composition. This alteration may damage reactor components. This research aims to develop advanced materials that can survive harsh conditions, prevent degradation, and improve reactor performance. This methodology is essential to understanding fusion reactor power facility development. To study how plasma affects material surface crystal structure, an atom-by-atom simulation model of tritium sputtering on beryllium reactor wall structures is needed. Molecular dynamics Multi Decision Criteria Analysis (MCDA) simplifies operational parameter understanding. Implementing these methods improves molecular dynamics simulation. Using molecular dynamic modeling parameters for sputtering Tritium plasma on a Beryllium-based model surface, we evaluated many MCDA approaches.

Keywords: MCDA in Tokamak Fusion Reactors · MCDA at Molecular Dynamics · MCDA at Plasma Material Interaction

1 Introduction

Utilizing Tokamak-type fusion reactors for energy production is highly efficient. Its absence of radioactivity and inexhaustible fuel supply render it significantly pertinent. Nuclear fusion in hydrogen-derived fuels generates energetic particles such as neutrons and helium. These phenomena manifest in the center plasma of fusion energy reactors. Nuclear fusion is the process in which two nuclei combine to create a single, bigger nucleus. Coulombic repulsion prevents two positively charged particles from approaching each other too closely. Their nuclear capabilities have diminished in efficiency, hence complicating fusion processes. An energy barrier arises from the extensive range of the repulsive Coulomb force.

C. Chira et al. (Eds.): InnoComp 2025, CCIS 2793, pp. 290–301, 2026.
https://doi.org/10.1007/978-3-032-12478-4_18

The short-range nuclear force, in contrast, scarcely encompasses the nuclear "surface". The magnitude of the Coulomb barrier is determined by the charges and radii of the interacting nuclei. Inherently, highly charged particles elevate the barrier. This configuration frequently serves as the foundation for plasma environments. In these circumstances, the interplay between temperature and density can profoundly influence the likelihood of overcoming the Coulomb barrier [1]. The increased likelihood of particles surpassing this energy barrier, along with their elevated kinetic energy, facilitates the successful occurrence of nuclear fusion. Plasma environments frequently exhibit this configuration. Ions and electrons in plasma exhibit exterior electrical neutrality and dissociate from one another. Ions and electrons in fusion reactors are compelled to follow a helical trajectory around the field lines in a magnetic field. The design of TOKAMAK fusion reactors is based on the principle of magnetic confinement. The selection of an appropriate material to sustain plasma exposure is a critical issue that commercial fusion energy must resolve. The impact of plasma ions on reactor wall surfaces has been the subject of numerous studies that have utilized molecular dynamics simulations to investigate the various aspects of this phenomenon. Through a computational approach, molecular dynamics simulations examine the physical movements of atoms and molecules. Atoms and molecules are permitted to interact. We provide a glimpse into the dynamic "evolution" of the system by enabling atoms and molecules to interact for specific durations. The trajectories of atoms and molecules are determined by dynamics, which numerically resolves Newton's equations of motion for a system of interacting particles. The interactions between particles and their potential energy are frequently determined using interatomic potentials or molecular mechanical force fields. Their assessment is facilitated by a comparative analysis of molecular dynamics experiments and MD simulation results. Scanning electron microscopy (SEM) and nuclear magnetic resonance (NMR) spectroscopy are two frequently implemented methodologies for this objective. Examination of the interactions between the plasma and the wall in the Tokamak nuclear fusion reactor in Fig. 1. [2] using molecular dynamics simulations. The design of a molecular dynamics simulation must consider the available computational resources, as a variety of simulation programs, including LAMMPS and Atomic Simulation Environment, are employed in these simulations. In order to guarantee that the computation is completed within the specified time frame, it is imperative to determine the simulation size (n = the number of particles), time step, and total duration. Nevertheless, the simulations' duration must correspond with the timescales of the observed natural phenomena. In order to obtain statistically significant results from the simulations, the simulated duration must be consistent with the kinetics of the natural process. Interatomic potential calculators, atomic kinetic energy, sputtering yield of the material, magnetic force of the impacting atom, and temperature are the primary factors that define the process in molecular dynamics simulations of material-plasma interactions. The results of molecular dynamics simulations are influenced by these parameters. The results of molecular dynamic simulations can be subjected to Multiple Criteria Decision Analysis techniques in order to enable rational evaluations of these criteria. This research investigates the simulation parameters that optimize the molecular dynamics simulation of tritium interacting with beryllium using four distinct MCDA methodologies.

Fig. 1. A Korea Tokamak Reactor.

2   Methods

Nuclear fusion reactors generate energy best. An ideal radiation-free application with an infinite fuel source. Fusion energy reactors produce helium and neutrons for hydrogen-based fuels from nuclear fusion events in the core plasma. A larger nucleus is formed when two nuclei fuse. The two positively charged particles cannot approach enough to engage the nuclear forces needed for fusion due to Coulomb repulsion. Nuclear force scarcely extends beyond the nuclear "surface", while the repulsive Coulomb force creates an energy barrier. Charges and radii of the two interacting nuclei determine this Coulomb barrier. The barrier is proportionally higher for highly charged particles. Plasmas often create this. Plasma is conductive and neutral outside, with ions and electrons moving independently. In magnetic field-immersed fusion reactors, ions and electrons follow a helical path along field lines and must advance parallel to the field. The magnetic confinement principle of TOKAMAK fusion reactors [3]. Molecular dynamics simulations model plasma-reactor wall interaction. High-energy ions sputter reactor wall materials during fusion. This study shows the beryllium material structure and H ions for magnetic high-energy tritium plasma. This study begins with molecular dynamics. The computational framework of molecular dynamics (MD) integrates Newton's second rule of motion to represent particle movement. Motion of the i-th atom as given below [3]:

$$m_i \ddot{x}_i = - \triangledown U_i \tag{1}$$

in Eq. (1), we identify the position, define the mass, and delineate an inter-atomic potential energy function that characterizes the interaction between each atom and its neighboring atoms. Potential energy functions are defined in traditional molecular dynamics by mathematical equations derived from atomic positions. Molecular dynamics simulations are performed with the Python programming language. To achieve this purpose, we utilize Spyder (Scientific Python Editor) version 6.0.3, which is based on the Anaconda framework. A Dell Precision 7680, featuring an Intel Core i7 13th Gen processor, Ubuntu 24.10 Linux, and Python compiler version 3.12.7, was utilized for the calculations. This work employs molecular dynamics simulations conducted in the Atomic Simulation Environment, implemented in Python. The Atomic Simulation Environment (ASE) is a collection of Python modules and tools designed to facilitate the preparation, execution, visualization, and assessment of atomic simulations [4]. The Velocity-Verlet algorithm is the second consideration in conducting a molecular dynamics simulation. The Verlet algorithm is one of the most often utilized time integration methods in molecular dynamics [5]. Essential factors for the demonstration encompassed the simulation's duration and the quantity of steps incorporated. In a valid inquiry, the simulation may function for a lengthy period. The hydrogen ion is subjected to the Lorentz force as a result of the two-Tesla magnetic fields. Thereafter, the magnetic force is manually applied using the molecular dynamics application. The parameters for setting up molecular dynamics simulations are outlined below. Incident energy in kiloelectronvolts (Energy keV) is 15. The magnetic field measures 2 T, and the beam energy is 1,602.18 J, corresponding to 1,000 kelvins. There are six hydrogen atoms and one unit of hydrogen charge. The hydrogen ion (H+) charge is located 2.5 Å above the crystal surface. The hydrogen charge is one unit multiplied by the starting energy of hydrogen, quantified as 35 keV, at a temperature of 300 K. Charge of H ion (Coulombs) Simulating n stages consisting of a total of 10 steps. Prior to executing the simulation of tritium plasma material contact with the beryllium surface, specific parameters must be configured to characterize it. Altering these parameters affects the kinetic energy, potential energy, and sputtering yield of the system. The parameters include the kinematic force of tritium incidents, the magnetic induction force of tritium, the interatomic potentials between the surface and tritium atoms, and the system's temperature. The parameters are systematically organized in a table that establishes various decision criteria regimes, and they are arranged utilizing the numerous decision criteria algorithms detailed in the subsequent paragraphs. Following table shows the molecular dynamic multiple criteria look up table given in Table 1.

Table 1. MCDA Input Table

DecCriNo#1	DecCriTyp#2	TotKinEngSys(eV)#3	TotPotEng(eV)#4	SputYield#5
1	TriEnergy = 15 keV MagField = 2 T Temp = 300 K IntAtPotCal = CHGNetCalculator MagForc = 1.17 N	3.223	−7.31	1
2	TriEnergy = 25 keV MagField = 2 T Temp = 300 K IntAtPotCal = CHGNetCalculator MagForc = 0.99 N	3.349	−7.37	0
3	TriEnergy = 35 keV MagField = 2 T Temp = 300 K IntAtPotCal = CHGNetCalculator MagForc = 6.6 N	3.372	−7.459	0
4	TriEnergy = 15 keV MagField = 3 T Temp = 300 K IntAtPotCal = CHGNetCalculator MagForc = 6.8 N	3.332	−7.399	0
5	TriEnergy = 15 keV MagField = 4 T Temp = 300 K IntAtPotCal = CHGNetCalculator MagForc = 1.39 N	3.706	−7.33	1
6	TriEnergy = 15 keV MagField = 2 T Temp = 400 K IntAtPotCal = CHGNetCalculator MagForc = 1.8625 N	3.4	−7.417	1
7	TriEnergy = 15 keV MagField = 2 T Temp = 500 K IntAtPotCal = CHGNetCalculator MagForc = 1.7255 N	3.401	−7.425	1
8	TriEnergy = 15 keV MagField = 2 T Temp = 500 K IntAtPotCal = MorsePotential MagForc = 1.43 N	3.495	−7.511	1
9	TriEnergy = 15 keV MagField = 2 T Temp = 300 K IntAtPotCal = MorsePotential MagForc = 0.95 N	3.507	−7.405	1
10	TriEnergy = 15 keV MagField = 2 T Temp = 400 K IntAtPotCal = MorsePotential MagForc = 0.965 N	3.549	−7.417	1
11	TriEnergy = 15 keV MagField = 3 T Temp = 300 K IntAtPotCal = MorsePotential MagForc = 5.9 N	3.394	−7.485	0
12	TriEnergy = 15 keV MagField = 4 T Temp = 300 K IntAtPotCal = MorsePotential MagForc = 1.32 N	3.364	−7.444	0
13	TriEnergy=25keV MagField = 2 T Temp = 300 K IntAtPotCal = MorsePotential MagForc = 1.04 N	3.36	−7.419	0
14	TriEnergy=35keV MagField = 2 T Temp = 300 K IntAtPotCal = MorsePotential MagForc = 5.98 N	3.355	−7.404	0

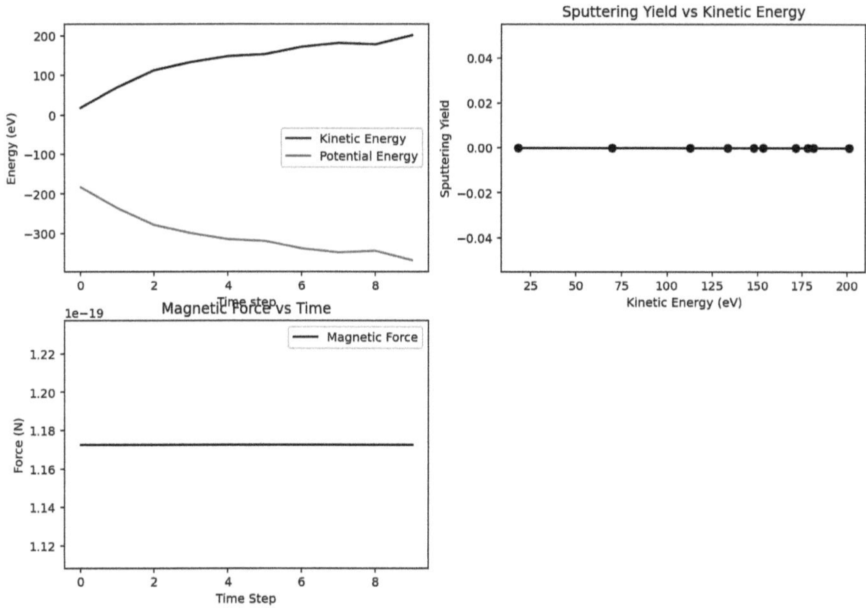

Fig. 2. MCDA Input Table Line 1 Energy Graph, Magnetic Force Graph and SputteringYield-KineticEnergy Graphs.

For Table 1 each of the multi decision critera has a graphical plot graphs given in the Fig. 2.

The Technique for Order of Preference by Similarity to Ideal Solution (TOPSIS) is a multi-criteria decision analysis method that was first developed by Ching-Lai Hwang and Yoon in 1981 [6]. Yoon and Hwang, Lai, and Liu subsequently improved the method in 1987 and 1993 [7]. Three The selected alternative must exhibit the minimal geometric distance from the positive ideal solution (PIS) and the maximal geometric distance from the negative ideal solution (NIS) in order to operate under the TOPSIS principle. This is a compensatory aggregation method that assesses a collection of alternatives by normalizing scores for each criterion and calculating the geometric distance between each alternative and the ideal alternative, which is defined as the optimal score for each criterion. The ordinal priority approach and the analytic hierarchy procedure can be employed to ascertain the weights of the criteria in the TOPSIS technique. The multi-attributive border approximation area comparison (MABAC) is the second multi-decision criteria analysis. In 2015, Pamučar and Ćirović introduced the MABAC approach, which has since been extensively employed to resolve a variety of real-world challenges. The MABAC multi-criteria framework offers the following primary advantages: (i) The outcomes generated by the MABAC approach are not influenced by the modification of the units of measurement for the criterion values of the alternatives. The MABAC approach does not alter the answers it provides when the criteria formulation type is changed, such as transitioning from a benefit type to a cost type. The mathematical calculations are not complicated by the inclusion of additional criteria and alternatives, which is why the MABAC method algorithm is highly effective for addressing multi-criteria problems with numerous criteria and alternatives [8]. The tertiary technique in the analysis is the Characteristics Objects technique (COMET). This method is predicated on the development of a rule basis that simplifies the evaluation of characteristic values [9]. The method's effective implementation in scenarios with imperfect data is facilitated by the method's assumptions, which are founded on fuzzy set theory [10]. Furthermore, the approach's ability to withstand ranking reversal phenomena is a substantial advantage [11]. The ranking mechanism remains unaltered by fluctuations in the number of potential outcomes. The criteria for evaluating choices are established by an expert by juxtaposing two or more Characteristic Objects (COs) and utilizing their knowledge [12,14]. Stable preference ordering toward ideal solutions (SPOTIS) is the ultimate strategy. The SPOTIS technique's fundamental concept is to ascertain the normalized distance dij (Ai, S*j) between each option Ai and the optimal solution S*j for each criterion Cj. Consequently, they compute the weighted average distance, which is a valid distance metric.

For comparing the stated methods Python "pymcda" [13,15] library is used in connection of the molecular dynamic simulation work. A versatile tool for resolving operations research multi-criteria problems is pymcdm. It possesses an extensive array of techniques for assessing choice alternatives and ascertaining the significance of qualities or normalization. This library includes a versatile module for visualizing the acquired findings. In this study, comparison of the 4 methods are given as chart, bar, trend line type of graph plots as given in the results and discussion section.

3   Results and Discussion

The many applications of the multi-decision criteria look-up table 1 assess the impact of tritium H on the beryllium crystal structure during molecular dynamics simulations. Subsequently, MD is executed, and as delineated in the methodology section, the results are analyzed with various multi-decision criterion analyses. The three-dimensional structure of the H tritium atom positioned above the beryllium surface functions as the input for the molecular dynamics simulation. The 3D output of tritium on the beryllium metal lattice is provided following the molecular dynamics simulation. In the given Fig. 3 white color atom denotes the H and the green colored atom shows the beryllium surface material.

The 3D resultant output molecular dynamic simulation of tritium on the beryllium metal lattice is provided in the following Fig. 4.

Based on the molecular dynamics performances by using the multi criteria decision inputs, the comparison of the TOPSIS, MABAC, COMET and SPOTIS bar chart result graphics are given in the following Fig. 5:

In Fig. 6, most polarized decisons are A6, A7 and A8 are similar in molecular dynamic results based on the comparison.

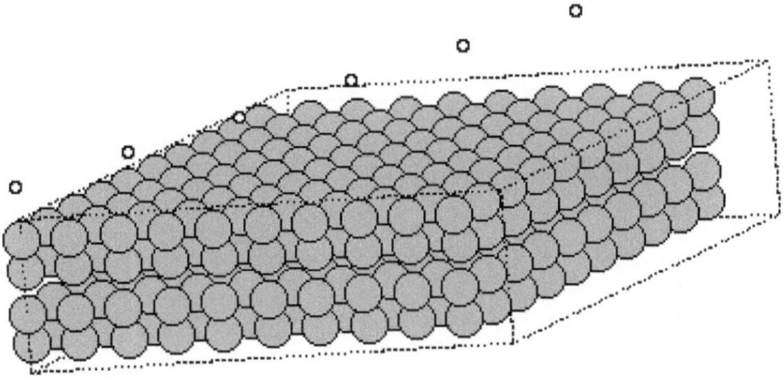

Fig. 3. Input system configuration where tritium H ions are placed on beryllium green atoms surface before the molecular dynamics simulation. (Color figure online)

In the following Fig. 7, ranking scatter is graphed for the decision criteria of the molecular dynamics. As from the graph it can be seen based on precedence A2 at the 14th ranking, A2 with the 13th ranking and A11 at 12th ranking decision criteria effective in molecular dynamics. The multi-decision criteria that are given in the ranking scatter graph and the bar charts graph show the same decision criteria selections MCDA algorithms are executed.ÂăThis means for the molecular dynamics of the input parameters given in the A14 criteria that involve The most effective simulation models use Tritium Energy: 25 keV, Magnetic Field: 2 T, Temperature: 300 K, and Interatomic Potential Calculator of Tritium: CHGNetCalculator. The second useful criterion in molecular dynamics is based on Tritium Energy of 15 keV, a Magnetic Field of 4 T, a Temperature of 300 K, and the Morse Potential for the interatomic potential of tritium. The third

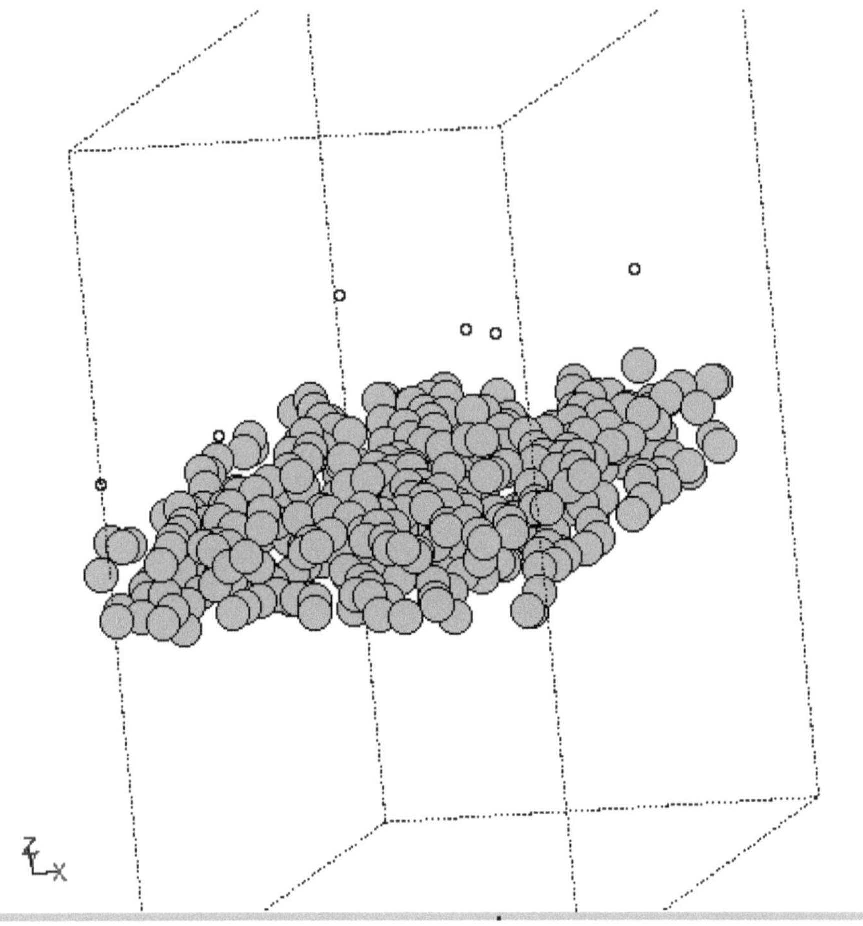

Fig. 4. Output system configuration after the molecular dynamics simulation with the sputtered beryllium surface metal and the H ions.

useful multi-decision criteria parameter for simulating the molecular dynamics of beryllium with tritium sputtering is Tritium Energy, which is set to 15 keV. Magnetic Field: 3 T, Temperature: 300 K, Interatomic Potential Calculator of Tritium: Morse Potential.

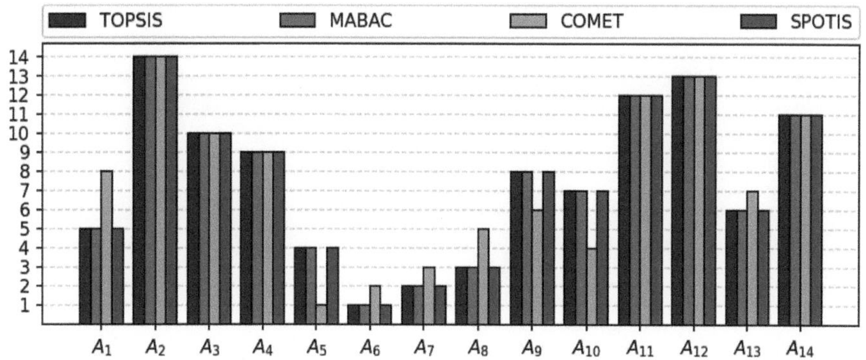

Fig. 5. Barchart comparison graphs of TOPSIS, MABAC, COMET and SPOTIS multi decision criteria analysis.

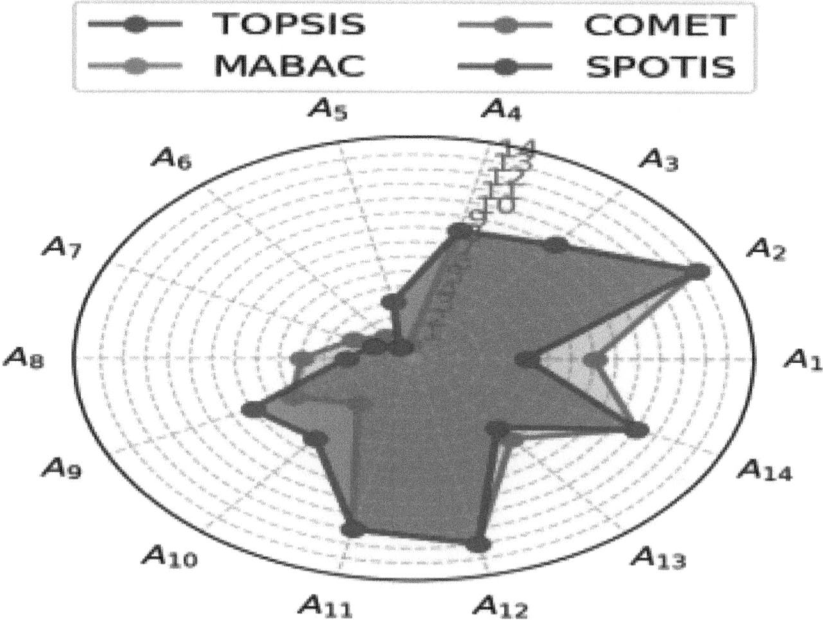

Fig. 6. Polar plot comparison graphs of TOPSIS, MABAC, COMET and SPOTIS multi decision criteria analysis.

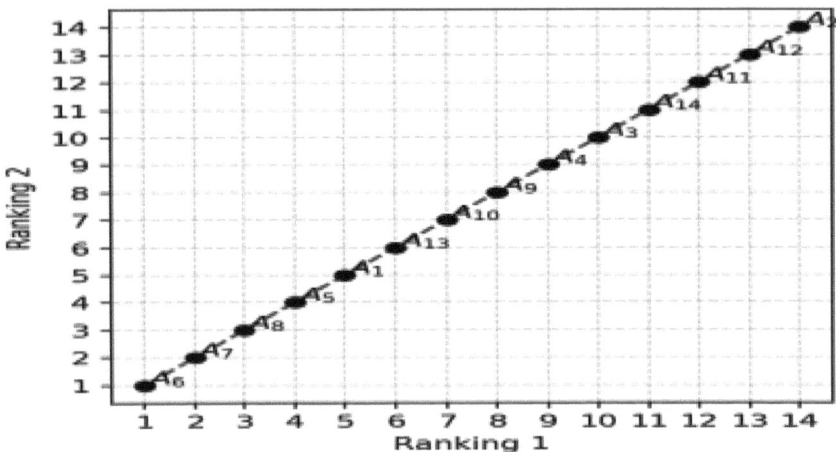

Fig. 7. Ranking scatter comparison graphs of TOPSIS, MABAC, COMET and SPO-TIS multi decision criteria analysis.

4   Conclusion

The simulation results and analyses help us understand how H works. Ions are involved in tritium sputtering within the structure of the beryllium crystal. The use of MCDA algorithms, like TOPSIS, COMET, MABACA, and SPOTIS, shows which parameters are effective in molecular dynamic simulation output trends, such as the accuracy and precision of simulation executions that are used in designing new Tokamak nuclear fusion reactor wall simulations. This paper suggests that the reliability lifecycle calculation of the tritium scattering effects on the various candidate materials for reactor structures could be further investigated, assuming that MCDA simulations of the fusion reactor wall structure molecular dynamics outputs enhance the understanding of the conditions in which the nuclear fusion reactor materials will be affected. This study shows molecular dynamic simulations can use the MCDA algorithms to understand the conditional requirements of the Tokamak-based nuclear fusion reactor that uses tritium material in the plasma core for sputtering, which affects the structures.

Acknowledgments. The authors would like to express their gratitude to HAVELSAN INC for their support in the establish of this study.

Disclosure of Interests. There was no prospective conflict of interest, whether financial or non-financial, during the course of the study. In addition, the study does not rely on informed consent from human participants or animals and adheres to the ethical standards of world-accepted research and literature.

References

1. AIAEA (International Atomic Energy Agency), Kikuchi M., Lackner K., Tran M.Q.: 1: Fusion Physics, Vienna, pp. 20-21 (2012)
2. Sparkes, M.: Korean nuclear fusion reactor achieves 100 million C for 30 seconds. NewScientist, Physics section (2022)
3. Rau, A., Jubin, S., Vella, J.R., Kaganovich, I.D.: Simulations of graphite boronization: a molecular dynamics study of amorphization resulting from bombardment. Front. Phys. **10**, 933494 (2022)
4. Larsen, A.H., et al.: The atomic simulation environment–a python library for working with atoms. Phys.: Condens. Matter. **29**, 273002 (2017)
5. Frenkel, D., Smit, B.: Understanding Molecular Simulation, 2nd edn. Academic Press (2002)
6. Yoon, K.: A reconciliation among discrete compromise situations. J. Oper. Res. Soc. **38**(3), 277–286 (1987)
7. Hwang, C.L., Lai, Y.J., Liu, T.Y.: A new approach for multiple objective decision making. Comput. Oper. Res. **20**(8), 889–899 (1993)
8. Torkayesh, A.E., et al.: A systematic literature review of MABAC method and applications: an outlook for sustainability and circularity, pp. 415 – 448 (2023)
9. Watróbski, J., Sałabun, W., Karczmarczyk, A., Wolski, W.: Sustainable decision-making using the comet method: an empirical study of the ammonium nitrate transport management. In: 2017 Federated Conference on Computer Science and Information Systems (FedCSIS), pp: 949–958. IEEE (2017)
10. Sałabun, W., Shekhovtsov, A., Kizielewicz, B.: A new consistency coefficient in the multi-criteria decision analysis domain. In: Paszynski, M., Kranzlmüller, D., Krzhizhanovskaya, V.V., Dongarra, J.J., Sloot, P.M.A. (eds.) ICCS 2021. LNCS, vol. 12742, pp. 715–727. Springer, Cham (2021). https://doi.org/10.1007/978-3-030-77961-0_57
11. Kizielewicz, B., Sałabun, W.: A new approach to identifying a multi-criteria decision model based on stochastic optimization techniques. Symmetry **12**, 1551 (2020)
12. Dezert, J., Tchamova, A., Han, D., Tacnet, J.-M.: The SPOTIS rank reversal free method for multi-criteria decision-making support. In: IEEE 23rd International Conference on Information Fusion (FUSION), Sun City, South Africa, pp. 1–8 (2020)
13. Kizielewicz, B., Shekhovtsov, A., Sałabun, W.: pymcdm-the universal library for solving multi-criteria decision-making problems. SoftwareX **22**, 101368 (2023)
14. Boggia, A., Massei, G., Pace, E., Rocchi, L., Paolotti, L., Attard, M.: Spatial multicriteria analysis for sustainability assessment: a new model for decision making. Land Use Policy **71**, 291–292 (2018)
15. Meigui, F., Yuan, S., Qu, Q., et al.: The synthesis and topochemical polymerization of o-carboranebased diacetylene macrocycles. New J. Chem. **47** (2023)

Exploring Natural Language Querying over Invoice Data: Experimental Insights on Vector and GraphRAG Approaches

Teodora-Cristiana Nemtoc[(✉)] and Ana-Maria Ghiran

Faculty of Economics and Business Administration, Babeş-Bolyai University,
Cluj-Napoca, Romania
`{cristiana.nemtoc,anamaria.ghiran}@econ.ubbcluj.ro`

Abstract. The expanding integration of AI techniques into daily organizational processes, has transformed the way business-related questions are addressed, significantly reducing the need for technical expertise. In this paper, we evaluate the extent to which responses to stakeholder-relevant questions can be addressed using Retrieval-Augmented Generation (RAG) and its variant, GraphRAG, for information extraction from financial documents such as invoices. This study assesses the performance of these approaches at varying levels of complexity and question categories, identifying how factors such as structure, complexity, and the processes involved in the query pipeline impact their effectiveness. Particular attention is given to how Graph-RAG broadens the scope of answerable questions by structuring documents through graph representations. Guidelines are also included on how each procedure can be integrated into financial workflows, highlighting their respective strengths and limitations compared to a SPARQL benchmark approach.

Keywords: Invoice processing · RAG · GraphRAG · Knowledge Graphs · Information Extraction

1 Introduction

The growing interest in natural language interfaces for enterprise data has driven the development of techniques that redefined the deployment and execution of business operations. As a result, tasks that once required specialized knowledge can now be performed through intuitive and accessible interfaces that allow non-technical stakeholders to interact and operate with enterprise data without intermediary support. The use of RAG (Retrieval Augmented Generation) and its extension, GraphRAG have opened up new opportunities for enabling natural language interfaces over structured business data, thereby revealing their applicability in enterprise scenarios involving domain-specific information retrieval from internal data sources.

In our previous study [1], we investigated and compared these two retrieval-based approaches: traditional RAG and GraphRAG in a business use case

C. Chira et al. (Eds.): InnoComp 2025, CCIS 2793, pp. 302–318, 2026.
https://doi.org/10.1007/978-3-032-12478-4_19

focused on extracting insights from raw invoice data. The traditional method uses a vector-based approach which retrieves information based on semantic similarity in the vector space, and treats each chunk node individually, while the graph-based alternative models entities and relationships, replacing the unstructured information with a Knowledge Graph, preserving the connections between nodes and thereby enabling reasoning. Using the LlamaIndex [15] framework, both Vector-Based and Graph-Based indexing techniques were implemented and evaluated based on the relevance of the results obtained.

We aim to empower non-technical users, such as general managers, financial officers, marketing professionals, and sales people, to interact with enterprise data in a fast and intuitive way [13]. Our initial work focused on outlining the problem state, describing the methodology, and the implementation of our approach. Although the technical pipeline was described in depth, the performance of the system from different user perspectives and question types received only a preliminary evaluation.

This follow-up work aims to continue the analysis of the results generated by the RAG [11] and GraphRAG [9] approaches in the processing of invoice related queries. After a theoretical introduction of the two RAG techniques, we shift our focus towards assessing the results obtained with each method. This analysis determined the appropriate application settings for each variant, identifying the potential constraints and shortcomings associated with each method. Beyond serving a comparative purpose, this work focuses on the GraphRAG technology with its role in enhancing the traditional method by integrating potential connections formed inside the invoice data, aiming to assess whether this structure determines a more contextually aware retrieval process.

The research objective is to investigate the practical applications of these methods in real-world contexts and understand how AI-enabled querying can facilitate daily business operations. Specifically, the study seeks to investigate the integration of RAG techniques to extract relevant information from financial documents such as invoices [12,14,16], thus streamlining various internal processes. The aim is to underscore the advancements that these methods can provide, while also recognizing the gaps that persist in responding to questions regarding common business operations.

The structure of the paper is as follows: the first section outlines the methodology underlying the two RAG techniques, describing their components, configuration parameters, and implementation requirements, followed by an extensive analysis of the experimental results obtained for multiple categories of questions targeting diverse stakeholders, covering a broad spectrum of scenarios and various levels of complexity. A comparative overview of the responses is conducted, highlighting key findings related to their performance and identifying their advantages and shortcomings. The limitations are discussed in a dedicated section, with an emphasis on the differences between the current approach for asking questions and obtaining results in natural language versus the old variant that demands translating into technical jargon and creating the queries. Future research to explore advanced variants of the RAG methodologies and to broaden the range of query types and addressed scenarios are proposed; the final chapter presents conclusions and actionable guidelines for the implementation of RAG and GraphRAG in domain-specific information extraction tasks from invoices.

2   Methodology

2.1   RAG and GraphRAG

In this study, we conduct a comparative analysis of two AI-driven approaches for extracting information from invoices, both utilizing RAG techniques, namely the traditional RAG and its extension, GraphRAG. [4,10]. The fundamental difference between these approaches lies in the nature of the data storage and representation mechanism. The traditional approach extracts information from a vector database, while GraphRAG extracts it from a Knowledge Graph. Both methods involve pre-processing unstructured information by segmenting it into manageable chunks and generating corresponding embeddings. However, GraphRAG additionally constructs a Knowledge Graph, based on the Document objects generated during the pre-processing stage in which detects entities and relationships mentioned in the chunks. Regarding the retrieval process, in the vector-based approach [9], similarity measures are applied to return relevant vectors in response to the user's query, while GraphRAG traverses the graph structure and detects nodes with the corresponding relationships that fit the query. The context generated with the retrieved information alongside the query is passed to the LLM, which generates the answer for the user, as illustrated in the operational framework presented in Fig. 1.

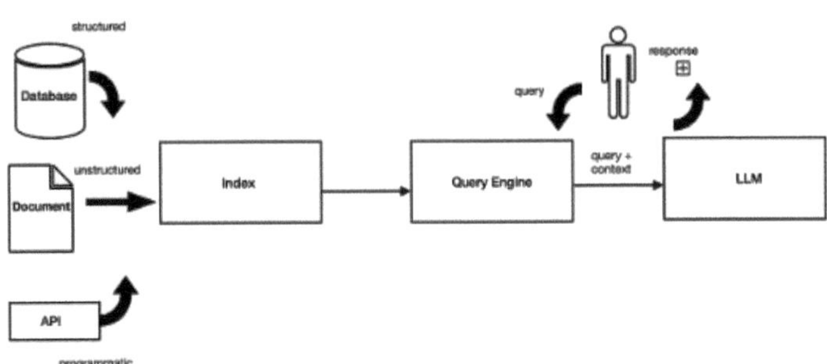

Fig. 1. Operational framework of LlamaIndex.

The experimental setup involved the LLamaIndex Framework. We adhered to LlamaIndex's default configuration, using GPT-3.5-turbo and text-embedding-ada-002 to ensure a reasonable compromise between efficiency and price.

LlamaIndex emerged as a versatile tool for supporting various RAG processes, providing multiple alternatives for data storage, organization, and retrieval, each tailored to specific data structures and scenarios. In this study, we explored two of its indexing techniques for implementing RAG pipelines: the VectorStoreIndex and the KnowledgeGraphIndex.

The first approach involves processing raw textual data by generating vector embeddings, transforming the input into smaller chunks that are then organized based on similarity within a high-dimensional space. Information retrieval using this index is guided by vector similarity measurements, enabling the identification of entities and concepts from the textual input depending on the formulation of precise and relevant queries. This method is best suited for tasks where contextual similarity is sufficient; the effectiveness of retrieval depends on the relevance and clarity of the query.

On the other hand, the KnowledgeGraphIndex converts the unstructured input into an organized structure in which data is organized into a Knowledge Graph, facilitating the understanding of connections between concepts as the entities are modeled as nodes and their corresponding relationships as edges. This method becomes suitable for scenarios that require interconnected data and graph traversals, where multiple paths are navigated to identify query answers, allowing for multi-hop reasoning and more complex retrieval.

2.2   Dataset

The invoices dataset used in this study is public and accessible via Hugging Face[1]. It is structured as 538 rows and 2 columns: one containing the raw invoice text and the other with the corresponding structured format in JSON. This dual structure provides versatility, enabling the application of NLP techniques on unformatted text, while also allowing for validation against the structured version. Each invoice lists 43 fields grouped into key categories: general invoice details, invoice summary, bank information, seller information, buyer information and product details. The entity and relationship extraction process applied to the raw invoices was validated using the structured JSON version, mitigating the risk of error propagation in the subsequent query responses. Although the structured format facilitated validation, data cleaning was still required. To evaluate the results, a SPARQL-based golden truth benchmark was constructed by transforming the JSON invoices into RDF triples loaded into Ontotext GraphDB. The JSON nested information was flatten into subjectâĂŞpredicateâĂŞobject triples and each new data entry received a unique subject identifier (e.g., Invoice1ProductDetails1, Invoice1ProductDetails2, Invoice2-ProductDetails1, etc.) ensuring that all extracted triples are traceable to their original source. Attribute cleaning was performed to clearly determine the proper data type, differentiating between numbers, monetary values, strings, dates, and additional types. In addition, blank nodes were created, where composite information was identified, such as monetary data that comprise both amount and currency. In creating the relationships, property semantics informed the mapping (e.g., BuyerDetails -> hasBuyerDetails, NetPrice -> hasNetPrice, etc.). The

---

[1] https://huggingface.co/datasets/lokaspire/invoice_dataset.

cleaned triples formed a well-structured Knowledge Graph structure, enabling SPARQL querying.

3   Results and Analysis

The results obtained with the two RAG approaches have revealed notable conclusions and potential application scenarios for each. The differences can be analyzed across several dimensions, such as the structure of the query, formulation complexity, and interpretability, or the number of subsequent steps required to produce an answer.

The prevailing tendency identifies the Vector Approach as particularly suitable for handling straightforward queries, particularly when the structure is simple and explicitly targets entities or attributes within the textual sources. When the formulation of the question becomes convoluted and requires nuanced interpretations and deductive reasoning, there is a noticeable decrease in the accuracy of the results. In such cases, GraphRAG proves to be more effective, increasing the flexibility of questions and managing queries that involve more complexity, ambiguity, multiple interpretations, or even hidden elements that are not directly mentioned. If the dataset were unstructured, lacking a clear format, but containing entities with multiple indirect connections and deeper hierarchical structures, the graph approach would provide a more robust framework for executing complex queries. Furthermore, its structural characteristics enable graph traversal operations, facilitating exploration of nodes located at varying hop distances [11].

To compare the two approaches and align them with relevant use cases, we formulated a set of targeted questions for company stakeholders, organized into distinct categories. The natural language queries and the SPARQL equivalent with the golden truth results are presented in Table 1 *Sample business questions and their SPARQL representation* from our previous research [1]. We included the results generated by each RAG variant to clearly identify their capabilities in providing answers compared to the standard SPARQL approach [8,17]. In the following section, we examine the results in detail and discuss the insights obtained for each approach.

3.1   Analysis of RAG Results for Business Stakeholder Queries

Table 1 presents questions targeting various business stakeholders and the results obtained with each method. Table 2 presents SPARQL responses that serve as the golden truth. Given that the majority of questions required retrieving specific elements from the Knowledge Graph on which various operations had to be performed, we selected, where applicable, the elements similar to those returned by the RAG approaches in order to analyze their content validity. For queries that involved extracting specific attributes, the complete SPARQL responses were retained.

Table 1. Comparison of RAG Approaches for Stakeholder Related Questions

ID	Stakeholder	Sample Question	Vector Approach	Graph Approach
1	General Manager	What could be a breakdown of purchases by supplier?	Shaw and Sons supplied Wall Mounted Upside Down Wine Bottle Rack Goblet Glass Holder Storage Organizer, Wine Openers Heavy Duty Stainless Steel Hinges Wine Key, EU Blichmann Riptide Brewing Pump, and Pier 1 Multi-color Cordial Liqueur Wine Glasses Set of 4.	Breakdown includes: Davis PLC (Chindi Rugs Carpet, Bohemian Yoga Mat, Kilim Counterpane, Christmas Rug, etc.), Invoice Numbers: 26675003, 86268868, 17548173, etc.
2	Financial Manager	Calculate the total amount for the items listed in the invoice with number 61356291.	$30,588.80.	$212.09.
3	Financial Manager	Which customers have the highest outstanding due amounts?	Shaw and Sons, Conrad, Smith and Watkins.	Bernard, Welch and Freeman, Chen-Ortega Estrada-Vasquez, Foster, Travis and Harris, Sims PLC Ochoa, Crane and Johnston.
4	Sales Manager	Which products are frequently bought together?	Dress Boy Lace Up Shoes Size 12 All Mattel Disney Pixar Model and Cars McQueen 1:55 Diecast Lot Choose Loose Kid Toy KushyShoo Children Kid Rain.	SONY Playstation2 Slim Console and Nintendo 64 N64 Console Black; Sega Saturn Console Gray and Nintendo Wii U Console Deluxe Bundle.
5	Sales Manager	What is the invoice number where Crawford, Acosta and Solomon are involved as sellers?	Invoice no: 94885760.	Invoice no: 94885760.
6	Chief Marketing Officer	What are the highest spending customers?	Chen-Ortega from Lisahaven, NY; Peterson-Long from East Mathew, OH.	Chen-Ortega, Parker Inc, Morris-Trujillo, Estrada-Vasquez, Bailey-Harris, David-Rhodes.
7	Chief Marketing Officer	What is the highest quantity bought for each product?	The Fountainhead (4 units), A Million Little Bricks (3 units) In the Footsteps of the Russian Snowman (25.87 units), Sega Saturn Console (3 units).	VINTAGE IRON WINE BOTTLE (22.5), Rabbit Wine Tool Kit (3), Open box Sealed Lolita Happy Retirement Wine (4) Nintendo Wii U Console Bundle (5).

Table 2. SPARQL Responses

ID	SPARQL Response
1	"Shaw and Sons" - "Wall Mounted Upside Down Wine Bottle Rack Goblet Glass Holder Storage Organizer, Wine Openers. Heavy Duty Stainless Steel Hinges Wine Key. 12Pcs/Random Color. New without boxes heavy duty plastic wine glasses lot of 16, EU Blichmann Riptide Brewing Pump - Hombrew Beer Wine Food Grade Stainless Steel, Pier 1 Multi-color Cordial Liqueur Wine Glasses Set of 4" "Davis PLC" - "Bohemian Garden Yoga Mat Indian Kilim Counterpane Xmas Christmas Rug Carpet, Chindi Rugs Carpet New Design, Cartoon Bedroom Kids Play Mat Soft Flannel Area Rugs, Bohemian Rag Rug- Woven Chindi Dari Living Room Rug / Hand-Woven Carpet Rug-Mats, Rug Beni Ourain, Moroccan Handmade 100% Wool Area Rug Berber Beni Ouarain Carpet, 40x120CM Christmas Door Mat Kitchen Floor Area Rug Bedroom Living Room Carpet"
2	"212.09"
3	"Parker Inc" "Dixon, Hall and Payne" "Mosley LLC" "Medina-Kennedy" "Wilson PLC"
4	"SONY Playstation2 Slim Console set" SCPH-90000/White " / TESTED Working 11192" - "Nintendo 64 N64 Console Black Japan Tested Working NUSHA" do wii U Console Deluxe/Handheld Bundle" - "Sega Saturn Console Gray HST-3200 HST-3210 Japanese" "Cars McQueen 1:55 Diecast Lot Choose Loose Kid Toy" - "The Children's Place Brown Dress Boy Lace Up Shoes Size 12 All Mattel Disney Pixar Model" "The Children's Place Brown Dress Boy Lace Up Shoes Size 12 All Mattel Disney Pixar Model" - "KushyShoo Children Kid Rain Boot Shoes Rubber, Monkey, Size Youth 1 Western Chief Kids' Boys Classic Rain Boots w/ Pull Handles Navy Blue Size 12" "Cars McQueen 1:55 Diecast Lot Choose Loose Kid Toy" - "KushyShoo Children Kid Rain Boot Shoes Rubber, Monkey, Size Youth 1 Western Chief Kids' Boys Classic Rain Boots w/ Pull Handles Navy Blue Size 12"
5	"94885760"
6	"Parker Inc"-228770.96;"Dixon, Hall and Payne"-163322.28;"Mosley LLC"-138277.46;"Medina-Kennedy"-117335.52;"Wilson PLC"-74307.34
7	"The Fountainhead: 50th Anniversary Edition" "4.0"; "A Million Little Bricks The Unofficial Illustrated History of the LEGO" "3.0"; "In the Footsteps of the Russian Snowman (Paperback or Softback)" "4.0"; "Sega Saturn Console Gray HST-3200 HST-3210 Japanese" "3.0"; "VINTAGE IRON WINE BOTTLE HOLDER CADDY HAND WROUGHT, CUT, SOLDERED" "5.0"; "Rabbit Wine Tool Kit 6 Piece Set. Open box Sealed" "3.0"; "Lolita Happy Retirement Wine Glass 15 Ounce GLS11-5534H" "4.0";"Nintendo wii U Console Deluxe/Handheld Bundle" "5.0"

Since we selected questions in a summary-oriented manner and the RAG approaches typically have a limit on the number of returned elements, evaluating answers using standard metrics becomes challenging in this context. Therefore, the evaluation has to be carried out manually by a qualified user. For future developments, an automatic evaluation metric will be established to concretely assess accuracy and F1 scores.

When queries specify a known predicate with its associated subject, object, or attribute and are expected to return the missing node that completes that triple, both graph-based and vector-based approaches return correct answers, as in the case of extracting the invoice number when the seller of an invoice is mentioned. The limitations of the techniques become noticeable when additional operations that require traversing the entire dataset are needed to answer the queries, such as aggregation or sorting. When the questions are formulated in a summary-oriented manner, as the expected output acts as a synthesized overview, both approaches return only the top specified k relevant nodes, as they are not designated to return all graph nodes before applying the requested operations. As a result, questions such as "What could be a breakdown of purchases by supplier?", "Who are the highest-spending customers?", and "What is the highest quantity bought for each product?" yield partial summaries, with each technique returning different items according to their embedded similarity measurements.

Common limitations lie in the aggregation and filtering operations, such as computing total values, followed by sorting requirements (e.g. descending). Although GraphRAG failed to correctly order the customers by their spending amounts, it successfully retrieved entities from the invoices with the correct seller role. In contrast, the vector-based approach identified only one actual customer, with the second returned entity having an incorrect customer role.

For the summary with the highest quantities per product, both methods generally provided accurate results by associating the correct products with their respective quantities. However, errors were present in both results, as unrealistic decimal quantities were returned for some items. For the complex case where building semantic bridges and entity association within the data is required, alongside the application of aggregation and sorting operations, the results remain notable. When querying for frequently purchased product pairs, both techniques correctly paired two items but omitted the frequency parameter as returned with SPARQL test.

It was noticed that RAG-based approaches relied on root-word similarity, as opposed to SPARQL which strictly matched the identifier of a product. For example, "Nintendo" identified by the graph-based approach, and "Cars McQueen" identified by the vector-based approach appeared in multiple transactions but with different details, potentially referring to distinct products, but because the description contained a part from the product, the RAG approaches counted them due to their text-similarities strategies. Despite these challenges in handling grouping and aggregation across the entire dataset, Graph-RAG approach performed well for scenarios in which queries were limited to specific entities and targeted known objects even if mathematical operations were required. For instance, in the query "Calculate the total amount for the items listed in the invoice with number 61356291" GraphRAG performed the summation and returned a value matching the ground truth, compared to the vector-based approach which used alternative computational logic and returned a different answer.

In synthesis, if the questions are formulated in an explicit manner and aim to return specific entities, both RAG approaches deliver highly accurate answers. The results are satisfactory when summaries of financial data are requested, as the techniques manage to return general analytical perspectives. In this way, the usefulness of the RAG methods was demonstrated, showcasing their role in assisting stakeholders who seek answers to financial queries without having to use tools that require prior programming knowledge. However, limitations arise when additional aggregation, grouping, or sorting operations are required across the entire dataset since RAG approaches are based on similarity algorithms for their retrieval pipeline.

For this reason, additional personnel to handle queries and manually extract information is still preferable because complex questions involving reasoning and alternative operations remain a challenge for RAG techniques [6]. However, if the scope of operations is narrowed by reducing the number of targeted entities, GraphRAG manages to apply complex calculations on the returned subgraph and provide accurate answers, eliminating the steps that would otherwise be necessary in a standard information extraction scenario. This positions the GraphRAG method as a more effective alternative for the traditional RAG approach for our information extraction use case within the context of the current study.

3.2 Pattern Identification and Strength Assessment of RAG Approaches for Simpler Queries

After examining the types of queries associated with business stakeholders and addressing specific cases of interest for each, given the relatively high complexity of RAG approaches, we intend to conduct an analysis of simpler questions that do not require aggregation operations, as represented in Table 3.

In this way, a clearer distinction between the vector-based and graph-based approaches can be captured. Thus, we expand the case of interest for the Sales Manager Stakeholder, where both methods successfully identified the invoice number based on known entities: Crawford, Acosta, and Solomon. Our objective is to explore how the formulation of questions and the requested information about the details of an entity influence the responses provided by the two methods [7]. Thus, when the query structure with the order of elements is changed or when key components of a triple are omitted, the role of the GraphRAG begins to stand out. For instance, when the term "what" is used to indicate the sellers in our dataset, the vector-based approach, which is not inherently centered on the relationship between nodes, identifies the entity broadly as a company, while the graph-based approach found the actual seller role within the dataset. When the query is more concrete, being modified with "What represents the entity", the vector approach returns a more comprehensive answer, while GraphRAG maintains its focus on delivering a more appropriate output. If the query requires retrieving an entire set of details rather than focusing on a specific property, the two approaches effectively return the key-value properties associated with the most relevant node identified.

Table 3. Comparison of RAG Approaches for Non-Complex Invoice Queries

Query	Vector Approach	Graph Approach
What is Crawford, Acosta and Solomon?	A business or company.	Crawford, Acosta and Solomon is the seller in the provided context.
What represent Crawford, Acosta and Solomon in these invoices?	Crawford, Acosta and Solomon represent the sellers in the invoices provided.	Crawford, Acosta and Solomon represent the seller in these invoices.
What are the details for the invoice where Crawford, Acosta and Solomon is the seller?	The details for the invoice where Crawford, Acosta and Solomon is the seller are as follows: - Invoice no: 94885760 - Date of issue: 08/22/2012 - Seller: Crawford, Acosta and Solomon - Client: Chen-Ortega - Seller's address: [...] - Seller's Tax Id: 995-99-2920 - Client's Tax Id: 974-99-3914 - Gross worth: $60.45	Identical as Vector Approach.

We proceed with the comparative analysis, aiming to uncover trends and patterns across different categories for the two approaches. Based on the analysis, we have reached the following conclusions:

GraphRAG Accurately Maintains Relational Structures Across Multiple Participants Representing a Single Entity

When multiple participants represent a single base entity, such as a Client, the GraphRAG method successfully captures the relationships and provides valid answers. This is achieved through the structured graph which retains the relational context of the target entity with its all relatives. In contrast, the conventional method fails to preserve these connections and returns incomplete or partially correct answers, or even hallucinated outputs.

Query: What are the roles of Brown, Rodriguez and Lopez?

RAG Answer: Brown is the seller, Rodriguez is the client, and Lopez is not mentioned in the context information provided.

Graph-RAG Answer: Brown, Rodriguez, and Lopez are clients.

Standard RAG Struggles with Simplistic or Fragmented Queries Unless Rephrased with More Context

If the question is overly simplistic and addresses only a fragment of the complete answer, the standard approach often fails to adequately respond to the query. It becomes essential to reformulate the question to provide additional information for improved accuracy, as this technique struggles to infer the broader context.

Query: Who is Rodriguez in this context?

RAG Answer: Rodriguez is not mentioned in this context.

GraphRAG Answer: Brown, Rodriguez and Lopez are the clients in this context.

GraphRAG Provides Richer Information for Attribute-Based Queries

When searching for a target attribute that can be inferred from unstructured information for an entity with a specified ID-type property, the results are more precise, even though differences may arise between the two methodologies. The vector-based approach provided partial correct information by accurately discovering the node linked with the invoice number. However, it also included the client in its response and not only the seller, as intended. GraphRAG distinguished between entities and isolated the seller.

Query: Who is the seller for the invoice with number 37292788?

RAG Answer: Bennett-Sanchez Riley-Miller.

Graph-RAG Answer: Bennett-Sanchez.

Standard RAG Is Sensitive to the Formulation and Specification of Predicates or Attributes to Search For

In some cases, if the relation to be identified is not explicitly stated as it is stored in the index, the vector engine encounters difficulties in identifying the source and the target. For example, as there are multiple columns that represent Date objects, a question regarding the invoice date, led to an accurate result only for GraphRAG which correctly found the date associated with the invoice, compared to the other method which failed to differentiate between the multiple date-related attributes.

Query: What is the invoice date of 49565075?

RAG Answer: The invoice date of 49565075 is not provided in the context information.

GraphRAG Answer: The invoice date of 49565075 is 10/28/2019.

GraphRAG's Flexibility in Similarity Matching Can Lead to Inaccurate Results in Subjective Queries

Limitations have also been identified for the GraphRAG approach. Due to its less restrictive similarity criteria during search process, for some queries, GraphRAG may fail to return complete answers or, in some cases, fails entirely to find the right answer. When dealing with queries that have subjective characteristics or are open to interpretation, such as searching for the tax id for a specific seller in an invoice where only its name is mentioned, the GraphRAG partially resolves the query as it identifies the correct invoice containing the name, and also identifies a value for the searched property, but the returned entity corresponds to the buyer rather than the seller. The vector-based method indicated the correct object from the triple, which likely picked the first value it encountered for the property in the store. GraphRAG enhances RAG by utilizing structural relationships for inferring the answers, but its flexibility in matching similarities can sometimes lead to inaccuracies in certain scenarios.

Query: What is the tax id for Jefferson Group?

RAG Answer: 980-76-7923.

GraphRAG Answer: The tax id for Jefferson Group is 937-75-3172.

RAG Techniques Face Limitations When Searching for Patterns Outside Relationship Contexts

When the context demands a broader scope that is not solely centered on the relationships formed but requires a comprehensive graph search for specific patterns, the two alternatives present limitations. They rely on retrieval mechanisms that differ from those employed by graph databases engines, which could serve as benchmarks for such tasks. The strength for this case is that both techniques managed to identify a common item denoted as correct, but the graph scenario stopped at a single returned one unit, limiting the completeness of its answer. Shortcomings also appear for GraphRAG as it found only one individual invoice that matches the searched criteria. Since vector-based technique is suitable for similarity search, particularly when the focus is not on the relationships between the entities, this type of indexing tends to be more accurate in its response by identifying a larger number of relevant nodes, leveraging its ability to measure semantic similarity.

Query: What are the invoice numbers for items containing Nike?

RAG Answer:

Invoice numbers for items containing Nike are 94885760 and 32656405.

GraphRAG Answer:

Invoice numbers for items containing Nike are 94885760.

Both RAG Techniques Perform well for Broader Queries

When the question allows for a broader response and does not require a highly specific answer, both methods perform well. The entire set of proper answers was detected with the vector strategy, the accuracy being high also for GraphRAG, even if it omitted one element.

Query: What are the financial details when items involve Marble Coffee?

RAG Answer: The financial details for items involving Marble Coffee are as follows:

- 48 Inches Marble Dining Table: Net worth \$9,741.60, VAT \$985.60, Gross worth \$10,727.20
- 14 Inches Black Marble Coffee Table: Net worth \$743.33, VAT \$74.33, Gross worth \$817.66
- 12" White Marble Coffee Table: Net worth \$650.61, VAT \$65.06, Gross worth \$715.67
- 12" Marble Coffee Side Table: Net worth \$995.12, VAT \$99.51, Gross worth \$1,094.63
- 15"x15" White Marble Coffee Table: Net worth \$2,303.96, VAT \$230.40, Gross worth \$2,534.36

GraphRAG Answer:

The financial details for items involving marble coffee are as follows:

- 14 Inches Black Marble Coffee: Net worth \$743.33, VAT \$74.33, Gross worth \$817.66
- 12" White Marble Coffee Table: Net worth \$591.46, VAT \$59.15, Gross worth \$650.61

- 12" Marble Coffee Side Table: Net worth $995.12, VAT $99.51, Gross worth $1,094.63

Both RAG techniques demonstrate strong performance for direct queries with clear context

Among the notable results observed for both types of indexing, there are cases where both procedures demonstrate strong performance, particularly in scenarios where queries are permissive or seek specific information while providing sufficient data for targeting elements capable of generating relevant responses.

Query: What is the client's name where the seller is Jefferson Group?

Common Answer: Parker Inc

RAG Techniques Excel in Extracting Complete Details when Restrictions on Extraction are minimized

Queries initiated from a concrete property and requesting all corresponding details akin to DESCRIBE query in SPARQL lead to identical results from both methods.

Query: What can you tell me about the invoice with number 37292788?

Common Answer: The invoice with number 37292788 was issued on 07/12/2013. The seller is Bennett-Sanchez and the client is Riley-Miller. The items listed on the invoice include Chindi Rugs Carpet New Design, Bohemian Garden Yoga Mat Indian Kilim Coverlet Fluffy Rug Carpet Non Slip Soft, Area Rugs Bathroom Room Round Floor Mat Washable Carton Tiger Print Area Rug, and Faux Skin Rug Hide Cute Carpet Floor Mat Home Decor. The total net worth of the items is $109.55, with a VAT of $10.96, resulting in a gross worth of $120.51.

3.3   Limitations

In general, the two approaches of RAG provide highly accurate results when the extraction process is simple. The vector approach easily identifies the sought elements based on textual similarity, while GraphRAG also permits the identification of more difficult-to-access elements due to its organized structure, preserving the connections between the graph elements. However, for real-world scenarios applicable to the financial domain regarding data extraction from invoices, potential stakeholders would formulate questions that would not only involve the straightforward identification of several entities but would require various operational techniques that should traverse the entire graph. The quality of the entity and relationship extraction process has a significant impact on the accuracy of the results, as any inconsistencies introduced at this stage can propagate through the pipeline and affect the final output.

Subsequently, potential mathematical functions would be required. The LLamaIndex implementation for RAG, in both its variants, engages in returning the graph nodes that most accurately respond to the formulated question. This degrades the results of the aggregation operations that should be performed across the entire graph and not just on the specified node list. However, this would imply changing the selection of the relevant information sent to the LLM for processing the final response.

Thus, either the formulation of the questions is adapted to be less restrictive, requesting alternatives that would assist in managing the financial context with more descriptive information rather than a mathematical result, or an overview for a filtered subset of entities is returned, along with the sought mathematical result. When the query involves mathematical operations that are not related to the context or to the textual meaning of the documents, both approaches fail in returning the answers, which is expected due to the characteristics they present. RAG is conceived for natural language questions that can reduce the need for writing rigorous queries that require technical capabilities but come with the limitation of restricting the questions to semantic constraints.

Given that the Vector-based approach relies solely on textual similarity and does not account for potential relationships and connections between elements, erroneous associations may arise between concepts that are not related by a direct relationship, simply because they are semantically close. This can significantly affect the returned response and reduce the confidence in the returned results. As the complexity of the questions increases [3], hallucinations become more frequent, especially when a direct answer cannot be derived purely from similarity measures, leading to responses that deviate significantly from the truth labels. Additionally, when constructing a response requires connecting multiple pieces of information from different text fragments and does not involve extraction from a single source, the absence of strong textual similarity often results in incomplete answers.

The GraphRAG approach also presents a set of limitations. Considering that the process starts from a Knowledge Graph, the quality of the results is directly linked to its correctness, which will influence the answers. There are scenarios where graph traversal logic for extracting the answer is correct, but inaccuracies in the identified entities and relationships propagate through to the final response. Additionally, in order to maintain a clear and repetitive structure for the graph identification, financial data must undergo cleaning processes to comply with the potential schema of the Knowledge Graph, which underpins the functionality of GraphRAG. The dependency on the Knowledge Graph becomes challenging when potential changes occur in the graph, as this implies restarting the indexing processes, an operation that can be hindered when working with large databases or frequently updated data, an issue relevant to the nature of the current domain involving financial information. Pertaining to the challenges of the domain, there are also LLM-specific limitations. One concern lies in the lack of interpretability in the generated responses that can raise challenges in domains requiring traceability and auditability. That is the case for the current financial domain as it is characterized by rigor and compliance requirements and also involves focusing on the segment of secure and reliable AI to develop trustworthy systems.

4   Future Directives

In light of the fact that the dataset consists of financial documents, which typically exhibit a well-defined structure due to the domain they originate from

[2] and the legal implications associated with them, the dataset does not fully leverage the differences between GraphRAG and traditional RAG. This occurs as the connections formed do not deviate significantly from the standard approach offered by vector-based search. In contrast, an unstructured dataset with deeper hierarchical structures would better showcase the advantages of a graph structure in uncovering complex or hidden connections. For upcoming development pathways, it is recommended to analyze the advancements that GraphRAG can offer across different types of datasets, starting from semi-structured inputs and progressing to completely unstructured documents. This approach will allow for a systematic evaluation of the associations between input formats and query types under which GraphRAG can provide increased performance, and effective outcomes will reveal feasible combinations between input formats and query types where GraphRAG outperforms the conventional approach. Moreover, since the Llama library comes with built-in functions for indexing and retrieval processes, user control remains quite limited. Thus, for future experiments, we aim to extend the techniques used for executing RAG, starting from a Knowledge Graph, allowing for greater influence over the implementation of algorithms used to preprocess, store, and retrieve data, extracting relevant parts passed to an LLM, trained using appropriate prompt engineering techniques to effectively respond to the queries. To better asses the overall quality of the generated answer, a standard evaluation protocol has to be followed. For questions that target specific elements or properties, metrics such as precision, recall, and F1-score can be easily calculated. However, in cases where the questions are formulated in a summary-oriented manner and require the return of large portions of the Knowledge Graph, the evaluation methods need to be adjusted. A standardized approach should be defined to suit the variety of query types used in this context and to reduce the manual effort needed to review and interpret the responses.

5   Conclusion

This study evaluates the performance of a RAG-solution for question-answering within business contexts, comparing it to SPARQL-based querying. By identifying a set of query types relevant to various stakeholders working with financial data, the responses returned by each approach were examined to identify trends and scenarios where each presents advantages or limitations. It was found that for simple questions that allow the extraction of entities and attributes of interest based on similarity, both methods demonstrated high accuracy. GraphRAG proves to be more useful when seeking to detect intricate patterns, entities connected through specific relationships that are not directly specified in the query, maintaining topic independence, and retrieving information located multiple hops away within the graph.

There are still tasks where RAG approaches fall significantly below the SPARQL standard. Complex questions, targeting everyday financial operations are challenging for both RAG techniques, given that they rely on similarity processes. Even though GraphRAG comes with improvements that register the connections formed between entities, returning a finite number of top matches becomes problematic for tasks that require logical reasoning across the entire graph. Thus, the area of applicability is narrowed, with manual input for verification still being necessary, placing AI technologies as a support for financial information extraction tasks, but not as a complete replacement.

References

1. Nemtoc, C., Ghiran, A.-M.: Natural language querying of invoice data using RAG and GraphRAG: leveraging LLMs for financial document insights. In: Proceedings of the 3rd International Workshop on Hybrid Artificial Intelligence and Enterprise Modelling for Intelligent Information Systems (HybridAIMS) at CAiSE 2025, Vienna, Austria (2025). https://doi.org/10.1007/978-3-031-94931-9_6
2. Abu-Salih, B.: Domain-specific knowledge graphs: a survey. J. Netw. Comput. Appl. **185**, 103076 (2021). https://doi.org/10.1016/j.jnca.2021.103076
3. Besta, M., et al.: Graph of thoughts: solving elaborate problems with large language models. In: Proceedings of the AAAI Conference on Artificial Intelligence, vol. 38, no. 16, pp. 17682–17690 (2024)
4. Sarmah, B., Mehta, D., Hall, B., Rao, R., Patel, S., Pasquali, S.: HybridRAG: integrating knowledge graphs and vector retrieval augmented generation for efficient information extraction. In: Proceedings of the 5th ACM International Conference on AI in Finance, pp. 608–616 (2024)
5. Lewis, P., et al.: Retrieval-augmented generation for knowledge-intensive NLP tasks. Adv. Neural. Inf. Process. Syst. **33**, 9459–9474 (2020)
6. Mavromatis, C., Karypis, G.: GNN-RAG: graph neural retrieval for large language model reasoning. arXiv preprint arXiv:2405.20139 (2024). https://arxiv.org/abs/2405.20139
7. Sanmartin, D.: KG-RAG: bridging the gap between knowledge and creativity. arXiv preprint arXiv:2405.12035 (2024). https://arxiv.org/abs/2405.12035
8. Rangel, J.C., Mendes de Farias, T., Sima, A.C., Kobayashi, N.: SPARQL generation: an analysis on fine-tuning OpenLLaMA for question answering over a life science knowledge graph. In: CEUT Workshop Proceedings. 3890 (2024)
9. Han, H., et al.: Retrieval-augmented generation with graphs (GraphRAG). arXiv preprint arXiv:2501.00309 (2024). https://arxiv.org/abs/2501.00309
10. Han, H., et al.: RAG vs. GraphRAG: A Systematic Evaluation and Key Insights. arXiv preprint arXiv:2502.11371 (2025). https://arxiv.org/abs/2502.11371
11. Gao, Y., et al.: Retrieval-augmented generation for large language models: a survey. arXiv preprint arXiv:2312.10997 (2023). https://arxiv.org/abs/2312.10997
12. Wang, D., Ma, Z., Nourbakhsh, A., Gu, K., Shah, S.: DocGraphLM: Documental Graph Language Model for Information Extraction. arXiv preprint arXiv:2401.02823 (2024). https://arxiv.org/abs/2401.02823
13. Wang, X., et al.: FinSage: A Multi-aspect RAG System for Financial Filings Question Answering. arXiv preprint arXiv:2504.14493 (2025). https://arxiv.org/abs/2504.14493
14. Wang, Y., Lipka, N., Rossi, R., Siu, A., Zhang, R., Derr, T.: Knowledge Graph Prompting for Multi-Document Question Answering. arXiv preprint (2023). https://arxiv.org/abs/2308.11730
15. LlamaIndex: Official Documentation. https://docs.llamaindex.ai/en/stable/understanding/. Accessed 31 Feb 2025

16. Yang, K., Al-Sabahi, K., Xiang, Y., Zhang, Z.: An integrated graph model for document summarization. Information **9**(9), 232 (2018). https://doi.org/10.3390/info9090232
17. Sima, A.C., Mendes de Farias, T., Anisimova, M., et al.: Bio-SODAUX: enabling natural language question answering over knowledge graphs with user disambiguation. Distrib. Parallel Databases **40**(2–3), 409–440 (2022). https://doi.org/10.1007/s10619-022-07414-w

Bayesian Ensemble Learning for Adaptive Time Series Forecasting

Mateusz Panasiuk[1,2]($\boxtimes$) (iD)

[1] AI Investments, Warsaw, Poland
`mat.panasiuk@gmail.com`
[2] Department of Electrical Engineering, Warsaw University of Technology, Warsaw, Poland

Abstract. We present a Bayesian approach to ensemble learning for time series forecasting, designed to address the challenge of model adaptation in non-stationary environments. Classical ensemble methods typically assume fixed model parameters and static combination rules, which limits their adaptability under temporal drift. Our method generalizes ensemble construction by framing both submodel weight assignment and parameter adaptation as problems of probabilistic inference. Using Bayes' theorem, we derive an online update rule that balances new observations with prior knowledge, enabling coherent uncertainty quantification. Submodel errors are transformed into ensemble weights via a parametrized function, with parameters governed by Gaussian priors and updated using a maximum a posteriori (MAP) approximation. This allows for fast, gradient-based updates without discarding accumulated experience. We evaluate the approach on a short-horizon directional forecasting task across 40 foreign exchange time series using eight neural network-based submodels. The proposed method consistently outperforms static ensembles, with small but statistically significant improvements in accuracy and increase in profits in simulated setting. Beyond empirical findings, this work contributes a tractable and theoretically grounded framework for adaptive ensemble learning with Bayesian guarantees.

Keywords: Ensemble Learning · Forecasting · Bayesian method

1 Introduction

Ensemble methods play a central role in modern machine learning, offering a principled means of combining predictions from multiple models to enhance predictive performance and robustness [2]. Techniques such as random forests and gradient boosting exemplify the success of this paradigm, where ensembles of weak learners are aggregated to form powerful predictors [1]. These approaches rely on the statistical benefits of model averaging, often yielding improvements in both accuracy and generalization.

© The Author(s), under exclusive license to Springer Nature Switzerland AG 2026
C. Chira et al. (Eds.): InnoComp 2025, CCIS 2793, pp. 319–330, 2026.
https://doi.org/10.1007/978-3-032-12478-4_20

However, classical ensemble methods typically operate under the assumption of stationarity - both in the data-generating process and in the structure of the ensemble itself [5]. That is, the manner in which constituent models are combined, as well as the parameters governing the ensemble, are fixed after training. While effective in many scenarios, such rigidity poses challenges in environments characterized by temporal drift, structural breaks, or evolving data distributions. Domains such as financial forecasting, health care analytics, and online recommendation systems frequently exhibit such dynamics, rendering static ensembles suboptimal or even misleading over time [4].

In this work, we propose a Bayesian approach to ensemble learning that explicitly addresses the challenges posed by non-stationary environments. By framing both the assignment of submodel weights and the adaptation of ensemble parameters as problems of probabilistic inference, we enable the ensemble to evolve in response to new observations in a principled manner. Concretely, we maintain Gaussian priors over the parameters that map submodel errors to weights, and use Bayes' theorem to update these beliefs as new data arrives. This formulation naturally accommodates temporal drift, allowing the model to balance historical information with recent evidence via the prior-likelihood trade-off intrinsic to Bayesian updating.

In contrast to traditional ensemble methods, which assume fixed combination rules or use heuristic reweighting strategies, our approach provides a coherent and mathematically grounded adaptation mechanism. Fixed-weight ensembles lack adaptability under distributional shift, while heuristic methods often lack transparency and require careful tuning [6]. Our framework, by contrast, offers a lightweight and tractable approximation to more general dynamic Bayesian models, such as state-space formulations or dynamic Bayesian networks, without the computational overhead typically associated with them. It also supports uncertainty quantification over ensemble parameters, which is essential in domains such as finance, where predictive confidence directly informs downstream decision-making. This combination of adaptability, tractability, and interpretability makes the proposed method particularly suitable for time series forecasting under non-stationarity.

We validate our approach empirically in the context of financial time series forecasting, using an ensemble of neural network-based models. Our results demonstrate statistically significant improvements in predictive accuracy compared to static ensemble baselines, highlighting the benefits of dynamic Bayesian model integration in volatile environments.

2   Ensemble Model

2.1   Model Structure

We consider an ensemble composed of m predictive submodels. Submodel weights are adapted dynamically based on recent predictive performance, with the intuition that models performing well in the recent past should receive higher weight. Specifically, we assume that the optimal weight allocation can be inferred from each submodel's recent error history.

Let $\bar{l}_t \in \mathbb{R}^m$ denote the vector of errors at time t, where each element corresponds to the mean error incurred by a submodel over a fixed-length sliding window of the past s observations. That is, for each model i, the scalar $l_{i,t}$ represents the average predictive error over the window $[t - s, t - 1]$. During the initial phase $(t < s)$, the ensemble is not evaluated until sufficient history is accumulated.

Let $\bar{\alpha}, \bar{\beta} \in \mathbb{R}^m$ denote learnable parameters with associated uncertainty $\sigma_\alpha, \sigma_\beta \in \mathbb{R}^m_{>0}$. These parameters govern the transformation from the normalized error vector $\bar{l}_t$ to the ensemble weight vector $\bar{w}_t \in \Delta^{m-1}$, where Δ^{m-1} denotes the $(m - 1)$-simplex.

The weight vector is computed by applying a softmax function to a linear transformation of the error vector:

$$\bar{w}_t = \mathrm{softmax}(\bar{l}_t \odot \bar{\alpha} + \bar{\beta}) \tag{1}$$

where $\odot$ denotes the element-wise (Hadamard) product.

The ensemble prediction $\hat{y}_t$ is then computed as a convex combination of submodel outputs:

$$\hat{y}_t = \sum_{i=1}^{m} w_{i,t} \cdot \hat{y}_{i,t} \tag{2}$$

This formulation ensures that submodels can be preferentially weighted based on their recent errors, while still allowing uncertainty-aware adaptation through Bayesian updating of $\bar{\alpha}, \bar{\beta}$ (Fig. 1).

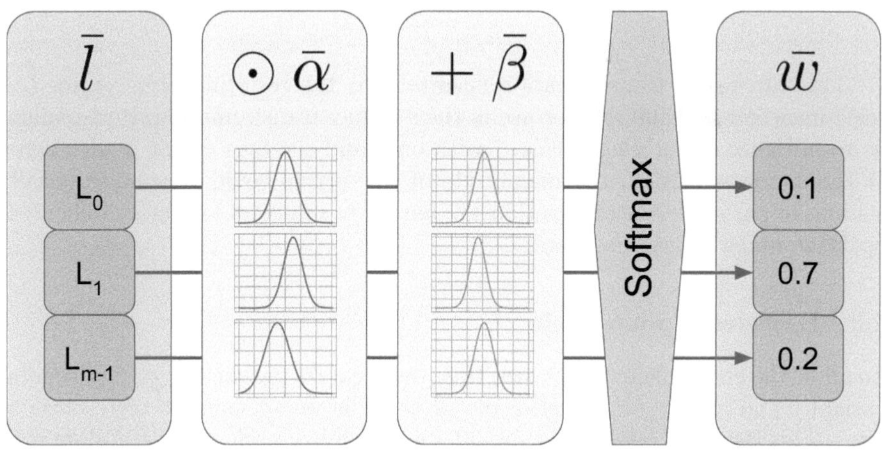

Fig. 1. Schematic illustration of the process generating ensemble weights from submodel errors.

It is possible to simplify the model and instead of relying on separate scaling parameters for each of the submodels - unify them. In such a setup vectors $\bar{\alpha}$ and

$\bar{\beta}$ are substituted by real values: $\alpha, \beta \in \mathbb{R}$, and so are the associated uncertainty parameters. $\sigma_\alpha, \sigma_\beta \in \mathbb{R}_{>0}$. In practice this simplification significantly speeds up the convergence and still yields good results, especially with low number of models used.

2.2 Error Representation

The ensemble relies exclusively on submodel errors to determine weights. To make these errors suitable for optimization and comparable across models, we transform them in three steps: averaging, centering, and variance normalization.

First, we compute the mean error incurred by each model i over the past s observations:

$$l_{i,t} = \frac{1}{s} \sum_{n=1}^{s} r_{i,t-n} \tag{3}$$

Next, we center the vector $\bar{l}_t = [l_{1,t}, \ldots, l_{m,t}]$ by subtracting its mean, yielding a zero-mean error vector $\tilde{l}_t$ with components:

$$\tilde{l}_{i,t} = l_{i,t} - \frac{1}{m} \sum_{j=1}^{m} l_{j,t} \tag{4}$$

Finally, we normalize the centered vector to have unit variance across submodels, defining the standardized error vector $\hat{\tilde{l}}_t$ as:

$$\bar{l}_{i,t} = \frac{\tilde{l}_{i,t}}{\sqrt{\frac{1}{m} \sum_{j=1}^{m} \tilde{l}_{j,t}^2}} \tag{5}$$

This three-step transformation ensures that the resulting error vector $\hat{\tilde{l}}_t$ is scale-invariant and suitable for use in the softmax transformation. It also allows us to initialize the scaling vector $\bar{\alpha}$ with ones and the bias vector $\bar{\beta}$ with zeros. Without normalization, the magnitude of raw errors could vary substantially, leading to parameter estimates that are harder to interpret and more difficult to optimize under Gaussian priors.

2.3 Bayesian Update Rule

To allow the ensemble to adapt over time, we treat its parameters $\bar{\alpha}, \bar{\beta}$ as random variables and infer their posterior distribution as we accumulate more observations. This Bayesian perspective provides a principled mechanism to update prior beliefs with incoming evidence, thereby enabling adaptation in non-stationary environments.

Formally, we place a Gaussian prior over the parameters and update them using Bayes' theorem:

$$p(\bar{\alpha}, \bar{\beta} \mid y) \propto p(y \mid \bar{\alpha}, \bar{\beta}) \cdot p(\bar{\alpha}, \bar{\beta})$$

Here, the likelihood reflects the fit to the most recent data, while the prior encodes accumulated experience.

To enable tractable inference, we adopt a Maximum A Posteriori (MAP) approximation, estimating the mode of the posterior by maximizing its logarithm. Letting $\theta = [\bar{\alpha}, \bar{\beta}] \in \mathbb{R}^{2m}$, the log-posterior becomes:

$$\log p(\theta \mid y) = \log p(y \mid \theta) + \log p(\theta)$$

The log-posterior over N most recent observations is given by:

$$\log p(y \mid \theta) = -\sum_{i=1}^{N} [y_i \log(\hat{y}_i) + (1 - y_i) \log(1 - \hat{y}_i)] \tag{6}$$

where $\hat{y}_i$ is the ensemble's predicted probability for instance i, and N denotes the number of labeled observations used for the update.

The prior is modeled as a Gaussian with diagonal covariance to model the parameters as independent:

$$\log p(\theta) = -\frac{1}{2} \sum_{k=1}^{2m} \frac{(\theta_k - \mu_k)^2}{\sigma_k^2} + \text{const} \tag{7}$$

The objective function used in optimization takes the form:

$$\mathcal{L}(\theta) = -\log p(y \mid \theta) - \log p(\theta) \tag{8}$$

We minimize $\mathcal{L}(\theta)$ using gradient-based optimization. Thanks to the low dimensionality of the parameter space and the smoothness of the objective, convergence is typically achieved in just a few iterations.

Conceptually, the prior acts as a memory of past experience, while the likelihood encodes the most recent experience. Bayesian updating provides a principled mechanism to balance these two sources of information, enabling adaptation in non-stationary environments.

3   Experimental Setup

3.1   Forecasting in Finances

To place emphasis of testing the adaptability of the ensemble, the experiment is placed in the realm of financial time series. They are known to be non-stationary and highly unpredictable what makes a good and challenging testing ground for our ensemble method.

The models are trained to predict direction of price change in the next 10 h. While the horizon of the forecast is short compared to conventional investment horizons it is widely used in algorithmic trading. Furthermore it allows for rapid adaptation of the ensemble model - as the effectiveness of the models can be assessed in a very short time. We get the information about the models accuracy with lag equal to the horizon, what allows for dynamic adjustment of weights that is close to real time.

3.2 Datasets

The model is tested on 40 time-series consisting of foreign exchange currency pairs. Each time-series is sampled with hourly period and contains data about open, high, low, and close ask and bid prices. Datasets were selected from all available foreign exchange time-series with history starting at 2010 or before to assure enough training data. Each dataset had at least 122 thousands of data points.

Prediction targets are computed using close price. Mean value between ask and bid close are considered close price.

$$m_t = C_{t+h} - C_t$$

where m_t denotes price change between t and t_h, C_t is close price at time t and C_{t+h} is close price at time $t + h$.

$$\hat{y}_t = \begin{cases} 1 & \text{if } m_t \geq 0 \\ 0 & \text{if } m_t < 0 \end{cases}$$

where $\hat{y}_t$ denotes correct prediction at time t.

Open, high and low prices are used to construct features used by neural networks, but are not used to compute target or used by the ensemble model.

3.3 Submodels

Four different neural networks were used to serve as submodels in the ensemble: Feedforward Network, Time Series Mixer [8], Deep Coupling [9] and Transformer [10]. All of them were altered to fit our task of binary forecasting and provided with the same features. The goal was not to provide strictly selected and fine-tuned models, but rather provide diverse set of forecasters for the ensemble.

For each time-series each of the four neural networks was trained in two distinct regimes to further bolster diversity. The final ensemble is made of eight different neural networks for each dataset.

To train the submodels datasets were split into training (70%), validation training (10%) and testing (20%). Only testing dataset - on which the final predictions were made - is used in further experiments with ensemble. Testing datasets have around 25 thousands observations with the shortest having 24.5 thousands.

To support model diversity, all submodels were trained independently from scratch for each dataset, using fixed architectures and consistent input features. No extensive hyperparameter tuning was performed, as the goal was not to optimize individual models, but to evaluate ensemble performance under heterogeneous yet plausible forecasters. This choice strengthens the case for the ensemble's adaptability rather than relying on highly optimized components. The optimization algorithm used for training the ensemble was Adam, applied to minimize the MAP-based loss defined in Sect. 3.3.

3.4 Submodels Effectiveness

Financial time-series are notoriously hard to predict. Even slight improvement over naive prediction constitutes a possible investment opportunity if it remains stable and reliable. By the effective market theorem no such opportunities should even be present if considering only data from inside the system. It is crucial to understand this to correctly interpret effectiveness of the models and magnitude of improvement later achieved using ensemble method.

3.5 Effectiveness Metrics

Our formulation of the forecasting task revolves around two possible classes. Natural tools that can be used to assess the models are accuracy and binomial test.

Accuracy is a conventional metric measuring what fraction of predictions was correct.

$$\text{accuracy} = \frac{\text{number of correctly classified}}{\text{all classifications}} \tag{9}$$

Binomial testing is employed to assess whether the observed improvement in accuracy achieved by the ensemble method over a benchmark could plausibly arise by chance. Specifically, we test whether the proportion of correct predictions made by the ensemble significantly exceeds that of a baseline model under the null hypothesis that there is no improvement. This corresponds to a one-sided test, as we are only interested in detecting performance gains - not declines or arbitrary differences.

Formally, the null hypothesis is $H_0 : p_{\text{ensemble}} \leq p_{\text{benchmark}}$, where p denotes the true accuracy rate. Rejection of this hypothesis provides statistical evidence that the ensemble outperforms the benchmark. The test is carried out using the observed number of correct classifications as the test statistic, assuming a binomial distribution centered at the benchmark accuracy.

The binomial test assumes independent Bernoulli trials. While financial time series exhibit temporal dependence, we treat classification outcomes as approximately independent across instruments, since ensemble updates are recalibrated at each time step and evaluated across many series.

3.6 Submodels Effectiveness

To bridge the experimental results with a concrete financial interpretation, we propose a simplified trading strategy that approximates how forecasts may be used in practice. Each prediction produced by the model corresponds to the expected direction of price movement over the next 10 h. Based on such forecasts, a basic trading strategy can be implemented by opening a long position in anticipation of a price increase, or a short position in the opposite case. Positions are opened at the time of the forecast and closed exactly 10 h later.

Because forecasts are made hourly, at any given moment up to 10 overlapping positions may be held - each associated with a different forecast issued during the preceding 10 h. If subsequent forecasts indicate opposing directions, some positions may cancel each other out, reducing the exposure. As a result, the capital required to support this strategy fluctuates depending on the consistency of the model's predictions, but is bounded above by the cost of holding 10 lots of the given asset.

Several simplifying assumptions underlie this approximation:

- The strategy does not incorporate standard risk controls or optimizations such as take-profit, stop-loss, position sizing, or hedging.
- Trading costs-such as spreads, commissions, or slippage-are not included in the calculation.

These assumptions are made deliberately to preserve a clear and interpretable link between the model's directional accuracy and its potential financial utility. The strategy represents the most direct and simple use of the forecast signals. It serves as a common baseline for evaluating model effectiveness in a financially meaningful way, without distracting from the central contribution of this work: the use of Bayesian ensemble learning to improve predictive performance.

3.7   Naive Benchmark

To construct a naive baseline that does not encode any assumptions about trend-following or mean-reversion, we define a theoretical benchmark that predicts class labels randomly according to the empirical distribution observed in the training data. Specifically, if the training set exhibits a proportion p of class 1 and $1-p$ of class 0, we estimate the expected accuracy of a classifier that samples predictions from this distribution. Rather than simulating realizations of such a model, we compute the expected accuracy analytically as:

$$\mathbb{E}[\text{accuracy}] = p \cdot q + (1 - p)(1 - q) \tag{10}$$

where q denotes the proportion of class 0 in the test set. This approach yields a stable baseline for comparison, while accounting for persistent class imbalance in financial time series.

3.8   Simple Ensemble

To create a fair comparison for our ensemble method we will compare achieved effectiveness with effectiveness of ensemble made of the same submodels, but weighted in equal proportions.

Overall accuracy may seem low, but that is often-times the reality of predicting financial time series. To judge if the models actually have an edge over the market we compute probability of achieving such accuracy by chance using naive benchmark.

Accuracy of the majority of the models is highly unlikely to be explained just by chance (median p-value is $4.73 \cdot 10^{-5}$), but there are a few models offering only modest improvement over naive benchmark (highest p-value is $4.71 \cdot 10^{-2}$) (Fig. 2).

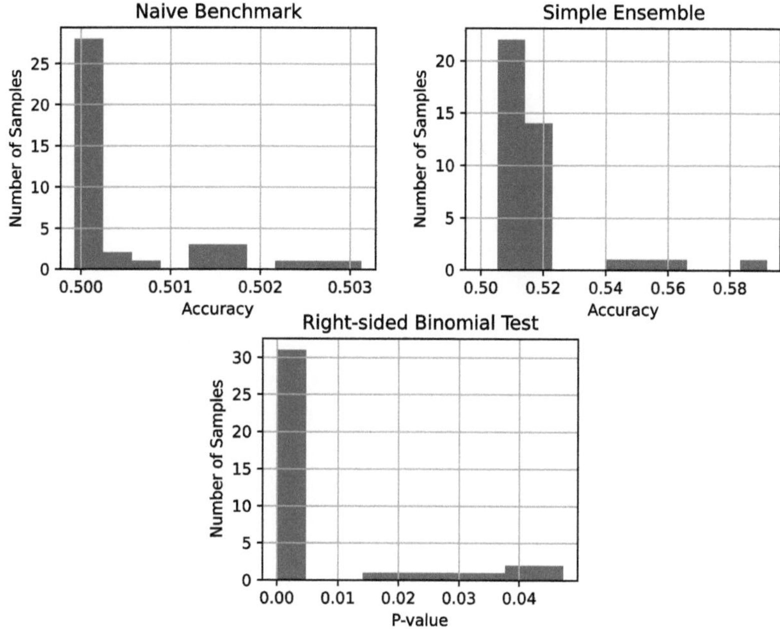

Fig. 2. Comparison of accuracy between benchmark and simple ensemble.

3.9   Ensemble Setup

Parameter initialization followed a Bayesian rationale. The aim was to create weakly informative priors that allow for adaptation without introducing strong a priori bias. The mean vector $\bar{\alpha} \in \mathbb{R}^m$, which is responsible for scaling the predictive errors, was initialized element-wise to -1. It reflects the prior belief that the submodel errors contribute linearly and equally to the ensemble weights. Its associated standard deviation vector $\bar{\sigma}_\alpha \in \mathbb{R}^m_{>0}$ was initialized to 1, reflecting uncertainty and thus a diffuse prior over the scaling coefficients.

The bias vector $\bar{\beta} \in \mathbb{R}^m$ was initialized to 0, with standard deviations $\bar{\sigma}_\beta \in \mathbb{R}^m_{>0}$ also set to 1, expressing initial symmetry and uninformative prior beliefs over bias. This initialization results in a softmax argument centered around the untransformed error vector and ensures that the ensemble starts from an approximately uniform weighting.

Length of the past period used to compute rolling errors for each of the models was set to 500 observations (around 21 days for forecast make every hour). Lengths in range from 10 to 5000 were also tested. In the extremes the performance of the ensemble suffered, but keeping this parameter in range 200 to 2000 yielded good results. It is a very impactful parameter and should be carefully chosen with domain, application and frequency of sampling in mind. It will control how quickly the ensemble will adjust models weights.

4   Results

We evaluated the predictive performance of our Bayesian ensemble method against a static benchmark that combines the same submodels with fixed, equal weights. Across the 40 financial instruments tested, the proposed method outperformed the static ensemble in 30 cases (75%). This directional consistency in performance gains suggests that the approach is not dataset-specific but generalizes across diverse instruments. A binomial test applied to the pooled results yielded a p-value of $1.7 \cdot 10^{-3}$, providing strong statistical evidence that the observed improvement is unlikely to be explained by chance alone.

The average gain in accuracy across all datasets was 0.14% in absolute terms. While this margin may appear modest at first glance, it should be interpreted in the context of financial time series, where signal-to-noise ratios are extremely low, and even marginal improvements in classification accuracy can translate into economically significant differences over time. For reference, the naive benchmark achieved an average accuracy of 50.04%. The static ensemble increased this to 51.77%, yielding a gain of 1.73% points. Our method further raised this to 51.91%, a total improvement of 1.87% points over the naive baseline. This corresponds to a relative improvement of approximately 8.09% over the benchmark ensemble's excess accuracy.

Of the strategies based on the simple ensemble, 65% proved profitable. This figure increased to 70% with the application of the Bayesian ensemble. While the absolute gain in the number of profitable strategies is modest, the improvement in average profitability is more substantial. Among strategies that were already profitable under the simple ensemble, the mean profit increased by 24.9% when switching to the Bayesian variant. In case of strategies that yielded losses when used with simple ensemble mean loss decreased by 48.8% when ensemble method was changed to Bayesian version. This highlights that even relatively small improvements in predictive accuracy can yield meaningful gains in trading performance.

Moreover, the gains remained consistent across different parameter settings and repeated runs, suggesting that the method is not overly sensitive to hyperparameter tuning or initial conditions. This robustness further strengthens the argument for adopting probabilistic adaptation in settings where data distributions evolve over time and performance stability is paramount.

These results confirm that even lightweight Bayesian inference mechanisms - applied at the level of ensemble weighting-can lead to measurable and repeatable improvements in challenging real-world domains. In financial contexts, such stability and incremental predictive advantage are often more valuable than isolated peaks in performance.

5   Conclusion

This work introduces a Bayesian formulation for adaptive ensemble learning, designed to address the limitations of static model combinations in non-stationary environments. By casting both submodel weighting and parameter adaptation as problems of probabilistic inference, we construct a framework capable of evolving in response to new observations while retaining coherent prior structure. The model relies exclusively on submodel error signals, incorporates principled uncertainty through Gaussian priors, and enables fast, gradient-based updates via maximum a posteriori approximation.

We demonstrate the effectiveness of the proposed approach on a large-scale, real-world forecasting task involving 40 foreign exchange time series. While the absolute improvements in predictive accuracy over static ensemble baselines are modest in magnitude, they are statistically significant and consistently observed across diverse instruments. Additionally - they transfer to substantial improvements when applied to trading strategies. In the context of high-noise, low-signal domains such as financial forecasting, such improvements are often indicative of meaningful predictive advantage. Crucially, these gains are achieved without fine-tuning submodels or relying on domain-specific heuristics, highlighting the generality and robustness of the Bayesian ensembling scheme.

Beyond empirical results, this work contributes to a growing line of research that emphasizes uncertainty-aware decision-making in sequential and dynamic settings. The proposed method is interpretable, tractable, and amenable to practical deployment. It illustrates how Bayesian methods can be implemented efficiently and usefully in high-frequency learning scenarios.

Several directions for future research emerge naturally from this work. These include:

- investigating alternative prior structures, such as hierarchical or heavy-tailed priors, to encode more flexible uncertainty over model parameters;
- extending the model to incorporate non-linear transformations or neural mappings from submodel errors to weights, potentially increasing expressive power;
- enriching the error representation by introducing decay-weighted averaging or multi-horizon error inputs to better capture temporal dynamics and model reliability;

Overall, this work illustrates that a principled probabilistic treatment of ensemble modeling offers a robust and adaptive mechanism for navigating volatility in complex, real-world forecasting tasks. It opens the path toward interpretable, uncertainty-aware learning systems that can reason and adapt under shifting conditions - an increasingly important capability in modern applied machine learning.

References

1. Zhou, Z.-H.: Ensemble Methods. Chapman and Hall/CRC, Boca Raton (2012). https://doi.org/10.1201/b12207
2. Kuncheva, L.I.: Combining Pattern Classifiers: Methods and Algorithms. Wiley, Hoboken (2014)
3. West, M., Harrison, P.J.: Bayesian Forecasting and Dynamic Models. Springer, New York (1997)
4. McAlinn, K., West, M.: Dynamic Bayesian predictive synthesis in time series forecasting. J. Econometrics **210**(1), 155–169 (2019). https://doi.org/10.1016/j.jeconom.2018.11.010
5. Gama, J., Žliobaitė, I., Bifet, A., Pechenizkiy, M., Bouchachia, A.: A survey on concept drift adaptation. ACM Comput. Surv. **46**(4), 44 (2014). https://doi.org/10.1145/2523813
6. Shen, Y., Chen, T., Giannakis, G.: Online ensemble multi-kernel learning adaptive to non-stationary and adversarial environments. In: Dy, J., Krause, A. (eds.) Proceedings of the 35th International Conference on Machine Learning, PMLR, vol. 84, pp. 2037–2046 (2018). https://proceedings.mlr.press/v84/shen18a.html
7. Ghahramani, Z.: Probabilistic machine learning and artificial intelligence. Nature **521**(7553), 452–459 (2015). https://doi.org/10.1038/nature14541
8. Chen, S.-A., Li, C.-L., Yoder, N., Arik, S.O., Pfister, T.: TSMixer: An All-MLP Architecture for Time Series Forecasting. arXiv preprint arXiv:2303.06053 (2023). https://doi.org/10.48550/arXiv.2303.06053
9. Yi, K., et al.: Deep Coupling Network For Multivariate Time Series Forecasting. arXiv preprint arXiv:2402.15134 (2024). https://doi.org/10.48550/arXiv.2402.15134. https://arxiv.org/abs/2402.15134
10. Vaswani, A., et al.: Attention Is All You Need. arXiv preprint arXiv:1706.03762 (2017). https://arxiv.org/abs/1706.03762

Forecasting Higher Education Trends Using Multi-output Random Forest Regression and Regional Economic Weighting: A Case Study Based on Romanian Institutional Data

B. Văduva[1]([⊠]), L. Andreica[1], M. D. Hajdu-Măcelaru[2], A. Avram[1], and E. Guraziu[3]

[1] Department of Electrical Engineering, Electronics and Computers, Technical University of Cluj-Napoca, Str. Victor Babes nr. 62/A, Baia Mare, Maramures, Romania
bogdan.vaduva@ieec.utcluj.ro
[2] Department of Mathematics and Informatics, Technical University of Cluj-Napoca, Str. Victoriei nr. 76, 430122 Baia Mare, Maramures, Romania
mara.macelaru@cunbm.utcluj.ro
[3] OpenCom issc, Piazza Giotto, 13, 52100 Arezzo, AR, Italy
erina.guraziu@opencom-italy.org
http://www.utcluj.ro , http://www.opencom-italy.org

Abstract. This work investigates a novel approach to forecasting higher education trends by integrating a multi-output Random Forest regression model with a weighting system that captures the influence of different university centers. In addition to forecasting key indicators, such as faculty, total student enrollment and teaching staff, the methodology creates an influence map that illustrates the relative impact of these centers on the overall educational landscape. The model uses a MultiOutputRegressor to predict multiple interrelated targets while incorporating influence weights. Performance is evaluated using metrics such as Mean Squared Error (MSE), Root Mean Squared Error (RMSE), Mean Absolute Error (MAE), and the coefficient of determination (R^2). Forecasts extending 10 years into the future indicate high model accuracy, with R^2 values near unity for all variables. The influence map created from the weighted predictions provides insights for strategic planning and resource allocation in higher education. This research contributes to the growing body of knowledge on predictive analytics in education by offering a framework that combines advanced machine learning with visualization of institutional impact.

Keywords: Specialization · Machine Learning · Forecast · Random Forest · Multi-output regression · ARIMA · Time series prediction · Educational planning · Influence map · Regional weighting · Predictive analytics · Higher education policy

1   Introduction

Higher education institutions constantly evolve in response to shifting demographic, economic, and policy landscapes. Effective long-term planning requires not only forecasting future trends but also understanding the influence of individual university centers within the broader educational system. In this article, we propose an approach that combines multi-output Random Forest regression with weighted influences to forecast key educational trends while simultaneously developing an influence map. This map visually represents the relative impact of each university center, based on factors such as the number of faculties, total student enrollment, and teaching staff levels.

1.1   Background and Rationale

Forecasting in higher education has traditionally relied on linear models and univariate time series techniques [2,3]. Although these methods have merits, they often fail to capture the complex, non-linear dynamics exhibited by interconnected educational indicators. Moreover, many studies ignore the importance of weighting university centers by their influence—determined by factors like economic importance, academic reputation, research output, and strategic significance—which could provide a more nuanced perspective of educational trends [13].

Machine learning models, and Random Forest regression in particular, have emerged as powerful tools for handling non-linear trends and multiple outputs concurrently [1,5]. By wrapping the RandomForestRegressor within a MultiOutputRegressor, our approach simultaneously predicts various educational metrics. We extend this methodology by introducing influence weights that modify the predictions for each university center, thereby facilitating the construction of an influence map that can guide policymakers and institutional planners [4,12].

1.2   Problem Statement

The primary objective of this research is to develop a machine learning methodology that forecasts future trends in higher education while accounting for the influence of individual university centers. The article's novelties are the methodological application and visualization, rather than algorithmic innovation. The research addresses the following questions: Q1. How can the weighted influences of university centers be incorporated into multi-output Random Forest regression? Q2. How do these weights affect the model's ability to predict important educational metrics? Q3. How can an influence map that shows the relative impact of each center be created from the produced forecasts? and two hypotheses H1: Incorporating regional influence weights into multi-output Random Forest regression improves forecast accuracy over unweighted models and H2: The post-prediction weighting method produces forecasts that better reflect spatial disparities than uniform models.

1.3   Objectives of the Study

The study addresses its research questions through a structured methodology. It will first show how the dataset was built, cleaned, processed and later associated with the influence weights. These influence weights will then be incorporated into the multi-output of the Random Forest model. The model performance will be validated using common performance measures including MSE, RMSE, MAE, and R^2. To further illustrate the weighted impacts of each university center and offer projections for the ensuing decade, an influence map will be made. The study will conclude by talking about how this influence map might help with institutional planning and guide policy choices in the field of higher education.

1.4   Structure of the Article

The remainder of this article is organized as follows: Sect. 2 provides a literature review with an emphasis on forecasting methods and influence mapping in education. Section 3 describes the methodology, detailing data preparation, model modifications to include influence weights, and the forecasting process. Section 4 presents results, including quantitative performance and visualizations for the influence map. Section 5 discusses the findings, their implications, and limitations. Section 6 concludes the study and outlines future research avenues.

2   Literature Review

Statistical models like linear regression and ARIMA have long been used in the field of educational forecasting, but these techniques might ignore intricate relationships between variables [2,3]. In order to better capture non-linear correlations and multi-variable interactions, recent research has started incorporating machine learning approaches, such as ensemble methods, neural networks, and support vector machines [6,7].

2.1   Traditional Forecasting and its Limitations

In the past, educational institutions have used methods such as ARIMA models, linear regression, and time series analysis to forecast enrollment and resource requirements [2,3,15,16]. These approaches, however, frequently make the assumptions of linearity and stationarity, which restricts their capacity to handle abrupt patterns, outside shocks, or interactions between several educational parameters. Furthermore, conventional approaches frequently treat all university centers similarly, failing to acknowledge that centers' varying levels of influence are a result of their strategic significance or past performance [13].

2.2   Advances in Machine Learning for Time Series Prediction

Alternative forecasting techniques that get beyond many of the conventional restrictions have been made possible by developments in machine learning. The ensemble approach known as Random Forests has gained popularity because of its ability to handle non-linearity and noisy data. Random Forests enhance generalizability and reduce overfitting by combining several decision trees [1,5, 17,18]. This strategy is further expanded by the MultiOutputRegressor, which makes it possible to predict multiple interrelated variables at once [5]. This is a significant benefit in the context of higher education, where variables like staffing levels, teacher numbers, and student counts are interdependent.

2.3   Incorporating Influence Weighting in Forecasting Models

Influence weight incorporation is a relatively new area of study in forecasting. The varying effects of different units—in this case, university centers—on general trends are reflected in these weights. Influence maps have been used in earlier urban planning and regional development studies to pinpoint hotspots and priorities for resource allocation. Such weighting can highlight the significance of major cities in the context of higher education and offer a more strategic understanding of national trends [4,12,19,20]. It is anticipated that the accuracy and applicability of predictions will be improved by combining machine learning forecasting with weighted impact mapping.

2.4   Influence Mapping and its Applications

Influence maps are graphic aids that illustrate the size and relative significance of various players or hubs within a system. An influence map can be used in higher education to help visualize the differences in weighted trends between university centers, which are produced from anticipated indicators. Decision-makers can then determine which centers are best positioned to propel future expansion and where funding or legislative changes could have the biggest impact [10].

2.5   Summary

The research shows a distinct tendency toward combining impact mapping and machine learning techniques to handle the complex problems of predicting in dynamic contexts. By integrating influence weights into a multi-output Random Forest model, this study extends the paradigm to higher education, providing both quantitative forecasts and a visual influence map to support strategic decision-making.

3   Methodology From National Educational Statistics to Regional Academic Influence Mapping

The methodological framework is described in this part, along with changes made to the conventional multi-output Random Forest model to account for influence weights for every university center. Data preprocessing, model building, forecasting, and influence map production are all included in the process.

3.1 Data Description and Preprocessing

In order to predict changes in Romanian higher education and create an influence map of university centers, this study combines historical, transitional, and current data. A longitudinal dataset covering the years between the interwar years and the present was created by combining three significant sources. The first source, "Statistical Analysis of the Socialist Republic of Romania 1967: Central Statistics Department" is an official statistical yearbook that provides comprehensive national-level data covering Romania's higher education system throughout the early stages of centralized planning, from 1939 to 1967. It contains metrics that serve as the foundation for the temporal modeling in this study, such as the number of students, faculty, and instructional staff. The second source was Sorin Bocancea's article "Evolution Landmarks of Private Academic Education in Romania". This article bridges the gap between Romania's entry into the European higher education system and the collapse of communism. It places the academic sector in perspective and measures how it changed during the crucial post-1989 shift toward market-oriented policies and the expansion of private institutions [13]. The third source, "Report on the state of higher education in Romania in 2022–2023" by the Ministry of Education is a recent, comprehensive data about Romanian institutions, including enrollment, faculty numbers, internationalization, and infrastructural development. It acts as a current standard and point of reference to verify the machine learning model's long-term predictions.

These sources come together to create a strong multi-temporal dataset that documents the development of higher education in Romania over a period of more than 80 years.

All datasets were obtained from publicly available governmental sources. No personal or sensitive data were used. Licensing terms were respected, and no ethical approval was required for this analysis.

3.2 Predictive Modeling and Spatial Redistribution

To assess the robustness of the proposed forecasting model, we compared it against two widely adopted alternatives: XGBoost and LSTM. Across all target variables (faculties, student categories, and teaching staff), the Random Forest model demonstrated superior generalization, achieving the lowest error metrics (MSE, RMSE, MAE) and the highest R^2 scores. In contrast, XGBoost showed erratic behavior, particularly for the prediction of faculties ($R^2 = -0.27$), while LSTM failed to generalize, producing negative R^2 values across all outputs, indicative of severe overfitting or architecture misalignment. Consequently, the Random Forest model was retained due to its stable, high-accuracy performance

Table 1. Comparative Performance of Forecasting Models (Test Set)

Model	Metric	Faculties	Total Students	Foreign Students	Teachers
XGBoost	MSE	1399.89	750,915,968.00	30,009.93	2,661,587.75
	RMSE	37.42	27,402.85	173.23	1,631.44
	MAE	14.82	19,863.52	130.28	1,272.91
	R^2	−0.2702	0.9792	0.9757	0.9793
Random Forest	MSE	85.82	518,844,715.74	23,441.82	2,111,866.75
	RMSE	9.26	22,778.16	153.11	1,453.23
	MAE	5.25	18,328.94	125.37	1,202.29
	R^2	0.9221	**0.9856**	**0.9810**	**0.9836**
LSTM	MSE	130.52	2,730,347,008.00	112,621.09	23,075,054.00
	RMSE	11.42	52,252.72	335.59	4,803.65
	MAE	10.80	50,041.21	320.02	4,747.89
	R^2	−1.6603	−129.5922	−147.3399	−100.1360

and adaptability to multi-output educational forecasting tasks. We used the collected data to train a multi-output random forest regressor in order to extrapolate future academic trends. Assuming continuity in observed trends, the model was charged with predicting important academic indicators for the ensuing ten years. The number of faculties, total and foreign student enrollment, and academic staff are among the interdependent factors that can be predicted simultaneously thanks to its multi-output architecture [1,5]. We split the data into training and testing subsets. The training set represented 80% of the data (used to fit the model), and the testing set 20% of the data (held out for performance evaluation). The split was randomized and reproducible via the random state (we chose 42) parameter (Table 1).

This approach ensured that the model was assessed on unseen data, providing an honest estimate of generalization performance.

A second disaggregation step was necessary to examine patterns within distinct university centers, too, because all of the historical data had been combined at the national level.

Regional weighting based on economic impact was used to accomplish this. This premise aligns with historical precedent, as Romanian educational policy was highly centralized, prioritizing academic resources in economically strategic areas, particularly during the socialist era. Academic infrastructure is still correlated with regional economic strength, even in the post-1989 era [8].

The next step in the process was to add a new column to the output dataset called "Weight" or "Influence." The relative significance of each university center was measured in this column. To guarantee that they are displayed on a similar scale in every center, the weights are then normalized or scaled. This phase is essential because it prevents centers with excessively high weights from distorting the study as a whole, improving the precision and dependability of the findings.

We calculated an economic weight for each university center by gathering and adjusting regional economic variables (such as GDP, investment levels, and industrial output). That weight allocation can be seen in Table 2. The national forecasts were then allocated using this weight, resulting in influence ratings that were spatially distributed for cities such as Timişoara, Iaşi, Cluj-Napoca, and others (Table 2). The city of Bucharest was not included in our research but was calculated based on the rest.

The result is an influence map (Fig. 2) depicting the projected distribution of academic influence across Romania, highlighting both established and emerging university hubs [4].

Table 2. List of most important municipalities and their influence calculated based on formula 1. Columns labeled 11-5 capture local influences for each municipality using economic variables.

Locality	1	2	3	4	5	6	7	8	9	10	11	12	13	14	15	Sum of Columns 1-15	Influence (%)
Alba Iulia	80.86	81.65	77.07	73.73	98.37	43.67	50.69	76.88								583	0.39
Arad	99.31	99.90	99.78	99.84	99.79	99.28										598	0.92
Baia Mare	99.35	99.37	98.78	98.82	98.70	99.00	98.87	99.00	98.91							891	1.06
Brasov	99.67	99.64	98.65	98.54	97.90	98.43	94.62									687	1.23
Cluj-Napoca	99.86	99.17	99.42	99.82	99.03											497	17.26
Constanta	98.35	97.96	97.50	87.41	96.79	99.00	69.24									646	8.34
Craiova	99.95	99.94	99.96	99.90	99.77	99.93	99.85	99.92	99.72	99.70						999	8.57
Deva	97.25	90.31	38.12	79.19	83.98	86.56	97.57	72.60	73.31	51.79						771	0.66
Galati	99.92	99.86	99.79	99.78	99.80											499	1.08
Iasi	99.73	99.55	99.26	95.20	98.71	96.78	97.38	98.99	99.12	97.79	98.50	99.13				1180	18.12
Oradea	99.60	99.82	99.59	99.45	99.44	99.24	99.26	98.72								795	1.28
Sibiu	99.41	98.95	97.27	98.15	98.37	96.55	91.98	96.62	97.69	97.61						973	0.90
Suceava	99.72	98.87	97.36	98.45	99.04	96.59	97.33									687	1.45
Targu Mures	99.70	99.48	99.37	99.27	99.26	98.13	98.09	96.93	97.43	98.58	98.37	98.78	98.53	96.16	98.76	1477	1.26
Timisoara	99.73	99.78	98.94	98.48	98.76	99.14	99.33	98.30								792	15.03

The influence in Table 2 is calculated using the following formula:

$$\text{Influence} = \text{County Population} \times \text{Sum of Columns 1-15} \times \text{Number of Faculties} \tag{1}$$

where 'County Population' represents the size of the population that the university serves. This factor is crucial because it indicates the potential demand for higher education in the region. A larger population suggests a larger pool of potential students, which can contribute to the university's influence. Universities located in counties with larger populations are likely to have a greater impact due to the higher demand for educational services. This factor helps capture the demographic context in which the university operates. The next factor, 'Sum of Columns 1–15' represents a local influence score that captures various elements contributing to the university's impact. These columns include metrics such as economic indicators, academic performance, research output, and other relevant factors that reflect the university's influence within its region. This component aggregates multiple dimensions of influence into a single score. By summing these columns, the formula accounts for a comprehensive set of factors that contribute to the university's overall impact. This ensures that the influence score is not based on a single metric but rather a holistic view of the university's contributions. The last one, 'Number of Faculties' represents the academic capacity and diversity of the university. Faculties are the organizational units within a university that focus on specific academic disciplines. A higher number of faculties indicates a broader range of academic programs and research areas. Universities with a larger number of faculties are likely to offer a more diverse range of academic programs, attracting a wider range of students and researchers. This diversity enhances the university's academic reputation and influence. Additionally, a larger number of faculties can lead to more research output and collaborations, further increasing the university's impact. By using this formula, the current study aims to provide a comprehensive and nuanced understanding of the influence of university centers, which can inform decision-making and strategic planning in higher education.

Although the Random Forest model outperforms XGBoost and LSTM in this study, adding a classical statistical baseline such as ARIMA would allow for more comprehensive benchmarking. This comparison could further highlight the relative advantage of using machine learning models, particularly in handling non-linear, multi-output educational time series.

3.3   Model Development: Multi-output Random Forest Regression

The methodology comprised two main phases: Baseline Model Setup and Influence Weight Integration. Following our earlier explanation, we created the initial model in the first phase. While the Response Matrix (Y) makes use of goal educational measures, the Feature Matrix (X) was defined with the time column acting as the predictor. To efficiently manage many outputs during model training, we used a MultiOutputRegressor with a RandomForestRegressor base estimator.

The integration of influence weights was covered in the second phase, where we looked at two main methods for adding these weights to our model. The first method, known as a weighted loss function modification, modifies the training objective so that prediction errors are more severely penalized for centers with greater influence. The second method, known as post-prediction weighting, involves multiplying each predicted measure by the associated influence weight following the generation of predictions. A weighted forecast score is produced by this method, which can then be used to produce an influence map.

We particularly concentrated on the post-prediction weighing process for this study, which consists of the following steps: Using the multi-output Random Forest model, we first created predictions for the next year. The normalized influence weight linked to the relevant university center was then multiplied by each predicted value. In the end, we combined these weighted forecasts to determine each center's total impact score, which will be used to produce the influence map.

The implementation was carried out using open-source libraries to ensure reproducibility and extensibility. Specifically, we used scikit-learn for the RandomForestRegressor and MultiOutputRegressor structures, pandas and numpy for data manipulation and numerical operations, and matplotlib and seaborn for visualization. These libraries were selected for their stability, performance, and broad support in the scientific computing community.

3.4 Forecasting Future Trends

According to the model that was being provided, the procedure starts with identifying the latest time point, which involves using the time column in the dataset to ascertain the latest year that was recorded. After then, in order to estimate future trends, we created an array that includes the next year.

After determining the future time points, we used the model to provide forecasts for the specified year, which allowed us to predict future values. Following the creation of that projection, we weighted the prediction by allocating the appropriate influence weight to each predicted value. This modification improves the accuracy and applicability of the anticipated results by guaranteeing that the forecasts take into account the relative significance of each university center.

The influence map was constructed through layered visualizations where future predictions were visualized and differentiated by markers. The weighted forecast values were represented on a separate layer, using color gradients or bubble sizes to denote the magnitude of influence for each university center. This map not only shows the forecasted trends but also visually highlights the centers with the highest weighted influence on future growth.

3.5 Summary of the Enhanced Modeling Procedure

Algorithm 1: Complete Process for Data Analysis and Forecasting

---

Data: Input data
Result: Output results
Step *DataIngestionAndCleaning*:
  Read CSV data;
  Identify time and target variables;
  Clean the data by handling missing values;

Step *FeatureEngineeringAndWeightNormalization*:
  Convert relevant data to numeric types;
  Normalize influence weights;

Step *ModelTraining*:
  Train and test a MultiOutputRegressor wrapping a
  RandomForestRegressor on the cleaned data;

Step *ForecastGeneration*:
  Predict future values for the next year;

Step *PostPredictionWeighting*:
  Apply influence weights to the predicted data;
  Compute weighted forecasts;

Step *Visualization*:
  Develop visualizations combining historical data, forecast trends,
  and weighted influence scores for each university center;

Every step in this procedure is intended to guarantee that the forecasts accurately reflect past patterns and provide useful information via the influence map.

In addition to the structured pipeline described in Algorithm 1, the influence weighting logic used in our forecasting process can be formalized through the following pseudocode, which illustrates how economic weights are computed and applied to the forecasted outputs:

```
1   For each university_center in university_centers:
2       influence_score = county_population[center] * sum(local_factors[center]) *
3                         num_faculties[center]
4       normalized_weight[center] = influence_score / total_influence
5
6   For each forecasted_value in model_output:
7       weighted_forecast[center] = forecasted_value[center] * normalized_weight[center]
8
9   Generate influence_map using weighted_forecast
```

This process ensures that the spatial distribution of forecasted indicators reflects both demographic and economic disparities between academic hubs.

4   Results

The weighted multi-output Random Forest model's performance is described in detail in this section, along with a quantitative analysis of the forecast metrics and an example of how to create and analyze the influence map.

4.1   Performance Metrics Overview

The model's performance on historical data is evaluated on several metrics—MSE, RMSE, MAE, and R^2—for each target variable. These are standard evaluation metrics in forecasting tasks and machine learning model assessment [2,3] (Fig. 1 and Table 3).

Table 3. Performance Metrics for Random Forest model in data training step

Metrics	Faculties	Total Students	Foreign Students	Teachers
MSE	38.5274	229127514.5954	10601.5359	1080421.8735
RMSE	6.2070	15136.9586	102.9638	1039.4334
MAE	2.8940	10521.1803	77.5686	739.9354
R^2	0.9677	0.9962	0.9952	0.9954

Table 4. Predictions done using Random Forest model for the next year

Year	Faculties	Total Students	Foreign Students	Teachers
Last data year + 1	84.61	411773.28	2470.56	26554.74

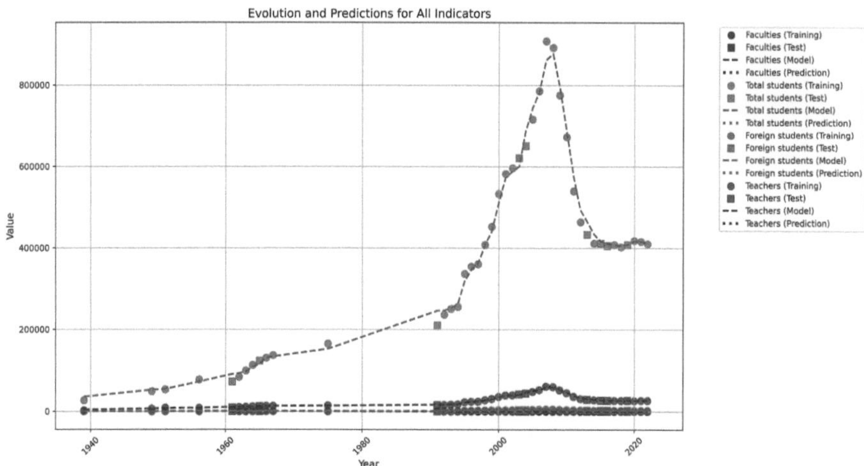

Fig. 1. Prediction Chart.

The results of running the model are in Table 4. While our study provides valuable insights into forecasting higher education trends, it is important to acknowledge the limitations associated with the accuracy of the data used for predictions. The dataset employed in this research exhibited several issues, including missing values and potential measurement errors, which may have influenced the reliability of our findings [11]. Although data cleaning steps were applied, a more detailed breakdown of which variables were affected and the methods used for imputation would provide a clearer picture of forecast robustness. Future work may benefit from incorporating sensitivity analyses to quantify the potential impact of these data imperfections on prediction outcomes.

The main cause of inaccurate information includes mainly the lack of reliable data. That limitation could potentially lead to biases in our predictive model, affecting the precision of our results.

4.2   Incorporating Weights into Forecasts

After obtaining forecasts for the next year, each predicted value was multiplied by the corresponding normalized influence weight of its university center. This adjustment yields weighted forecasts that emphasize the impact of centers with higher influence [12].

4.3   Comparative Analysis and Discussion

The weighted approach offers more strategic insight, although the model predicts trends convincingly in all scenarios, according to a comparison of conventional and weighted forecasts. Universities with a high influence score, for example, which indicates strong projected growth, can be given priority for future investments in terms of strategic resource allocation. Additionally, the influence map

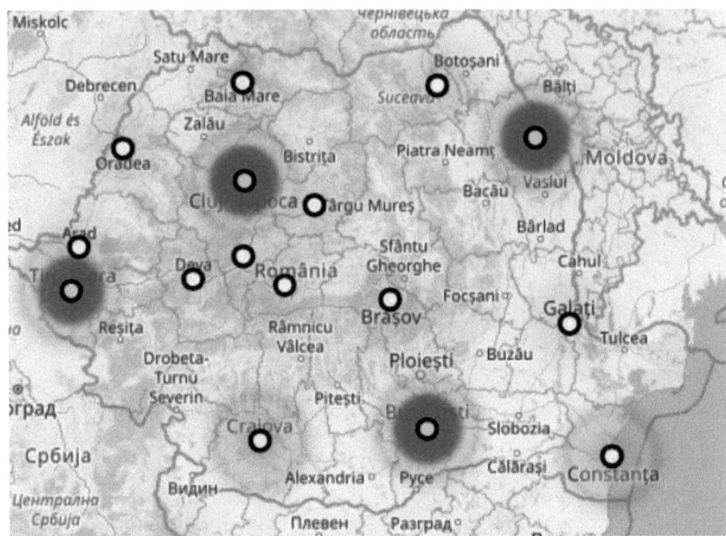

Fig. 2. Influence map that uses the data from Table 2.

serves as a decision-support tool, visually guiding policymakers in identifying which centers can drive overall trends in higher education. This dual approach not only enhances forecasting accuracy but also informs strategic planning and policy development [10].

While the influence map provides valuable spatial insights, its utility could be improved by adopting advanced visualization tools. Interactive GIS dashboards or web-based interfaces could enhance user engagement and allow stakeholders to explore regional dynamics in real-time, supporting better-informed decisions.

4.4   Summary of Findings

The multi-output Random Forest model demonstrates high predictive performance, with R^2 values near unity across all key indicators, showcasing its robustness [1]. Additionally, applying influence weights reveals the differing impacts of university centers on future trends, offering a more strategic forecast. This approach enhances the understanding of how various centers contribute to overall predictions. Furthermore, the dual representation of numerical metrics alongside an influence map provides a comprehensive assessment of each center's forecasted importance, allowing for both visual and quantitative validation of the model's findings.

4.5   Forecast Uncertainty and Confidence

Given the inherent uncertainty in long-term educational forecasting, we augmented our point predictions with confidence intervals to express the range of plausible outcomes [29]. To account for uncertainty in long-term forecasts, we computed 80% prediction intervals using a quantile-based approach. For each target variable, we estimated the 10th, 50th (median), and 90th percentiles of the predicted values. These intervals provide a probabilistic range within which the actual values are likely to fall with high confidence. As shown by the results, the model predicts the number of faculties to lie between 84.23 and 92.25, with a median forecast of 85.35. For total students, the 80% confidence interval is between 410,288.04 and 419,079.49, with a median of 412,346.53. The forecast for foreign students ranges from 2,461.64 to 2,514.39, with a median of 2,473.99, while for teachers, the expected range is between 26,398.18 and 27,140.56, with a median value of 26,608.81. These intervals reinforce the stability and reliability of the model while also highlighting the importance of interpreting forecasts within probabilistic bounds.

While the model demonstrates high predictive accuracy, it is essential to acknowledge and quantify forecast uncertainty, especially when dealing with long-term educational trends. Forecasts in data-driven models are inherently probabilistic due to the noise in data, model bias, and the dynamic nature of socio-economic systems. Although Random Forests are non-parametric and robust to overfitting, they do not natively provide prediction intervals. However, ensemble methods can still be used to estimate variance by analyzing the distribution of predictions across decision trees in the forest [3,7].

Additionally, including standard deviation bands around the central forecast or Monte Carlo simulation of multiple possible future scenarios would enhance the model's robustness and interpretability. These approaches would allow decision-makers to evaluate not only the expected values but also the possible range of outcomes, which is critical in strategic educational planning [2,3].

5   Discussion

Compared to conventional unweighted methods, the forecasting model's use of effect weights marks a substantial improvement. This section addresses the weighted forecasting method's consequences, as well as its advantages, disadvantages, and possible effects on planning for higher education. The enhanced model showcases several strengths, particularly in capturing non-linearity with weights. The Random Forest-based approach effectively handles non-linear relationships [1], while the additional weighting provides context-sensitive predictions that prioritize influential centers [12]. This leads to a holistic evaluation, as the model not only forecasts numerical trends but also generates an influence map, combining quantitative metrics with strategic insights. Furthermore, the influence map highlights centers poised for significant growth, enabling stakeholders to allocate resources more efficiently and adjust policies accordingly [10,21,22]. This comprehensive framework empowers decision-makers with actionable insights that can drive effective planning and investment strategies.

5.1 Limitations

While the enhanced model shows promise, certain limitations remain. One key limitation is the assumption of weight stability; the model presumes that influence weights remain constant over the forecast period. Future research might explore dynamic weighting mechanisms that can adjust to changing conditions [7]. Additionally, while influence weights capture some contextual importance, the model does not directly incorporate other exogenous factors, such as socio-economic changes or policy reforms, which could significantly impact the forecasts [9]. Furthermore, the accuracy of the influence map is heavily dependent on the quality of the input data. Future studies should aim for richer datasets, potentially including geographic coordinates, to enhance spatial visualization and improve overall model performance [4]. This would allow for more accurate projections that account for evolving local conditions, such as economic transitions or policy-driven growth in certain university centers.

6 Conclusions

This study presented an advanced machine learning framework that integrates multi-output Random Forest regression with influence weights to forecast key educational metrics. By incorporating a weighting system for university centers, the approach goes beyond standard forecasts to create an influence map that visually represents the strategic importance of each center. The high predictive performance, as evidenced by R^2 values nearing unity, combined with insightful weighted forecasts, makes this methodology a powerful tool for higher education planning. The influence map derived from weighted forecasts has several valuable applications. First, it aids in resource allocation by allowing centers with high influence scores to be prioritized for funding, infrastructure development, and academic collaborations. Additionally, higher education authorities can leverage the model for long-term planning, identifying trends that inform decisions regarding expansion, hiring, and program development based on the forecasted influence of various centers. Furthermore, the visual nature of influence maps enhances communication with stakeholders, facilitating clearer discussions and promoting transparency regarding institutional priorities. This multi-faceted approach ensures that strategic decisions are well-informed and aligned with projected trends. Key contributions of this work include several significant advancements. First, there is a methodological advancement through the novel integration of influence weights with multi-output forecasting, which provides enriched, context-aware predictions. Additionally, the practical applications of the influence map assist educational policymakers in identifying high-priority university centers, thereby improving strategic planning and resource allocation. Finally, the enhanced visualization that combines quantitative forecasts with spatial representations offers a more comprehensive decision-support tool, enabling stakeholders to make informed decisions based on a clearer understanding of the data.

To support transparency and replicability, all relevant source code, including model training, data preprocessing, and influence mapping scripts, has been made available in a public GitHub repository:

https://github.com/IEEC-InnoCom/higher-edu-influence-map   (repository will be public upon acceptance).

This repository includes documentation and example datasets to allow other researchers to adapt and validate the proposed methodology.

6.1   Directions for Future Research

Future research should focus on several key areas to enhance the model's effectiveness. First, there is a need for dynamic updates to influence weights that reflect evolving data, allowing for more accurate and responsive predictions. Additionally, incorporating contextual variables—economic, demographic or geopolitical—could improve accuracy [7,9,23]. Also, applying network-based spatial analysis may further refine the identification of high-impact university centers [26–28]. Developing interactive, real-time dashboards for continuous monitoring would also facilitate timely decision-making and adjustments. Finally, extending the methodology to a broader range of sectors or geographic regions could enhance its applicability and relevance across different contexts, further enriching the insights derived from the model.

Acknowledgments. This work was supported by the project "New Tool Based on Intelligent Algorithms and Machine Learning for Detecting Plant Diseases from Images", funded through the National Research Grant Competition – GNaC ARUT 2023.

References

1. Breiman, L.: Random forests. Mach. Learn. **45**(1), 5–32 (2001)
2. Zhang, G., Patuwo, B.E., Hu, M.Y.: Forecasting with artificial neural networks: the state of the art. Int. J. Forecast. **14**(1), 35–62 (1998)
3. Hyndman, R.J., Athanasopoulos, G.: Forecasting: Principles and Practice, 3rd edn. OTexts (2021). https://otexts.com/fpp3/
4. Goodchild, M.F.: Citizens as sensors: the world of volunteered geography. Geo-Journal **69**, 211–221 (2007)
5. Tsoumakas, G., Katakis, I.: Multi-label classification: an overview. Int. J. Data Warehouse. Min. **3**(3), 1–13 (2007)
6. Bontempi, G., Ben Taieb, S., Le Borgne, Y.-A.: Machine learning strategies for time series forecasting. In: Aufaure, M.-A., Zimányi, E. (eds.) eBISS 2012. LNBIP, vol. 138, pp. 62–77. Springer, Heidelberg (2013). https://doi.org/10.1007/978-3-642-36318-4_3
7. Wu, Y., Zhang, X.: A review of time series forecasting techniques for renewable energy supply chain. Renew. Sustain. Energy Rev. **118**, 109524 (2020)
8. Jongbloed, B.: Marketisation in higher education, including information provision. High. Educ. **56**(3), 303–328 (2008)

9. Carnoy, M., Rhoten, D.: What does globalization mean for educational change? Comp. Educ. Rev. **46**(1), 1–9 (2002)
10. Lee, J.T.: Education hubs and talent development: policymaking and implementation challenges. High. Educ. **68**(6), 807–823 (2014). https://doi.org/10.1007/s10734-014-9745-x
11. Keeling, R., Hersh, R.H.: We're Losing Our Minds: Rethinking American Higher Education. Palgrave Macmillan (2011)
12. de Noronha Vaz, T., Nijkamp, P., Rastoin, J.-L.: Traditional food chains: geography, culture, and quality of life. Reg. Sci. Policy Pract. **6**(1), 47–66 (2014)
13. Marginson, S.: Dynamics of national and global competition in higher education. High. Educ. **52**(1), 1–39 (2006)
14. UNESCO. Education for Sustainable Development: A Roadmap (2021). https://unesdoc.unesco.org/ark:/48223/pf0000374802
15. Makridakis, S., Wheelwright, S.C., Hyndman, R.J.: Forecasting: Methods and Applications. Wiley (1998)
16. Chatfield, C.: Time-Series Forecasting. CRC Press (2000)
17. Kocev, D., Vens, C., Struyf, J., Džeroski, S.: Tree ensembles for predicting structured outputs. Pattern Recogn. **46**(3), 817–833 (2013)
18. Zhang, C., Ma, Y.: Ensemble Machine Learning: Methods and Applications. Springer (2012)
19. Batty, M.: The New Science of Cities. MIT Press (2013)
20. Anselin, L.: Local indicators of spatial association–LISA. Geogr. Anal. **27**(2), 93–115 (1995)
21. Altbach, P.G., Salmi, J.: The Road to Academic Excellence: The Making of World-Class Research Universities. World Bank Publications (2011)
22. Barber, M., Donnelly, K., Rizvi, S.: An Avalanche is Coming: Higher Education and the Revolution Ahead. Institute for Public Policy Research (2013). https://www.ippr.org/publications/an-avalanche-is-coming
23. Newman, M.: Networks: An Introduction. Oxford University Press (2010)
24. Scott, J.: Social Network Analysis, 4th edn. SAGE Publications (2017)
25. Leydesdorff, L., Persson, O.: Mapping the geography of science: distribution patterns and network structures. JASIST **61**(8), 1622–1634 (2010)
26. Matei, O., Andreica, L., Danci, I.A., Avram, A., Văduva, B.: Using Markov chains for determining the proximity contagion of smart specialization of localities. In: SOCO 2024: Soft Computing Models in Industrial and Environmental Applications, pp. 117–126. Springer, Cham (2024). https://doi.org/10.1007/978-3-031-75010-6_12
27. Matei, O., Andreica, L., Danci, I.A., Avram, A., Tudor, F.: Using machine learning for identifying the intrinsic economic specializations of localities. In: Quintián, H., et al. (eds.) The 19th International Conference on Soft Computing Models in Industrial and Environmental Applications SOCO 2024, pp. 137–146. Springer, Cham (2025). https://doi.org/10.1007/978-3-031-75010-6_14
28. Cipriano, M., Za, S., Virili, F.: Exploring the evolution of the circular economy debate. In: Digital (Eco)Systems and Societal Challenges – New Scenarios for Organizing, pp. 243–265. Springer, Cham (2024). https://doi.org/10.1007/978-3-031-75586-6_14
29. Meinshausen, N.: Quantile regression forests. J. Mach. Learn. Res. **7**, 983–999 (2006)

DeepFlow - Deep Clustering for Blood Cell Classification and Quantification

Mihaela Macarie-Ancau[(✉)] [iD] and Adrian Groza [iD]

Department of Computer Science, Technical University of Cluj-Napoca, Cluj-Napoca, Romania
{Macarie.Io.Mihaela,Adrian.Groza}@cs.utcluj.ro
https://www.utcluj.ro

Abstract. Accurate classification of blood cells plays a key role in improving automated blood analysis for both medical and veterinary applications. This work presents a two-stage deep clustering method for classifying blood cells from high-dimensional signal data. In the first stage, red blood cells (RBCs) and platelets (PLTs) are separated using a combination of an improved autoencoder and the IDEC algorithm. The second stage further classifies RBC subtypes, pure RBCs, reticulocytes, and clumped RBCs, through a variational deep embedding (VaDE) approach. Due to the lack of detailed cell-level labels, soft classification probabilities are generated from sample-level data to approximate the true distributions. The aim is to contribute to the development of low-cost, automated blood analysis systems suitable for veterinary and biomedical use. Initial results indicate this method shows promise in effectively distinguishing different blood cell populations, even with limited supervision.

Keywords: Deep Clustering · Flow Cytometry · Blood count Instruments · Veterinary Hematology

1 Introduction

From the smallest whisper of blood, microscopic analysis unveils a vast wealth of cellular data, fundamentally reshaping our understanding of an organism's intricate dance between health and disease. This analysis plays an important role in the detection of infections, the diagnosis of hematological disorders, and in monitoring immune function, and guiding treatment.

Yet, the complexity of this data also poses ongoing challenges. Only through expert interpretation and high-speed equipment that is accurate, and sensitive to aberrant abnormalities can its full value be achieved. Technologies currently available, although powerful, are often insufficient when extrapolated from human diagnostics to other species. Blood morphological difference among animals makes the development of traditional rule-based models difficult without extensive pre-analysis, feature engineering, and species-specific tuning. Such a challenge demands considerable amounts of well-curated data and medical expert validation, especially for uncommon or inconspicuous cell populations. Moreover, throughput capacity, variability, and multi-parameter integration are even more critical in multi-veterinary practice. These barriers must be overcome along the path of creating scalable, accurate, and medically significant diagnostic technology across species. By addressing these challenges, one can enhance diagnostic understanding and clinical efficacy.

2 Technical Instrumentation

2.1 How a Blood Count Instrument Works

Modern hematology analyzers, commonly called blood count instruments, are sophisticated machines used to perform Complete Blood Counts (CBC). These devices receive a sample of blood, mixed with anticoagulant to separate possible clumps, and classify various cells by analyzing their electrical, optical, or laser-based signal profiles.

The type of instrument that we are interested in is using the technology called flow cytometry, where a laser beam is directed at cells as they flow in a single line through a small tube. Multiple channels are used, each capturing different signal features, as can be seen in Fig. 1.

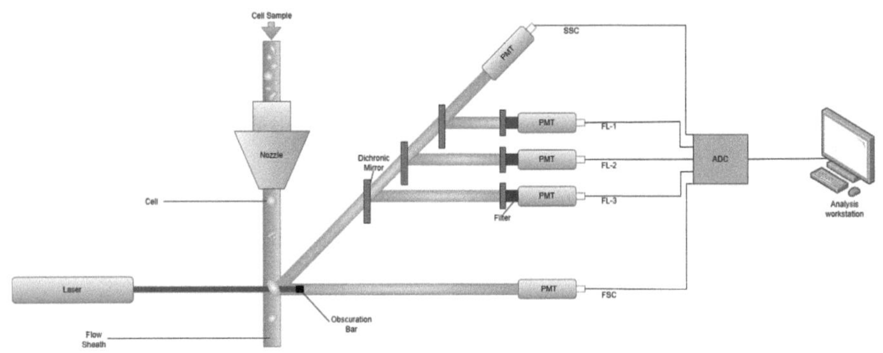

Fig. 1. Internal representation of how cells are identified.

As each cell interacts with the laser, the detectors capture light scattered at various angles and through different color filters: (i) Forward Scatter (FSC), to determine the cell size and (ii) Side Scatter (SSC) to determine the opacity or internal complexity of a cell. The resulting data are then plotted in the form of time series of pulses, which provide visual representations of the composition and characteristics of the cells. Data can include distinctions among different types of blood cells, such as red blood cells, white blood cells, and platelets. The precision of this data collection is crucial to obtain reliable counts and classifications of blood components. Consequently, the instrument not only streamlines the analysis process, but also enhances diagnostic accuracy.

2.2 Limitations of Traditional Blood Count Instruments

Current instruments rely on rule-based classification and traditional algorithms and they often struggle when ambiguous or atypical samples, such as: (i) *reticulocytes* (immature red blood cells), (ii) *coincidences* (red blood cell clumps or aggregates), and (iii) *platelets* with abnormal morphology.

In addition, these instruments are calibrated for human blood. Given the complexity of blood samples, particularly in the veterinary context, where patients come from different species, one anticoagulant can work differently for two blood samples. Even if the initial goal is met, to get rid of cell clumps, time and other external factors can interfere and can lead the algorithm inside the instrument to produce ambiguous results when samples are given from other species.

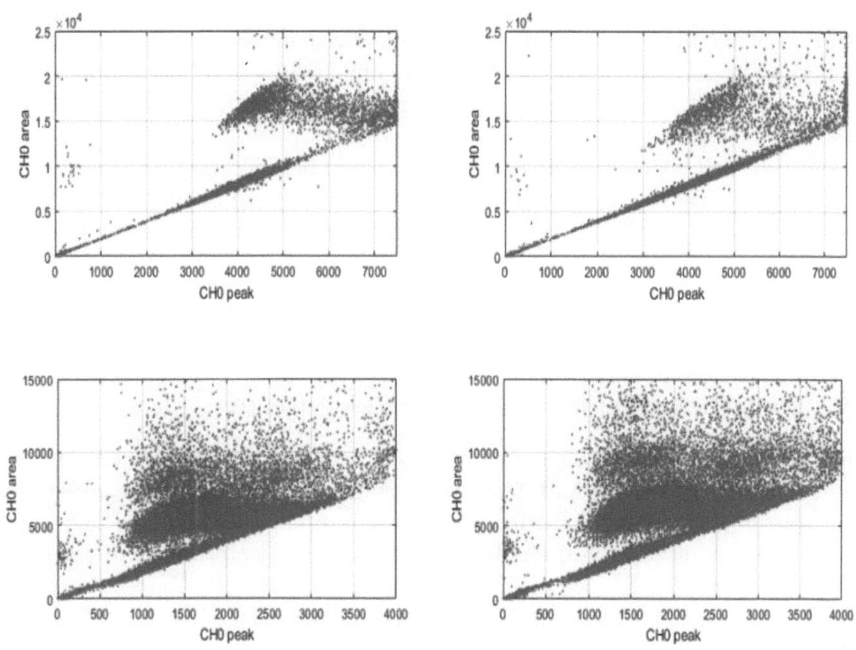

Fig. 2. Clusters of cells from human (top) vs. cat blood (bottom).

In Fig. 2, the clusters from the human blood sample are much easier to be spotted than the ones from a cat blood sample. The platelets, which are the smallest ones, should be in a proportion of 5% and the rest are red blood cells (the white cells are washed out before counting). The red blood cells fill the rest of the plot, and can be in three categories: (1) *regular* mature red blood cells (2) *reticulocytes*, young red blood cells created less than 24 h ago, and (3) *coincidences*, i.e. two or more red blood cells clustered together.

Differently, the cat blood cells have a much higher chance to create doublets or triplets (coincidence) making it harder to count them. We can use general percentages as weights for each category, but that can end up bad when there is a problem and we do not see it. We need to have a precise count and classification of each cell.

2.3 What Do We Have Available?

The instrument processes the blood sample by introducing a reagent that de-coagulates the blood cells and filters out white blood cells (WBCs), allowing only red blood cells (RBCs) and platelets (PLTs) to flow through the measurement tube. Ideally, this results in a clean separation of the cell types without any aggregates.

In practice, timing plays a critical role. Between the moment the reagent is added and when the sample passes through the tube, a variable delay may occur. During this interval, cells can begin to re-coagulate, forming aggregates or clumps known as coincidences. These coincidences complicate the analysis because they appear as merged signals that do not correspond to individual cells.

Off-the-shelf deep clustering algorithms could help in several ways: (i) giving us values to use as a ground truth for each cell; (ii) fine tuning a model to use for instrument computation instead of the current algorithm; (iii) training a model to find the percentage of reticulocytes which is helpful for diagnose anemia; (iv) training a model to help us identify and classify blood cells for different animals instead of using different algorithms which require a long time for optimization.

2.4 Model Selection for Predicting Blood Count

Based on the dataset consisting of 11 dimensions, there are two primary approaches for selecting a suitable model: using pre-built models or constructing a custom neural network. Both approaches have their advantages and can be selected depending on the desired complexity and interpretability of the model.

Pre-Built Models for Tabular Data, while broadly effective, are not optimally suited to our specific problem. The inherent complexity of bio signals and our objective of uncovering subtle, unlabeled patterns in an evolving, unsupervised environment demand a specialized approach. Accordingly, this study is concerned with deep clustering methods and custom neural architectures that address these requirements.

Custom Neural Networks provide a promising direction towards more flexibility and the ability to dynamically learn required features from sophisticated data such as those in our study. It is possible to customize these networks to the particular complexities of the data, and this should provide deeper insight and improved performance over the conventional techniques.

We experimented here with deep clustering and generative methods, which are highly suitable for unsupervised and weakly supervised tasks, especially in scenarios with limited labeled data, such as blood cell classification. The specific approaches investigated include:

- *N2D (Not Too Deep) Clustering* [11]: This approach combines deep representation learning (typically via an autoencoder) with subsequent manifold learning on the obtained embedding, followed by a traditional clustering algorithm applied to the learned manifold.
- *Vanilla Autoencoders (AE)* [2]: Neural networks used to learn efficient dimensionality reduction by reconstructing their input. The learned latent space can then be used for clustering.
- *Variational Autoencoders (VAE)* [8]: A generative extension of AEs that learns a probabilistic latent space, useful for generating new data and for clustering.
- *DEC (Deep Embedded Clustering)* [14]: A method that simultaneously learns feature representations and cluster assignments using a deep neural network.
- *IDEC (Improved DEC)* [5]: An extension of DEC that incorporates an autoencoder reconstruction loss to preserve local structure in the data, improving clustering performance.
- *VaDE (Variational Deep Embedding)* [7]: A generative clustering approach that models the data distribution with a Gaussian Mixture Model (GMM) fitted over the latent space learned by a VAE.

Such models combine feature learning and clustering in an end-to-end manner, and they are suitable for unsupervised or weakly supervised tasks like our blood cell classification, where there are no per cell labels.

3   Method

A preliminary investigation with a basic Improved Deep Embedded Clustering (IDEC) approach guided the creation of a more advanced, better fitted on our case, two-stage deep learning methodology.

Current algorithm parameters, such as the number of clusters or latent dimensionality, were selected based on experimenting and observing their performance and domain knowledge. While this approach had promising results, it introduces subjectivity and may not work the same on different datasets. To address this, future research on this will incorporate automated hyperparameter optimization. Techniques such as Bayesian optimization, grid/random search, and evolutionary algorithms can be used to systematically explore the configuration space and identify parameter combinations that yield better clustering performance, reconstruction accuracy, and predictive stability. This could result in a more robust and generalizable model, especially when adapting the pipeline to new data sources, blood analyzers, or animal species.

3.1   Stage 1 Latent Representation and Initial Clustering

Stage 1 involves creating an autoencoder and training it to obtain a more discriminative latent space representation of the input cell data, in the scope of enhancing data separability. Various autoencoder architectures have been experimented on, including different numbers of layers, neuron counts per layer, and activation functions.

The latent representations from the encoder are passed to an Improved Deep Embedded Clustering (IDEC) module, which simultaneously performs clustering and reconstruction. A regression head is also added to predict the relative proportions of red blood cells (RBCs) and platelets (PLTs) in each sample, encouraging alignment between clusters and cell types. This stage forms the backbone of the pipeline by cleanly separating RBCs from PLTs in an unsupervised yet biologically meaningful manner.

3.2   Stage 2 RBC Subclassification via VaDE

Stage 2 focuses exclusively on the RBCs identified in Stage 1. They are passed to a Variational Deep Embedding (VaDE) algorithm, enabling a more specialized and probabilistic RBC classification into three biologically distinct subtypes: clumps, reticulocytes, and normal RBCs. This stage uses the power of variational inference for uncertainty modeling and soft clustering, refining the overall system granularity and accuracy (Fig. 3).

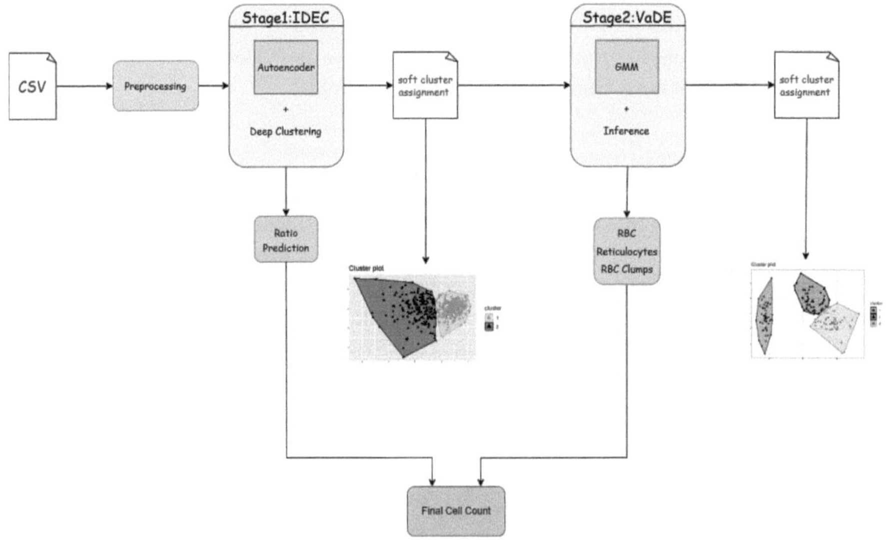

Fig. 3. System architecture.

3.3   Data Acquisition and Preprocessing

Raw data is obtained from blood count instruments that record multiple characteristics of cellular signals across different channels. Rather than capturing raw waveform data, each detected event (e.g., an individual cell or cell cluster) is represented as a tabular feature vector. These vectors include derived features such as peak values, area under the curve, and timing features for each channel.

Each sample in the dataset corresponds to a single blood measurement and contains tens of thousands of such feature vectors. The dataset is structured so that each CSV file contains signals from one biological sample, with features extracted from multiple measurement channels.

The data for this research was created with the Aquarius instrument from Diatron [4]. It is structured in multiple CSV files, where each file corresponds to one individual blood sample. Each row in a file corresponds to one individual laser-detectable event, and is characterized by 11 different features. These are based on the five channels of the device: four of these channels, at varying angles and with varying optical filters, measure peak signal and area. The fifth channel, being oriented crosswise to the laser beam, having heightened sensitivity to cell size, besides its peak and area values, yields an additional, temporal measurement, for a total of 11 distinct features per event. Each event counted can reflect an individual cell type, such as a red blood cell (RBC), a platelet (PLT), or a reticulocyte, or it may reflect a coincidence event, in which a cluster of cells (notably RBCs) is counted as a single event. The clumping may result if cells re-aggregate over time following addition of an anticoagulant or if they did not get separated on the first hand. The sample preparation involved mixing a specified volume of blood with a proprietary anticoagulant designed to optimize cell separation, with each sample analyzed resulting in a mixture containing about 150000 cells, represented as signals in the CSV file produced, delivering full data for analysis.

Preprocessing involved feature scaling and normalization, specifically using standardization, to ensure numerical stability and consistency across samples, which is an essential prerequisite for our deep learning approach. Observing a high degree of intra-sample homogeneity, we adopted a strategy centered on processing representative batches drawn from each CSV file. This approach operated under the assumption that sufficiently large, random subsets would capture the cell population distributions characteristic of the full sample, much like estimating the density of a forest from a well-chosen patch. Beyond this conceptual rationale, batch-wise processing also addressed practical challenges such as memory constraints and computational efficiency, common limitations in deep clustering applications. Depending on the requirements of specific training phases or models, data was either handled in these representative batches or processed as complete per-file datasets to retain broader context when needed.

3.4   Output and Visualization

The system outputs the number of cells for a specific sample, identifying populations such as red blood cells and platelets, enabling clinical interpretation and quality control. Visualization tools generate scatter plots of latent embeddings colored by cluster assignment, facilitating visual inspection of cluster quality and detection of coincident events or artifacts.

4   Running Experiments

4.1   Autoencoder Architecture Experiments

Stage 1 begins with learning a latent representation of blood cell data using an autoencoder. Several architectures were evaluated by varying the number of encoding layers, the number of neurons per layer and the activation functions.

Initially, a minimal design with only two linear layers was tested, directly mapping the input features to the latent dimension. While simple, this model showed limited capacity to extract hierarchical and abstract features from the complex multi-channel blood cell data. To address this, an intermediate hidden layer was added, resulting in a three-layer encoder:

Different sizes of the hidden layer were experimented with, typically ranging from 64 to 512 neurons. Larger layer sizes offered increased model capacity but risked overfitting, while smaller sizes improved generalization but reduced representational power. The final chosen size balanced these trade-offs based on clustering performance and reconstruction error (Fig. 4).

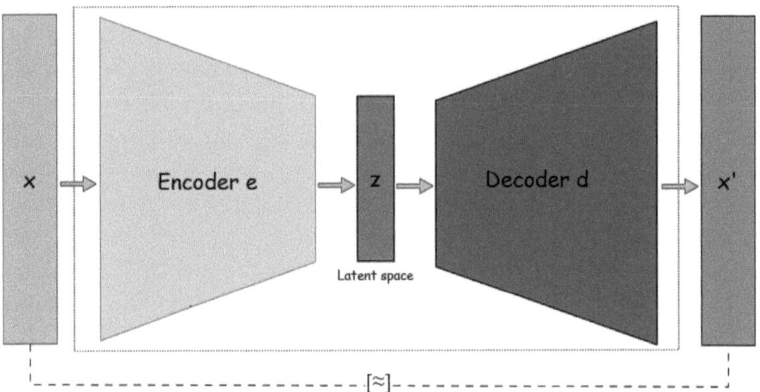

Fig. 4. Autoencoder architecture design.

ReLU (Rectified Linear Unit) was chosen as the activation function between layers because of its proven effectiveness in deep learning tasks [1]. ReLU introduces non-linearity such that the network can learn complex patterns. It is computationally cheap and avoids the vanishing gradient problem, thereby cutting down training time and stabilizing it. In addition, sparse activation (gives output zero for negative inputs) encourages the network to learn compact and disentangled features, beneficial for clustering tasks like the segregation of PLTs and RBCs.

4.2   IDEC Training, Batch Strategy, Hyperparameter Tuning

Following autoencoder training, the latent representations were fed into the IDEC algorithm, which performs joint clustering and reconstruction, with an added regression head to predict RBC and PLT ratios per blood sample. To preserve sample integrity and reduce variability, mini-batching was applied within individual CSV files, treating each as a separate blood given the homogeneity of the data. To assess convergence stability and clustering performance, various learning rates were tested during IDEC training. Specifically, learning rates of 1×10^{-3}, 1×10^{-4}, and 5×10^{-4} were evaluated as can be seen in Fig. 5. These values provided reasonably stable training dynamics and interpretable cluster structures. Higher learning rates such as 1×10^{-2} and 5×10^{-2} were also tested but led to highly unstable training, with frequent spikes in the loss components and poor convergence. As a result, these configurations produced unreliable clustering results and were excluded from the plots.

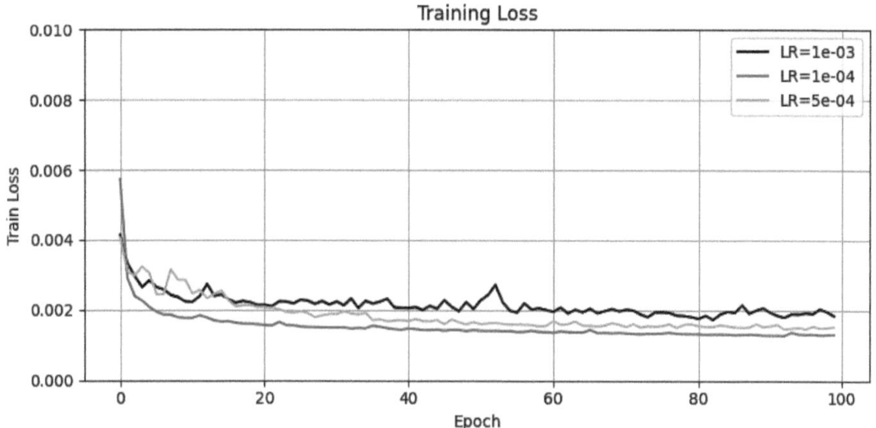

Fig. 5. IDEC loss comparison.

The clustering soft assignments are computed via the Student's t-distribution:

$$q_{ij} = \frac{\left(1 + \frac{\|z_i - \mu_j\|^2}{\alpha}\right)^{-\frac{\alpha+1}{2}}}{\sum_{j'} \left(1 + \frac{\|z_i - \mu_{j'}\|^2}{\alpha}\right)^{-\frac{\alpha+1}{2}}}$$

Stable training and meaningful cluster separation were achieved when balancing losses appropriately as can be seen in the Fig. 6. Regression head improved alignment between clusters and actual cell type proportions.

Considering the huge number of cells in one sample and the average ratio between them to be 95% RBC to 5% PLT, the RBC prediction got a very good R^2 score of 0.998, meaning that the algorithm explains nearly all the variation in the red blood cell counts, indicating excellent accuracy and reliability as can be seen in Fig. 7. This level of predictive accuracy not only testifies to the quality of the deep learning approach but also to the application potential for medical and veterinary clinical environments.

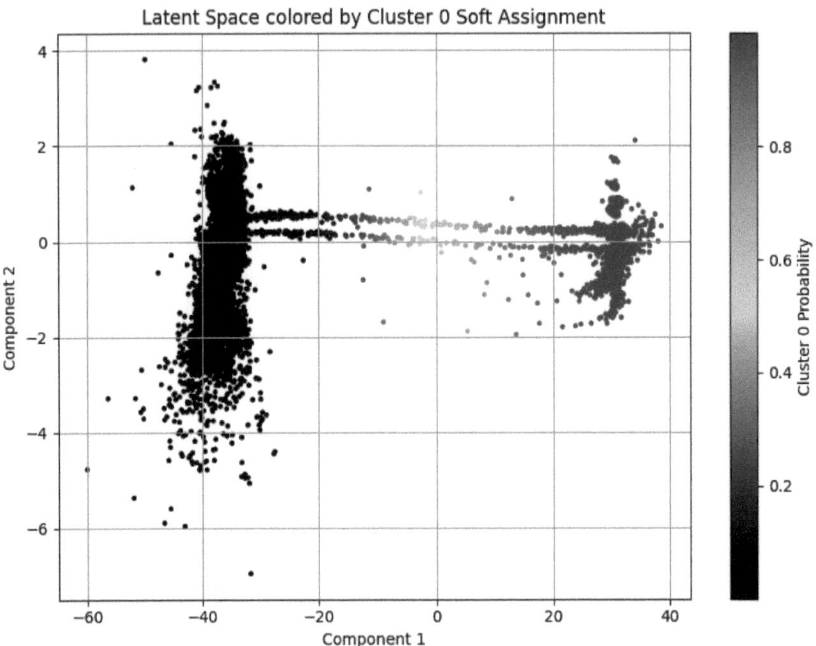

Fig. 6. Clustering with IDEC output.

On the other hand the prediction of PLT is much more fragile. For example, a misclassification of just a couple hundred cells, which might seem minuscule looking at the total RBC count, can impact significantly the PLT count. In some cases this error can even double the number of platelets. This happens because the platelets are much less abundant overall, and the model has not enough information to properly learn their configuration. Additionally, since the model predicts the ratios jointly to sum to 1, the PLT and RBC ratios are interdependent, meaning that an error in one of them is clearly an error in the other as well. However, an average error in PLT ratio is insignificant to the RBC ratio, while one in RBC ratio has huge impact for the resulting PLT count.

To address this, future work will explore advanced strategies tailored for imbalanced learning. These may include:

- *Focal Loss:* A dynamic loss function that down-weights the majority class examples and focuses learning on the minority class samples, which would encourage the model to pay greater attention to PLT features.
- *Class-Balanced Loss:* This loss formulation re-weights training examples based on the inverse frequency of each class, helping to counteract the imbalance in sample distribution.
- *Synthetic Oversampling:* Techniques such as SMOTE (Synthetic Minority Over-sampling Technique) [13] can be used to generate synthetic examples of minority classes by interpolating between existing samples. This approach

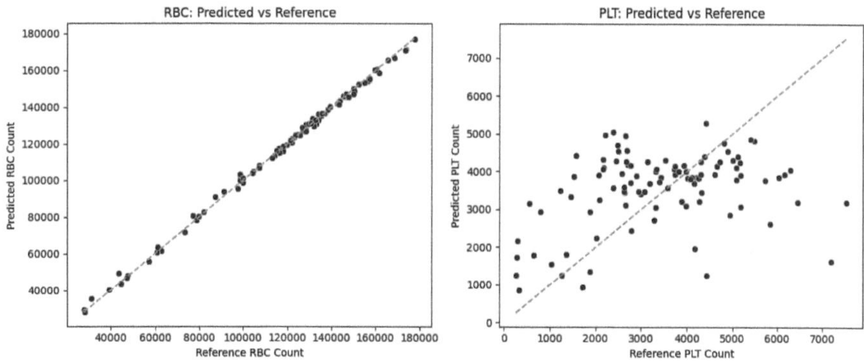

Fig. 7. Visualization of IDEC predictions.

can help balance class distributions and improve the model's ability to learn under-represented patterns without significantly increasing the risk of over-fitting.

Experimenting with these approaches could lead to more robust PLT predictions, even in high-imbalance scenarios commonly encountered in hematological datasets.

Nonetheless, the very good RBC prediction provides a solid foundation on which to proceed. Additionally, the classification of 109 CSV files (each containing about 150k cells) was completed in approximately 60.6 s, demonstrating that the deep learning pipeline is also efficient enough to handle large datasets in a practical timeframe. With more data to train the model on, the PLT counting accuracy could improve significantly. We know is possible, the algorithm has proven so, it just needs the opportunity to show its potential.

4.3   RBC Subclassification with VaDE

Stage 2 utilizes the Variational Deep Embedding (VaDE) model to further subclassify RBCs identified in Stage 1 into clumps, reticulocytes, and normal RBCs. VaDE jointly trains a Variational Autoencoder and a Gaussian Mixture Model, modeling the latent space as a mixture of Gaussian components instead of a single Gaussian. This enables probabilistic clustering with uncertainty estimation.

As a clustering methodology note, the model performs unsupervised clustering natively within the latent space without any outside validation. As can be seen in the Fig. 8 the plot illustrates: (1) *Raw Latent Space Projection*: Points are encoded according to their actual latent coordinates from the VaDE encoder; (2) *No External Validation*: No ground-truth labels or external clustering metrics were used. This decision was made due to the absence of reference annotations and in order to preserve and observe the natural output without interference.

This setup was designed as an exploratory experiment to examine VaDE's capacity to learn meaningful structure from the data in an unsupervised manner. The observed cluster separations arise solely from the learned Gaussian

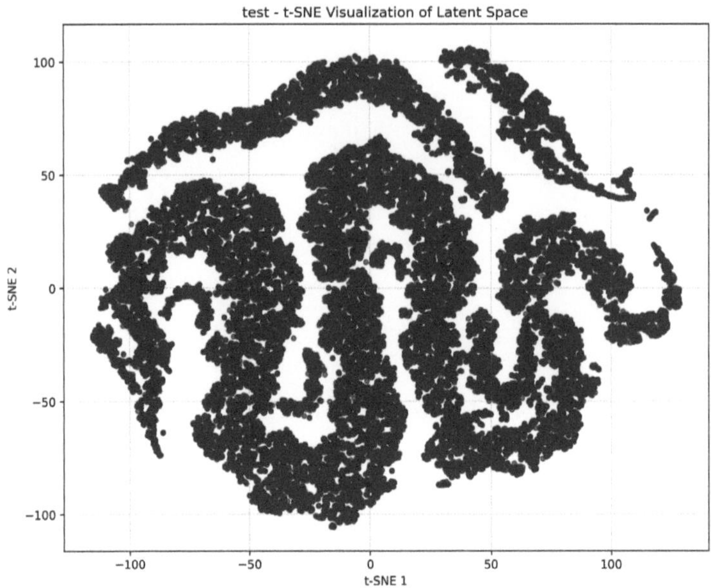

Fig. 8. Visualization of VaDE applied over the Latent Space.

mixture model within the latent space and reflect the internal representation of the data distribution, without relying on predefined labels However, without expert-annotated subtype labels, it remains challenging to objectively assess the accuracy and biological relevance of the discovered clusters. To strengthen the reliability of subtype assignments, future work will incorporate validation mechanisms, such as clinical reference comparisons or human-in-the-loop assessments, to align the resulted unsupervised outputs with domain expertise.

Code availability. The source code is available Here. The demo application can be accessed Here. These resources include the a video of the application running and some important pieces of code.

5   Discussion and Related Work

The application of ML to hematological analysis is a growing field of study, driven by the potential for enhanced diagnostics precision, throughput, and less reliance on labor-intensive manual workflows. Supervised deep learning has proven to be effective, particularly in the context of image-based analysis. For instance, researchers have shown that it's possible to accurately classify human white blood cells using stain-free imaging flow cytometry data, highlighting how these models can learn meaningful features from complex cell morphology without relying on traditional staining methods [10]. This achievement highlights a major trend: the move towards automated systems that can extract clinically relevant information directly from raw instrumental data. However, one significant bottleneck for supervised learning is its requirement for big, painstakingly annotated datasets. In many biomedical applications, like examining data from high-throughput blood counters, it is often not feasible or very expensive to obtain precise, cell-by-cell labels. This has instigated active interest in unsupervised learning methods, which have the potential to detect intristic patterns in data without explicit labels. One of the most promising paradigms among them is deep clustering, since it combines the strength of deep neural networks in learning rich, hierarchical features with the task of data clustering.

One of the early significant contributions in this area is Deep Embedded Clustering (DEC), proposed by Xie et al. [14]. The main novelty of DEC was to use a ML method that learns a mapping from the data space to a low-dimensional feature space and, at the same time, updates cluster assignments in that space. This is achieved by adding a clustering specific loss function that encourages the learned embeddings to form clusters around cluster centroids. Innovative as it is, one potential pitfall of the original DEC formulation is that the clustering loss can, during training, increasingly distort the learned feature space, undermining the autoencoder's ability to preserve the local structure of the data. To alleviate this, Guo et al. introduced Improved Deep Embedding Clustering (IDEC), a simple yet effective variant of DEC approach [5]. IDEC elegantly resolves the feature space distortion problem by incorporating the autoencoder's reconstruction loss into the total objective function, where the feature learning components are trained jointly. In this manner, the representation in the latent space is ensured to be faithful to the original data's local structure while being simultaneously optimized for cluster separability.

The first part of our study, which performs the initial, rough classification of Red Blood Cells (RBCs) and Platelets (PLTs), is directly inspired by the principles of the IDEC framework. For use cases requiring more than just discriminative clustering, generative models offer a more sophisticated solution by learning the actual data distribution itself. The Variational Autoencoder (VAE), proposed by Kingma & Welling, is a building block of the modern generative modeling [8]. The VAE is a learnable probabilistic mapping to a structured latent space from which new data points can be sampled. Building on this generative ability, Jiang et al. proposed Variational Deep Embedding (VaDE), a framework where a Gaussian Mixture Model (GMM) is employed as a prior in the VAE's latent space [7]. In VaDE, the data generation process is posed as first selecting a cluster from the GMM parameters end-to-end, VaDE learns a powerful generative model for each cluster. The second stage of our project, which aims to perform a more granular sub-classification of the RBC population, uses the VaDE approach for its ability to model complex, multi-modal distributions.

The practical relevance of this work is grounded in the technology of modern automated hematology analyzers, such as Aquarius series from Diatron [4]. These instruments generate a high-dimensional signal data that serves as the input for our algorithms. Our two-staged pipeline, which combines the discriminative power of IDEC with the generative ability of VaDE, is an approach to modeling this instrumental data in the low-resource scenario of sample-level reference counts. By recasting these high-level counts as soft probability targets, out algorithm has a semi-supervised nature, demonstrating an effective framework for applying state-of-the-art deep clustering techniques to real-world biomedical data challenges.

While this study focused on custom deep clustering architectures, future work will evaluate their performance relative to other hybrid or classical approaches using public datasets. Additionally, it is important to assess the pipeline's computational scalability and deployment feasibility for high-throughput lab settings. This includes evaluating the performance across various hardware configurations, as well as analyzing memory usage, and latency.

The current research examined a deep clustering algorithm for blood cell analysis via a two-stage pipeline with IDEC and VaDE models. The first stage utilized an IDEC-based model to detect red blood cells (RBCs) and platelets (PLTs) from a mixture of cell data. Although the model was trained in an semi-supervised manner (knowing only the number of cells per file), the integration of a regression head allowed for the prediction of sample-level RBC/PLT ratios, which helped guide the interpretation of clustering results. This method provided a way to assign per-cell labels that better matched known proportions from reference data.

A key observation during development was the importance of aligning model outputs with expected biological ratios. While semi-unsupervised clustering may often return arbitrary or non-symmetric groupings, post-processing with predicted ratios provided more interpretable and useful results. Using soft cluster probabilities and ranking them in order of likelihood was a handy workaround for having no ground truth cell-level annotations.

Stage two utilized a VaDE approach on the candidate set of cells verified as RBCs in an attempt to further separate into subtypes, such as pure RBCs, reticulocytes, and RBC clumps. The hierarchical architecture allowed the pipeline to be selective enough to process only relevant cells for subsequent clustering.

Two limitations of this study are: First, the algorithms were trained and tested on non-annotated cell-level data, which limits ultimate classification accuracy assessment. While the current approach uses sample-level counts and soft clustering, the addition of even partial cell-level labels, either manually annotated or synthetically generated from a developed simulator, could substantially improve model validation. Future work will concentrate on exploring ways of obtaining and using such labels. Second, some design choices, such as the number of VaDE clusters or latent feature interpretation, remain somewhat heuristic and would be an interesting topic of investigation.

Despite these limitations, the work highlights the potential for deep clustering algorithms in biomedical applications. By blending representation learning with soft probabilistic assignments and post-processing, the pipeline demonstrated that valuable conclusions can actually be learned from unlabeled data. With improving annotation quality and greater accessibility, this approach could be a suitable scalable foundation for more advanced and accurate blood cell analysis systems.

A future direction would be the integration of explainable artificial intelligence (XAI). In line with [3] and [6], explainability could be achieved through methods such as feature attribution, latent space visualization or analysis of individual sample influence on cluster formation. Additionally, methods like SHAP (SHapley Additive exPlanations) [12] and LIME (Local Interpretable Model-Agnostic Explanations) [9] could be explored to provide transparency in model decisions. Visualizing contributions of each feature to latent representations or clustering output would increase trust and interpretability in clinical environments.

As deep clustering methods continue to gain traction in biomedical data analysis, it becomes essential to enhance transparency, scalability, and clinical alignment. By addressing current limitations such as the lack of annotations, class imbalance, and manually selected parameters, and by integrating interpretability into the pipeline, future work aims to develop robust, efficient, and clinically trustworthy blood analysis systems that can generalize across different species and instruments.

6   Conclusion

This study presented a two-stage deep clustering pipeline aimed at detecting and classifying types of blood cells using unsupervised and semi-supervised learning methods. IDEC was used in the first stage to distinguish RBCs from platelets, while VaDE was used in the second stage to explore potential RBC subtypes.

The use of a latent space obtained from training an autoencoder with regression and clustering provided a generalizable framework for processing high-dimensional, unlabeled biological data. Although the algorithms themselves were based on relatively simple architectural components, they were able to produce cell-level classifications, generally in agreement with sample-level reference values.

This is especially valuable in real biomedical datasets, where class imbalance is common and often breaks standard models. The model's strong RBC predictions, show its ability to generalize well. Just as important, processing over 100 large samples in about a minute highlights its practical efficiency. While platelet predictions remain more sensitive due to their low representation, the system is well-positioned to improve. By capturing meaningful features in this latent space, the model can identify and quantify cell populations, even when it is faced with high class imbalance.

Disclosure of Interests. The authors declare no competing interests.

References

1. Agarap, A.F.: Deep Learning Using Rectified Linear Units (ReLU). arXiv preprint arXiv:1803.08375 (2019). https://arxiv.org/abs/1803.08375

2. Bank, D., Koenigstein, N., Giryes, R.: Autoencoders. arXiv preprint arXiv:2003.05991 (2021). https://arxiv.org/abs/2003.05991
3. Bhatia, K., Dhalla, S., Mittal, A., Gupta, S., Gupta, A., Jindal, A.: Integrating explainability into deep learning-based models for white blood cells classification. Comput. Electr. Eng. **110**, 108913 (2023)
4. Diatron: Aquarius 3. https://www.diatron.com/aquarius-3. Accessed 14 June 2025
5. Guo, X., Gao, L., Liu, X., Yin, J.: Improved deep embedded clustering with local structure preservation. In: Proceedings of the 26th International Joint Conference on Artificial Intelligence (IJCAI-17), pp. 1753–1759 (2017). https://doi.org/10.24963/ijcai.2017/243
6. Islam, O., Assaduzzaman, M., Hasan, M.Z.: An explainable AI-based blood cell classification using optimized convolutional neural network. J. Pathol. Inform. **15**, 100389 (2024)
7. Jiang, Z., Zheng, Y., Tan, H., Tang, B., Zhou, H.: Variational Deep Embedding: An Unsupervised and Generative Approach to Clustering. arXiv preprint arXiv:1611.05148 (2017). https://arxiv.org/abs/1611.05148
8. Kingma, D. P., Welling, M.: Auto-Encoding Variational Bayes. arXiv preprint arXiv:1312.6114 (2022). https://arxiv.org/abs/1312.6114
9. Ribeiro, M.T., Singh, S., Guestrin, C.: "Why Should I Trust You?": Explaining the Predictions of Any Classifier. CoRR, abs/1602.04938 (2016). http://arxiv.org/abs/1602.04938
10. Lippeveld, M., et al.: Classification of human white blood cells using machine learning for stain-free imaging flow cytometry. Cytometry A **97**(3), 308–319 (2020). https://doi.org/10.1002/cyto.a.23920
11. McConville, R., Santos-Rodriguez, R., Piechocki, R. J., Craddock, I.: N2d:(not too) deep clustering via clustering the local manifold of an autoencoded embedding. In: 2020 25th International Conference on Pattern Recognition (ICPR), pp. 5145–5152 (2021)
12. Lundberg, S.M., Lee, S.-I.: A unified approach to interpreting model predictions. CoRR, abs/1705.07874 (2017). http://arxiv.org/abs/1705.07874
13. Bowyer, K.W., Chawla, N.V., Hall, L.O., Kegelmeyer, W.P.: SMOTE: Synthetic Minority Over-sampling Technique. CoRR, abs/1106.1813 (2011). http://arxiv.org/abs/1106.1813
14. Xie, J., Girshick, R., Farhadi, A.: Unsupervised Deep Embedding for Clustering Analysis. arXiv preprint arXiv:1511.06335 (2016). https://arxiv.org/abs/1511.06335

Forecasting Technical Debt in Software Projects with Limited Historical Data

Oskar Picus[(✉)][iD]

Department of Computer Science, Faculty of Mathematics and Computer Science,
Babeş-Bolyai University, 1 M. Kogălniceanu, Cluj-Napoca, Romania
oskar.picus@ubbcluj.ro

Abstract. Technical debt refers to the compromises made during software development that enable rapid delivery but may hurt the long-term health of a software system. Recently, forecasting technical debt has become a topic of interest to researchers and techniques for it were used successfully for mature software projects. This work aims at analyzing if existing techniques in forecasting technical debt are still applicable and effective for software projects with limited historical data. We have partially applied the methodology of a reference study, using the sliding-window method and various Machine Learning algorithms, including a boosting one, which was not used before for forecasting technical debt. These techniques were applied on a data set from three projects, 62% to 74% smaller than the one in the reference study. We have obtained comparable results, however, we observed a general inability of existing Machine Learning techniques of coping with small and irregular data sets. This research contributes to the ongoing effort to improve software maintainability and offers insights to practitioners on the methods for mitigating the adverse effects of technical debt. Our results partially replicated the original study's, emphasizing the need for further research aimed at forecasting technical debt for software projects with minimal data available.

Keywords: Technical debt · Time series · Regression · Forecasting

1 Introduction

As software size and complexity grows, the pressure for delivering features at a rapid pace for the ever-eager users becomes a constant within software engineering. However, in this desire to deliver as much and as fast as possible, internal quality attributes of the software such as reliability and especially maintainability tend to be overlooked. Therefore, this phenomenon, called technical debt, refers to technical compromises made to the health of the software project for various short-term benefits [6].

Managing technical debt is not a straightforward process, as it takes into account the nature of the debt (for instance, intentional or unintentional) or

C. Chira et al. (Eds.): InnoComp 2025, CCIS 2793, pp. 363–376, 2026.
https://doi.org/10.1007/978-3-032-12478-4_23

the developers' understanding of this concept and the software project they are working on [9]. Various methodologies and tools were proposed, each supporting one or more activities related to the process of managing technical debt [1].

However, there is a general consensus that managing software projects should be done proactively, not reactively [3]. Extrapolating this perspective to the field of technical debt, it is more desirable to prevent the occurrence of this phenomenon, rather than mitigate it.

A possible method for approaching this is to predict the future value of technical debt, assisting stakeholders into making decisions regarding the evolution and the maintainability of the software project. An empirical study is performed in [20], where the authors have evaluated various Machine Learning algorithms for predicting future values of technical debt. They have modeled the problem as a forecasting one [14] by creating a data set of weekly snapshots for various software projects, each annotated with the amount of technical debt at that point in time using SonarQube[1]. As a result, they have created a time series, a univariate statistical model where historical data indicates the value of a certain characteristic, in this case, the quantity of technical debt [19].

Technical debt, including its various aspects such as characterization and management, has been a focal point for various researchers. For instance, the authors of [11] analyzed historical data for various open-source software projects and concluded that each project has its own technical debt particularities and that it evolves differently, according to the project's lifecycle. In their endeavor, they have created a data set that quantifies technical debt for each version of the three software projects they have studied.

In this paper, we partially apply the methodology described in [20], utilizing the data set in [11] to forecast technical debt in future software releases, with feature extraction performed using the `tsfresh` package [5]. This research addresses data scarcity by demonstrating that such approach does not necessarily remain effective when applied to a smaller dataset. In addition, this contribution further demonstrates that each software project has its own technical debt particularities, given that the same methodology, applied to different projects, yields different results. Thus, the problem that our work attempts to solve can be summarized in the following research question (RQ):

RQ: *Do methods for forecasting technical debt remain effective for software projects with limited historical data?*

The remainder of this paper is structured as follows. The methodology of our work, including a description of the used dataset, is presented in Sect. 2. Section 3 reveals the obtained results and discusses their implications, with Sect. 4 detailing the limitations of our study. Related work in the field of technical debt forecasting is presented in Sect. 5. The paper is concluded in Sect. 6, which also highlights planned future work. The replication package for obtaining our results is available at [15].

---

[1] https://www.sonarqube.org/.

2   Methodology

2.1   Formalisation

We formalise the problem as follows. Forecasting models intend to predict future values of a target y_t at a time point t, based on past observations [10]. Therefore, one-step ahead forecasting models with no associated static metadata take the form of $\hat{y}_{t+1} = f(y_{t-k:t}, x_{t-k:t})$, where $\hat{y}_{t+1}$ is the model, $y_{t-k:t} = \{y_{t-k}, \ldots, y_t\}$ and $x_{t-k:t} = \{x_{t-k}, \ldots, x_t\}$ are observations for the target and input variables over a look-back window k and f is the model's prediction function. This paper focuses on one-step ahead forecasting, however the formalisation can be extended to predict the future values for a horizon h. In addition, we consider as target variable the technical debt value, while the input variables represent the considered features, as detailed in Sect. 2.4.

2.2   Data Set

The data set described in [11] presents historical data for three Java software projects: FreeMind[2], jEdit[3] and TuxGuitar[4]. From this, for each release of the considered projects, we retrieve two variables: *Date*, the time variable indicating when the respective project version was released and *Existing Debt*, the numerical variable representing the amount of technical debt already existent in the current version.

Table 1 presents some descriptive statistics for each project, in the context of the *Existing Debt* variable. On average, each project had 30 releases, with varying amounts of technical debt, ranging from a minimum 0 (at the start of each project) to a maximum average of 88156.93. All projects present high population standard deviations, indicating that the existing amounts of technical debt are spread around a wide range of values.

FreeMind exhibits a positive skew score, suggesting that most releases have less technical debt, while a small number of releases have significantly higher quantities of technical debt. For instance, the 26th of February 2006 introduced a great amount of technical debt (281708), while the previous and the next release only introduced 25342 and 31006, respectively. This presents itself as an opportunity of analyzing projects with non-uniform data, allowing for a deeper understanding of rare events, such as a software release introducing a magnitude of software quality issues. On the other hand, jEdit and TuxGuitar present negative, but close to 0, skew scores, suggesting nearly symmetrical data.

Figure 1 presents the evolution in time of the quantity of technical debt for the three studied projects. While FreeMind presents a spike in technical debt (the 26th of February 2006 release), followed by a sudden drop, suggesting successful application of technical debt management techniques, jEdit and TuxGuitar accumulate in time significant, unresolved debt.

---

[2] http://freemind.sourceforge.net/.
[3] http://jedit.org.
[4] https://www.tuxguitar.app/.

Table 1. Descriptive statistics for the existing technical debt in each studied software project.

	FreeMind	jEdit	TuxGuitar
Number of releases	38	46	30
Time period	2000-07-09 2016-02-05	2000-01-29 2020-09-03	2005-11-18 2022-04-28
Mean	65258.894737	102047.434783	97164.466667
Median	81596.0	116101.0	131900.5
Standard deviation	47345.780111	36403.605012	65632.703511
Skewness	2.2419	-0.763869	-0.185158
Minimum	0	0	0
Maximum	281708.0	146965	189791

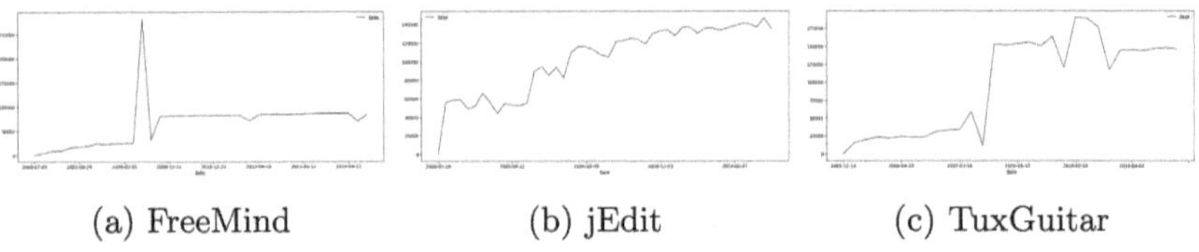

(a) FreeMind   (b) jEdit   (c) TuxGuitar

Fig. 1. Technical debt distribution for the three studied projects.

This data set is significantly smaller than the one used in [20]. In terms of software size, FreeMind had at maximum 65603 lines of code, jEdit, 96912 lines of code and TuxGuitar, 107650 lines of code. Compared to the projects used in [20], FreeMind is 40% smaller, jEdit, 46.6% smaller and TuxGuitar, 53.3% smaller. This was obtained by computing the percentile rank for the three projects, relative to the lines of code metrics of the compared projects. In addition, depending on the project, the authors of [20] have compared 100 to 150 versions of the same code base, a significantly greater value than the average of 38 releases present in the data set from [11].

2.3 Data Preprocessing

The data set was already curated by the authors of [11]; as such, no steps for cleaning the data (such as filling missing values) were needed.

Machine Learning approaches do not work, by default, with time series and forecasting problems. That is because in the training phase of the model, multiple examples are needed to train and test. Thus, an approach where only the time series until today is used and the value for tomorrow needs to be determined in order to have a target is not satisfactory.

Replaying the history is a possibility that can be used to extend the data set and take advantage of the fact that Machine Learning approaches support features with multiple inputs. We have used the *sliding-window* technique [7], which transforms the data set such as a single sample contains multiple prior time steps as inputs (the past information) to predict future time steps as output (the future information). Choosing the *window* size requires both experimentation with multiple values and consideration for the data set size. Given the relatively small size of our data set and the results obtained by using different *window* dimensions, we have decided to use a *window* size of 2.

2.4   Feature Engineering

In this step, the raw data from the data sets were transformed into more informative features for better understanding of the data and capturing more patterns, all leading to better predictions. For this, we have used the `tsfresh` [5] package, which is capable of computing a large number of time series characteristics, including statistical, temporal-specific, frequency-based and others. These features were later provided to the Machine Learning algorithms that we have employed. Then, we have filtered these characteristics, retaining only those that are statistically relevant, thus obtaining 37 features for both FreeMind and jEdit and 27 for TuxGuitar.

Software releases are usually performed at irregular intervals, with this aspect also being reflected in our data set, where each project followed different release calendars. In order to capture the temporal variability, we have incorporated the time in days since the previous release as an additional feature, allowing our models to capture the natural pacing of the development activity.

2.5   Analysis Techniques

Four regression algorithms were employed, in order to ensure a comprehensive evaluation and to increase the robustness and reliability of our results:

- Multivariate Linear Regression (MLR), a statistical technique which models the relationship between multiple independent variables and a single dependent variable
- Support Vector Regression (SVR), based on the principles of Support Vector Machines, aiming to find a function that approximates the underlying relationship in the data while ensuring that most of the data points are within a specified margin of error
- Random Forest (RF), built on the principles of ensemble learning. It combines multiple decision trees to produce more accurate and robust predictions
- AdaBoost (Adaptive Boosting), another ensemble learning method that combines multiple weak regressors to create a stronger one.

MLR, SVR and RF were all used in [20]; we are also using them in order to obtain comparable results. We have also used AdaBoost, a boosting algorithm, in an effort to complete the family of Machine Learning techniques used in forecasting technical debt. Despite not being used before in technical debt prediction, AdaBoost has proven to be successful in other software engineering problems, such as refactoring [12].

2.6 Performance Measurements

For measuring the performance of the regressors, the following metrics were used, coinciding with the evaluation methodology used in [20]:

- Mean Absolute Error (MAE), which computes the average of the absolute differences between the predicted values and the actual values
- Root Mean Squared Error (RMSE), which computes the square root of the average of the squared differences between predicted and actual values
- Mean Absolute Percentage Error (MAPE), particularly useful in forecasting and time series prediction tasks, measures the average absolute percentage difference between predicted values and actual values

2.7 Experimental Design

For training our models, using the algorithms specified in Sect. 2.5, we have followed the classical 80-20 train test split; 80% of the data is used for training, while 20% is used for evaluating it.

Aiming to improve the robustness of our evaluation, we have used the Walk-forward Train Test validation method [18], which is inspired by k-fold cross validation. Using 3 splits, we report the minimum of the performance measurements discussed in Sect. 2.6, alongside a 95% confidence interval of the mean.

3 Results and Discussion

Table 2 presents the obtained confidence intervals of the metrics for each of the three studied projects and for each chosen Machine Learning algorithm, following the execution of our experiment. Table 2 also presents a comparison between our results and those reported in [20]. Furthermore, Fig. 2 displays a comparison between the actual values of technical debt for each project (in blue) and the predicted ones (in orange).

MLR performs poorly for our data set, with a minimum MAE of 11919.05, a minimum RMSE of 21981.55 and a minimum MAPE of 8.5%. One of the reasons why MLR underperforms stems from its assumption of the data's linearity and its sensitivity to outliers. For example, as explained in Sect. 2.2, FreeMind presents a spike in technical debt, followed by a sudden drop; this is an example of an outlier, to which MLR is very sensitive. In addition, none of the three projects present a linear distribution of technical debt.

Comparing the results in [20] and the minimum scores that we obtained for MLR, we see a 256.13% increase in MAE, a 470.45% increase in RMSE and a 28.99% increase in MAPE, highlighting this technique's disadvantages for a data set of this kind.

SVR's scores are more favourable, with a minimum of 2.31% for MAPE for jEdit and comparable results for the other projects (3.85% for FreeMind, 8.19% for TuxGuitar). This is a result of SVR's ability to work with smaller data sets

Table 2. Obtained experimental results, reported with a 95% confidence interval (CI) and compared to [20]'s.

Data set	Regressor	Measurement	CI	Minimum	[20]'s results
FreeMind	MLR	MAE	8235127.97 ± 35319302.63	23862.85	3346.77
		RMSE	9065197.71 ± 38849399.8	30499.93	3853.36
		MAPE (%)	100.57 ± 431.37	28.77	6.59
	SVR	MAE	13112.28 ± 41556.21	2856.99	3428.83
		RMSE	14750.83 ± 39138.59	5575.30	3970.34
		MAPE (%)	0.16 ± 0.50	3.85	6.7
	RF	MAE	19724.92 ± 58330.70	2862.82	4262.78
		RMSE	21587.21 ± 54994.57	4334.63	4767.65
		MAPE (%)	0.24 ± 0.71	3.72	8.66
	AdaBoost	MAE	50008.76 ± 201279.24	2869.04	N/A
		RMSE	56745.01 ±223856.76	4318.63	
		MAPE (%)	0.61 ± 2.45	3.72	
jEdit	MLR	MAE	35972.59 ± 74695.71	11919.05	3346.77
		RMSE	42683.31 ± 73205.31	21981.55	3853.36
		MAPE (%)	0.28 ± 0.56	8.5	6.59
	SVR	MAE	6974.39 ± 10833.46	3179.50	3428.83
		RMSE	8454.67 ± 11640.27	4764.58	3970.34
		MAPE (%)	0.05 ± 0.1	2.31	6.7
	RF	MAE	15177.12 ± 21289.52	5339.36	4262.78
		RMSE	16773.32 ± 20016.004	7657.32	4767.65
		MAPE (%)	0.12 ± 0.19	3.78	8.66
	AdaBoost	MAE	17002.36 ± 24086.0002	5942.5	N/A
		RMSE	18289.96 ± 25243.61	6841.6	
		MAPE (%)	0.13 ± 0.21	4.24	
TuxGuitar	MLR	MAE	1247649.62 ± 5112303.56	47666.61	3346.77
		RMSE	1492475.27 ± 6043500.6	60073.69	3853.36
		MAPE (%)	9.205 ± 37.84	32.71	6.59
	SVR	MAE	43519.014 ±94646.29	11941.87	3428.83
		RMSE	46828.31 ± 98098.34	12307.39	3970.34
		MAPE (%)	0.46 ± 1.36	8.19	6.7
	RF	MAE	45082.31 ± 112740.54	9867.2	4262.78
		RMSE	47286.54 ± 117114.23	10658.14	4767.65
		MAPE (%)	0.44 ± 1.34	6.76	8.66
	AdaBoost	MAE	43893.02 ± 92425.32	15778.9	N/A
		RMSE	46041.78 ± 91116.39	19039.74	
		MAPE (%)	0.502 ± 1.509	10.85	

Table 3. Top 5 averaged feature importance scores from RF.

Data set	Feature name	Importance score
FreeMind	Days since last release	0.327
	Quantile at 0.6 for *Existing Debt*	0.074
	Sum of *Existing Debt* over the time series	0.067
	Quantile at 0.3 for *Existing Debt*	0.066
	Ricker wavelet for *Existing Debt* at points = (2, 5, 10, 20), coefficient = 0, width = 2	0.065
jEdit	Mean of *Existing Debt*	0.176
	Ricker wavelet for *Existing Debt*, at points = (2, 5, 10, 20), coefficient = 2, width = 20	0.161
	Ricker wavelet for *Existing Debt*, at points = (2, 5, 10, 20), coefficient = 1, width = 20	0.161
	Ricker wavelet for *Existing Debt*, at points = (2, 5, 10, 20), coefficient = 1, width = 10	0.129
	C3 statistic for *Existing Debt* with lag = 1	0.103
TuxGuitar	Ricker wavelet for *Existing Debt*, at points = (2, 5, 10, 20), coefficient = 1, width = 10	0.256
	Largest fixed point of dynamics for *Existing Debt*, with order of polynomial = 3, number of quantiles for averaging = 30	0.195
	Ricker wavelet for *Existing Debt*, at points = (2, 5, 10, 20), coefficient = 2, width = 10	0.19
	Ricker wavelet for *Existing Debt*, at points = (2, 5, 10, 20), coefficient = 1, width = 2	0.131
	Sum over the squared values of *Existing Debt*	0.129

and its robustness to outliers. We have also obtained favourable results for the other two considered metrics (a minimum of 2856.99 in MAE and 4764.58 in RMSE).

In comparison to [20], we have obtained a 20% increase in RMSE and decreases in both MAE and MAPE (16.67% and 65.48%, respectively). These results indicate that SVR is an adequate option when working with smaller data sets for predicting technical debt.

Approaching the problem of forecasting technical debt with Random Forests resulted in fairly consistent results for FreeMind and jEdit; however, it underperforms for TuxGuitar, obtaining a MAE of 9867.2 and a RMSE of 10658.14 due to overfitting. Random Forest Regressor predicts by averaging the outputs of individual decision trees, thus they tend to regress towards the mean. Because TuxGuitar presents extreme highs and lows (0 technical debt on the 18th of November 2005 or 11489 on the 3rd of March 2008, followed by 152938 technical debt on the 11th of March 2008), Random Forest tends to move toward the center of these values. This is also noticeable in Fig. 2, where Random Forest's predictions tend to the mean of the previous quantities of technical debt.

Compared to our reference study, we have obtained decreases on all three considered metrics: 32.84% for MAE, 9.08% for RMSE and 57.02% for MAPE.

Table 3 reports the top 5 feature importance scores in the Random Forest models, with scores averaged across the cross-validation. With the exception of TuxGuitar, simpler features such as the mean or the sum of the technical debt values drive the predictions, being capable of best capturing the characteristics of the time series. Notably, for FreeMind, its most important feature is related to the difference in days between the releases, implying that the release schedule for software projects is a strong predictor for technical debt.

Conversely, the majority of the most influential features are complex ones (e.g. the Ricker wavelet [16], the largest fixed point of deterministic dynamics [8] or the C3 statistic [17]), with them best describing the time series for TuxGuitar. This variance in the chosen features further supports the fact that each project has its own technical debt characteristics, highlighting the need for personalized techniques in predicting it.

Regression for technical debt using AdaBoost also proves to be a viable option; the obtained results for jEdit are comparable to those obtained in our reference study, with a MAPE of 4.24%. However, the noise in the data for both FreeMind and TuxGuitar determined AdaBoost to focus on the outliers, instead of trying to generalise for the other data points present in the data set. This is most noticeable with FreeMind's spike in technical debt, related to the 26th February 2006 release, which was mostly paid back in the following release. As mentioned earlier, [20] did not use in their study AdaBoost; thus, we were unable to compare our results to theirs.

Answer to RQ: Methods for forecasting technical debt are sensitive to the size of the data set but remain effective for software projects with a smaller number of releases. We have observed that with a decrease between 62% and 74% in the number of data points (corresponding to a decrease in the available historical data for the software projects), no consistent trend was observed, since we have obtained both increases and decreases in the three selected metrics. The results suggest that the investigated Machine Learning techniques are not able to fully fit the technical debt trend that some software projects display, with most techniques struggling to fit the spikes and drops of the quantities of technical debt.

4   Limitations

While this study presents promising results and confirms the findings in [20], several limitations must be acknowledged. Firstly, we have only used the default parameter combinations, as specified in the `sklearn`[5] documentation, for the regression algorithms mentioned in Sect. 2.5. As a result, we were unable to

---

[5] https://scikit-learn.org/stable/.

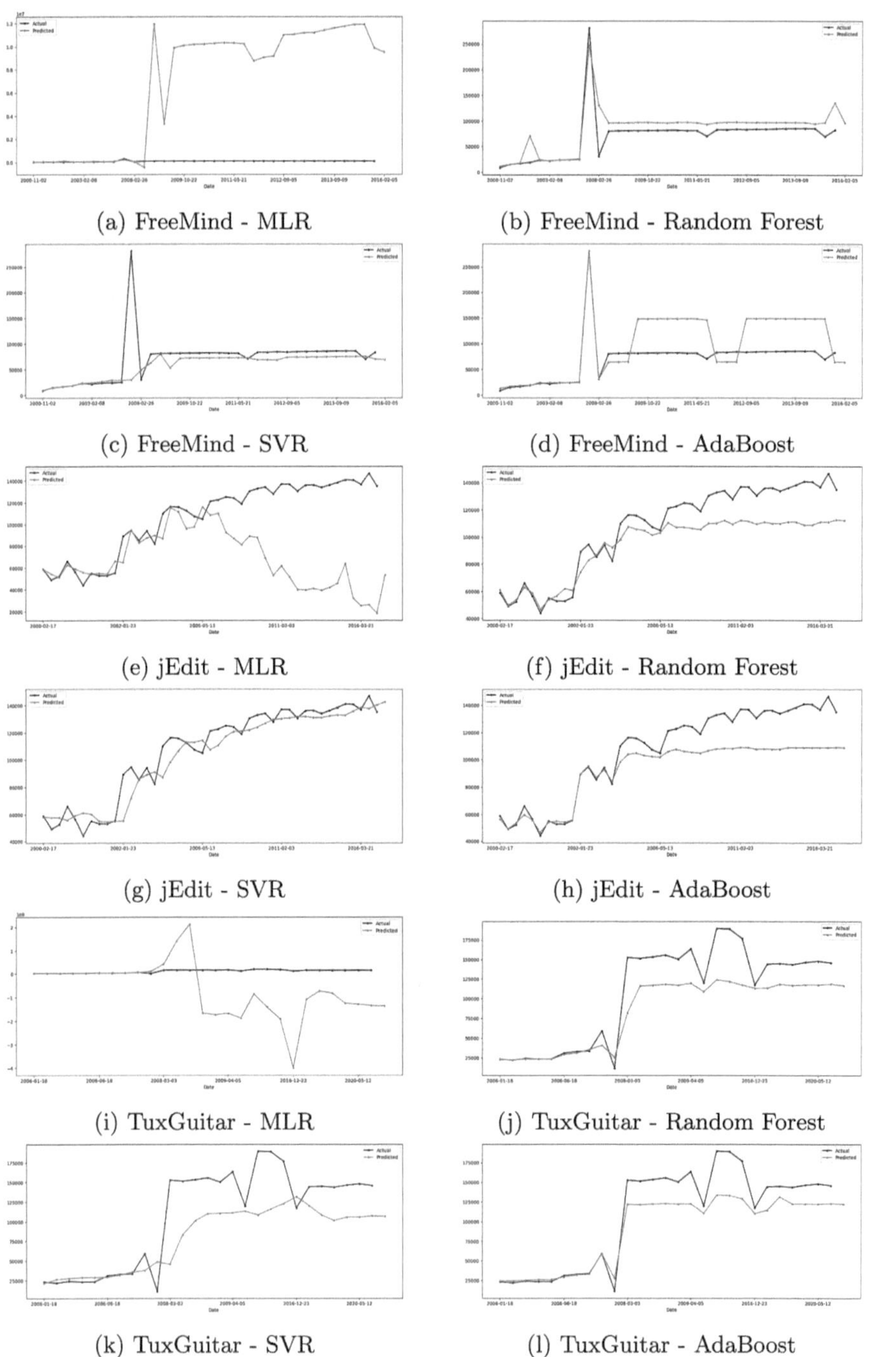

(a) FreeMind - MLR

(b) FreeMind - Random Forest

(c) FreeMind - SVR

(d) FreeMind - AdaBoost

(e) jEdit - MLR

(f) jEdit - Random Forest

(g) jEdit - SVR

(h) jEdit - AdaBoost

(i) TuxGuitar - MLR

(j) TuxGuitar - Random Forest

(k) TuxGuitar - SVR

(l) TuxGuitar - AdaBoost

Fig. 2. Distribution of the actual and predicted technical debt for each employed algorithm and for each studied software project.

experiment with and analyze the differences between these parameters, which might have led to improved results. We intend to explore this further in future work.

Secondly, our findings are context-dependent and may not be directly transferable to other software projects. As we have demonstrated in Sect. 3, the analyzed methods for forecasting technical debt are sensitive to the particularities of the data set, i.e. the way technical debt evolves in the software project. This impacts the generalizability of our results, given that the analysis was conducted on only three open-source Java projects, which may not be representative of other software projects written in other programming languages, or even in Java itself, given the small sample size. Further validation on diverse and larger datasets is necessary to assess the robustness and broader applicability of our findings.

We do not consider the size of the data set to be a limitation, since the aim of our study was specifically analyzing the topic of technical debt forecasting in the context of data scarcity.

5   Related Work

Technical debt has been a topic of great interest to the research community, with numerous contributions to this domain being made in the last 10 years [13]. In particular, forecasting technical debt has started to gain attention in 2020, notably with the work in [20]. In this section, we explore the existing contributions in the field of technical debt forecasting.

The authors of [20] have constructed a data set of weekly snapshots for 15 Java projects of various sizes. They have then predicted the future technical debt values by using both project-level metrics, computed by SonarQube and various object-oriented class-level metrics, aggregated at project-level. They employed multiple Machine Learning algorithms (MLR, Ridge and Lasso regression, SVR and RF) and performed cross-validation using the Walk-forward TrainâĂŞTest method [18]. They conclude that their approach captures technical debt values with high accuracy, while being best suited for projects with a relatively long history.

In [2], the authors studied technical debt forecasting on 8 Java projects at the commit level. They have used as features the past debt values, computed using SonarQube and the metrics from the Chidamber and Kemerer suite [4]. Using MLR, Decision tree (DT), Bagged decision tree (BDT) and RF, with RMSE as the only performance indicator, their results are promising, obtaining a 95% accuracy for RF.

The study in [22] focuses on technical debt in JavaScript applications, further emphasizing the need mentioned in [21] for quantifying debt with respect to the particularities of the chosen programming language. Therefore, the authors used a diverse collection of predictors, which includes external indicators (e.g. the number of open issues), the technical debt value and source code, JavaScript-specific and maintainability-specific metrics. They have used Regression Analysis and Autoregressive Integrated Moving Average (ARIMA) on 19636 releases from 105 JavaScript projects, concluding that the information from just 1 or 2 releases is generally sufficient for predicting the future technical debt value.

In relation to existing approaches, our work makes valuable contributions. We analyze if forecasting technical debt demonstrates strong performance for smaller data sets as well. Furthermore, we distinguish our work from previous efforts by experimenting with a new class of Machine Learning algorithms, namely boosting, while also using an innovative feature engineering process, namely with the `tsfresh` *package.*

6   Conclusions and Future Work

In this study, we aimed to replicate the findings in [20] in the context of forecasting technical debt and limited availability of historical data for software projects. Our results partially replicated the original study's conclusions, with our contribution highlighting the importance of the project's historical data in regards to software releases. Specifically, we have found that forecasting techniques for technical debt are sensitive to the amount and variations in historical data, yet we were able to obtain favourable results using SVR and AdaBoost.

As future work, we aim to analyze different parameter combinations for the Machine Learning algorithms that we have selected. In addition, we also intend to offer an even more comprehensive view on the techniques that can be used for forecasting technical debt, by experimenting with different Machine Learning algorithms and comparing the results with those obtained in this study.

Acknowledgments. Grateful appreciation is extended to Simona Motogna for the invaluable guidance, support and insightful feedback throughout this research.

Disclosure of Interests. The author has no competing interests to declare that are relevant to the content of this article.

References

1. Alves, N.S., Mendes, T.S., de Mendonça, M.G., Spínola, R.O., Shull, F., Seaman, C.: Identification and management of technical debt: a systematic mapping study. Inf. Softw. Technol. **70**, 100–121 (2016). https://doi.org/10.1016/j.infsof.2015.10.008
2. Aversano, L., Bernardi, M.L., Cimitile, M., Iammarino, M., Montano, D.: Forecasting technical debt evolution in software systems: an empirical study. Front. Comp. Sci. **17**(3), 173210 (2022). https://doi.org/10.1007/s11704-022-1541-7

3. Castelli, V., et al.: Proactive management of software aging. IBM J. Res. Dev. **45**(2), 311–332 (2001). https://doi.org/10.1147/rd.452.0311

4. Chidamber, S., Kemerer, C.: A metrics suite for object oriented design. IEEE Trans. Software Eng. **20**(6), 476–493 (1994). https://doi.org/10.1109/32.295895

5. Christ, M., Braun, N., Neuffer, J., Kempa-Liehr, A.W.: Time series feature extraction on basis of scalable hypothesis tests (tsfresh – a Python package). Neurocomputing **307**, 72–77 (2018). https://doi.org/10.1016/j.neucom.2018.03.067

6. Cunningham, W.: The WyCash portfolio management system. SIGPLAN OOPS Mess. **4**(2), 29–30 (1992). https://doi.org/10.1145/157710.157715

7. Dietterich, T.G.: Machine learning for sequential data: a review. In: Caelli, T., Amin, A., Duin, R.P.W., de Ridder, D., Kamel, M. (eds.) SSPR /SPR 2002. LNCS, vol. 2396, pp. 15–30. Springer, Heidelberg (2002). https://doi.org/10.1007/3-540-70659-3_2

8. Friedrich, R., et al.: Extracting model equations from experimental data. Phys. Lett. A **271**(3), 217–222 (2000). https://doi.org/10.1016/S0375-9601(00)00334-0

9. Li, Z., Avgeriou, P., Liang, P.: A systematic mapping study on technical debt and its management. J. Syst. Softw. **101**, 193–220 (2015). https://doi.org/10.1016/j.jss.2014.12.027

10. Lim, B., Zohren, S.: Time-series forecasting with deep learning: a survey. Phil. Trans. R. Soc. A **379**(2194), 20200209 (2021). https://doi.org/10.1098/rsta.2020.0209

11. Molnar, A., Motogna, S.: Characterizing technical debt in evolving open-source software. In: Proceedings of the 17th International Conference on Evaluation of Novel Approaches to Software Engineering - ENASE, pp. 174–185. INSTICC, SciTePress (2022). https://doi.org/10.5220/0011073600003176

12. Motogna, S., Berciu, L.M., Moldovan, V.A.: Artificial intelligence methods in software refactoring: a systematic literature review. In: 2024 50th Euromicro Conference on Software Engineering and Advanced Applications (SEAA), pp. 309–316 (2024). https://doi.org/10.1109/SEAA64295.2024.00055

13. Murillo, M.I., López, G., Spínola, R., Guzmán, J., Rios, N., Pacheco, A.: Identification and management of technical debt: a systematic mapping study update. J. Softw. Eng. Res. Dev. **11**(1), 8:1–8:20 (2023). https://doi.org/10.5753/jserd.2023.2671

14. Petropoulos, F., et al.: Forecasting: theory and practice. Int. J. Forecast. **38**(3), 705–871 (2022). https://doi.org/10.1016/j.ijforecast.2021.11.001

15. Picus, O.: Replication package for "Forecasting technical debt in software projects with limited historical data" (2025). https://doi.org/10.6084/m9.figshare.28805942.v2

16. Ricker, N.: Wavelet functions and their polynomials. Geophysics **9**(3), 314–323 (1944). https://doi.org/10.1190/1.1445082

17. Schreiber, T., Schmitz, A.: Discrimination power of measures for nonlinearity in a time series. Phys. Rev. E **55**, 5443–5447 (1997). https://doi.org/10.1103/PhysRevE.55.5443

18. Stone, M.: Cross-validatory choice and assessment of statistical predictions. J. Roy. Stat. Soc.: Ser. B (Methodol.) **36**(2), 111–133 (1974). https://doi.org/10.1111/j.2517-6161.1974.tb00994.x

19. Tsoukalas, D., Jankovic, M., Siavvas, M., Kehagias, D., Chatzigeorgiou, A., Tzovaras, D.: On the applicability of time series models for technical debt forecasting. In: 15th China-Europe International Symposium on Software Engineering Education (2019). https://doi.org/10.13140/RG.2.2.33152.79367

20. Tsoukalas, D., Kehagias, D., Siavvas, M., Chatzigeorgiou, A.: Technical debt forecasting: an empirical study on open-source repositories. J. Syst. Softw. **170**, 110777 (2020). https://doi.org/10.1016/j.jss.2020.110777
21. Tsoukalas, D., Siavvas, M., Jankovic, M., Kehagias, D., Chatzigeorgiou, A., Tzovaras, D.: Methods and tools for TD estimation and forecasting: A state-of-the-art survey. In: 2018 International Conference on Intelligent Systems (IS), pp. 698–705 (2018). https://doi.org/10.1109/IS.2018.8710521
22. Zozas, I., Bibi, S., Ampatzoglou, A.: Forecasting the principal of code technical debt in JavaScript applications. IEEE Trans. Software Eng. **49**(4), 2498–2512 (2023). https://doi.org/10.1109/TSE.2022.3222318

Author Index

FSC
www.fsc.org
MIX
Papier aus verantwortungsvollen Quellen
Paper from responsible sources
FSC® C105338